GLUCOCORTICOID ACTION

BASIC AND CLINICAL IMPLICATIONS

ANNALS OF THE NEW YORK ACADEMY OF SCIENCES
Volume 1024

GLUCOCORTICOID ACTION

BASIC AND CLINICAL IMPLICATIONS

Edited by Tomoshige Kino, Evangelia Charmandari, and George P. Chrousos

The New York Academy of Sciences
New York, New York
2004

∞ *The paper used in this publication meets the minimum requirements of the American National Standard for Information Sciences—Permanence of Paper for Printed Library Materials, ANSI Z39.48-1984.*

Library of Congress Cataloging-in-Publication Data

Glucocorticoid action : basic and clinical implications / edited by Tomoshige Kino, Evangelia Charmandari, and George P. Chrousos.
p. ; cm. — (Annals of the New York Academy of Sciences ; v. 1024)
"This volume is the result of a conference entitled 'Basic and Clinical Implications of Glucocorticoid Action—Focus on Development,' held on June 17–18, 2003 in Bethesda, Maryland"—Contents p.
Includes bibliographical references and index.
ISBN 1-57331-557-5 (cloth : alk. paper) — ISBN 1-57331-558-3 (pbk. : alk. paper)
1. Glucocorticoids—Physiological effect—Congresses. 2. Glucocorticoids—Therapeutic use—Congresses.
[DNLM: 1. Glucocorticoids—physiology—Congresses. 2. Glucocorticoids—therapeutic use—Congresses. 3. Homeostasis—physiology—Congresses. 4. Receptors, Glucocorticoid—physiology—Congresses. WK 755 G5667 2004] I. Kino, Tomoshige. II. Charmandari, Evangelia. III. Chrousos, George P. IV. Series.
Q11.N5 vol. 1024
[QP572.G54]
500 s—dc22
[612.4/

2004012476

GYAT / PCP

Printed in the United States of America
ISBN 1-57331-557-5 (cloth)
ISBN 1-57331-558-3 (paper)
ISSN 0077-8923

ANNALS OF THE NEW YORK ACADEMY OF SCIENCES

Volume 1024
June 2004

GLUCOCORTICOID ACTION

BASIC AND CLINICAL IMPLICATIONS

Editors and Conference Organizers
TOMOSHIGE KINO, EVANGELIA CHARMANDARI, AND GEORGE P. CHROUSOS

This volume is the result of a conference entitled Basic and Clinical Implications of Glucocorticoid Action—Focus on Development, held June 17–18, 2003 by the National Institutes of Health, Bethesda, Maryland.

CONTENTS

Financial assistance was received from:

- NATIONAL INSTITUTE OF CHILD HEALTH AND HUMAN DEVELOPMENT
- OFFICE OF RARE DISEASES, NATIONAL INSTITUTES OF HEALTH
- TAKEDA PHARMACEUTICALS NORTH AMERICA

Preface

TOMOSHIGE KINO, EVANGELIA CHARMANDARI, AND
GEORGE P. CHROUSOS

Pediatric and Reproductive Endocrinology Branch, National Institute of Child Health and Human Development, National Institutes of Health, Bethesda, Maryland 2089-1583, USA

The Pediatric and Reproductive Endocrinology Branch, National Institute of Child Health and Human Development, National Institutes of Health (NIH), organized a conference entitled Basic and Clinical Implications of Glucocorticoid Action—Focus on Development, which was held at the NIH on June 17–18, 2003. The reviews that follow cover most of the topics presented at the conference and discuss the most recent advances in basic and clinical aspects of glucocorticoid action.

The book is divided into five parts: the first three parts focus on the glucocorticoid receptor (GR) mechanisms of action, its interaction with other transcription factors, as well as with factors that modulate GR function. The last two parts present an update on the effects of glucocorticoids on the immune system and the clinical implications of glucocorticoid action.

The conference was organized in memory of the late Dr. Yukitaka Miyachi (1939–2003), former Professor and Chairman of the Department of Internal Medicine, Toho University School of Medicine, Tokyo, Japan.

This conference was supported by the National Institute of Child Health and Human Development and the Office of Rare Diseases, National Institutes of Health, Bethesda, Maryland, with the aid of an educational grant from Takeda Pharmaceuticals North America, Wilmington, Delaware.

Ann. N.Y. Acad. Sci. 1024: ix (2004).
doi: 10.1196/annals.1321.016

Yukitaka Miyachi, M.D., Ph.D.
1939–2003

In Memoriam

Yukitaka Miyachi graduated from the Medical School of the University of Tokyo, Japan, in 1963 and received his Ph.D. degree from the Graduate School of the same university in 1967. He subsequently completed a residency in medicine and a fellowship in endocrinology and metabolism at the Third Department of Internal Medicine, University of Tokyo Medical School. In 1971, Dr. Miyachi joined the National Institutes of Health as a Visiting Fellow and continued his research in the field of hypothalamic-pituitary-adrenal axis for two years, under the supervision of the late Dr. M.B. Lipsett, before returning to his native country as an Associate Professor of Internal Medicine.

Dr. Miyachi became Chief and Director of Internal Medicine at Shizuoka General Hospital, Shizuoka, Japan, in 1983; Associate Professor and then Professor of Internal Medicine at the University of Toho in 1987 and 1990, respectively; and Chairman of the Department of Internal Medicine at the Toho University School of Medicine, Tokyo, in 1998.

Yukitaka Miyachi was a leader of the Japanese Society of Endocrinology for many years and founded the Japanese Society on Hormonal Steroids. Dr. Miyachi made major research contributions to the fields of peptide and steroid radioimmunoassays and performed a number of excellent studies in clinical endocrinology. He always maintained close contact with his former mentor and other colleagues at the National Institutes of Health, and he collaborated with scientists worldwide.

Colleagues will remember Dr. Miyachi for his ebullient personality, his humor, his dedication to research and the clinical practice of endocrinology, his kindness, and his generosity. He trained many young researchers and clinicians who will always be grateful for his unwavering support, encouragement and guidance, and for the opportunity to undertake excellent clinical and/or research training in endocrinology under his supervision.

—E.C.

Glucocorticoids and Their Actions

An Introduction

EVANGELIA CHARMANDARI, TOMOSHIGE KINO, AND GEORGE P. CHROUSOS

Pediatric and Reproductive Endocrinology Branch, National Institute of Child Health and Human Development, National Institutes of Health, Bethesda, Maryland 20892-1583, USA

KEYWORDS: **glucocorticoid receptor (GR); immune response; HPA axis**

Glucocorticoids regulate a variety of growth, metabolic, developmental, and immune functions and play a pivotal role in preserving basal and stress-related homeostasis.[1–3] They also represent one of the most widely prescribed drugs worldwide. During the last 50 years, pharmacologic doses of glucocorticoids have been used in the treatment of inflammatory, autoimmune, and lymphoproliferative diseases, and in the prevention of graft rejection, while substitution doses have been employed in the management of adrenocortical insufficiency states.[1–3]

Glucocorticoids exert their effects through the glucocorticoid receptor (GR), which belongs to the superfamily of nuclear receptors that function as ligand-dependent transcription factors.[4,5] Alternative splicing of the human (h) GR in exon 9 generates two highly homologous receptor isoforms, hGRα and hGRβ, which are identical through amino acid 727, but differ at their carboxyl-termini. hGRα is ubiquitously expressed in almost all human tissues and cells and represents the classic GR that functions as a ligand-dependent transcription factor, whereas hGRβ does not bind glucocorticoids and inhibits the transcriptional activity of hGRα in a dose-dependent manner.[6–8]

In the absence of ligand, hGRα resides mostly in the cytoplasm of cells as part of a large multiprotein complex, which consists of the receptor polypeptide, two molecules of heat-shock protein (hsp)-90, and several other proteins. Upon ligand binding, hGRα dissociates from the hsps and translocates into the nucleus of cells, where it modulates the transcriptional activity of glucocorticoid-responsive genes in either of two ways: by binding to specific sequences in the promoter region of target genes, the glucocorticoid-response elements (GREs), or through protein–protein interactions with other transcription factors, such as nuclear factor (NF)-κB, activator protein-1 (AP-1), and several signal transducers and activators of transcription (STATs).[9–11] Newly characterized nuclear receptor coregulators (coactivators or corepressors) may also enhance or attenuate glucocorticoid signal transduction

Address for correspondence: Evangelia Charmandari, M.D., National Institutes of Health, Building 10, Room 9D42, 10 Center Drive MSC 1583, Bethesda, MD 20892-1583. Voice: 301-496-5800; fax: 301-402-0884.
charmane@mail.nih.gov

Ann. N.Y. Acad. Sci. 1024: 1–8 (2004).
doi: 10.1196/annals.1321.001

through different pathways, further accounting for the gene-, cell- and tissue-specific actions of glucocorticoids.[12,13]

Circulating glucocorticoid concentrations are largely under the control of the hypothalamic-pituitary-adrenal (HPA) axis. Corticotropin-releasing hormone (CRH), the principal hypothalamic regulator of the HPA axis, stimulates the secretion of adrenocorticotropic hormone (ACTH) from the anterior pituitary, which in turn stimulates cortisol secretion from the adrenal cortex. Arginine-vasopressin (AVP), although a potent synergistic factor of CRH, has very little ACTH secretagogue activity on its own. Glucocorticoids exert a negative feedback effect on the secretion of CRH and AVP by the hypothalamus, on the secretion of ACTH by the anterior pituitary, and on suprahypothalamic centers that control the activity of the HPA axis. The inhibition of CRH/AVP and ACTH secretion serves to limit the duration of the total tissue exposure to glucocorticoids, thus minimizing the catabolic, lipogenic, anti-reproductive, and immunosuppressive effects of these hormones. Therefore, glucocorticoids play a key regulatory role in maintaining basal and stress-related homeostasis and preserving normal physiology.[1–3]

The last few decades have witnessed an upsurge in our understanding of the action of glucocorticoids and their involvement in human pathophysiology. Glucocorticoids are involved in every organ system of the human organism and in almost every physiologic, cellular, and molecular network. However, alterations in the activity of the HPA axis and/or tissue sensitivity to glucocorticoids may lead to disease if they are not followed by "compensatory" adaptive central nervous system (CNS) and peripheral responses, which are activated in an attempt by the organism to maintain homeostasis.[14–16] This adaptive response of an individual is determined by multiple genetic, environmental, and developmental factors. Lack of adequate compensation, whether excessive or deficient, may lead to "allostasis" and target tissue pathology, and may define vulnerability to endocrine, developmental, psychiatric, or immunologic disorders[16–19] (FIG. 1).

Prolonged or chronic activation of the HPA axis may lead to a number of disorders that arise as a result of increased and prolonged secretion of CRH and/or glucocorticoids[14,16,20–22] (TABLE 1). *Growth and development* may be affected because of suppression of growth hormone (GH) secretion and inhibition of the effects of insulin-like growth factor (IGF)-I and other growth factors on target tissues. These suppressive effects on growth may be direct, due to increased glucocorticoid concentrations, or indirect, due to CRH-induced increases in somatostatin secretion. Anxiety or melancholic depression, psychosocial short stature, malnutrition, and the "inhibited child syndrome" are examples of the effects of HPA axis hyperactivity on growth and development. Increased HPA axis activity also results in suppression of *thyroid function*, as evidenced by the decreased production of thyroid-stimulating hormone (TSH) and inhibition of peripheral conversion of thyroxine to triiodothyronine.

Gonadal function may also be inhibited at many levels. CRH suppresses the secretion of gonadotropin-releasing hormone (GnRH) either directly or indirectly, via stimulation of arcuate pro-opiomelanocortin peptide–secreting neurons. Furthermore, glucocorticoids exert an inhibitory effect on the GnRH neuron, pituitary gonadotroph and the gonads, and render target tissues of gonadal steroids resistant to these hormones. Suppression of gonadal function caused by chronic activation of the HPA axis has been demonstrated in highly trained runners of either sex and ballet dancers.[14,16,20–22]

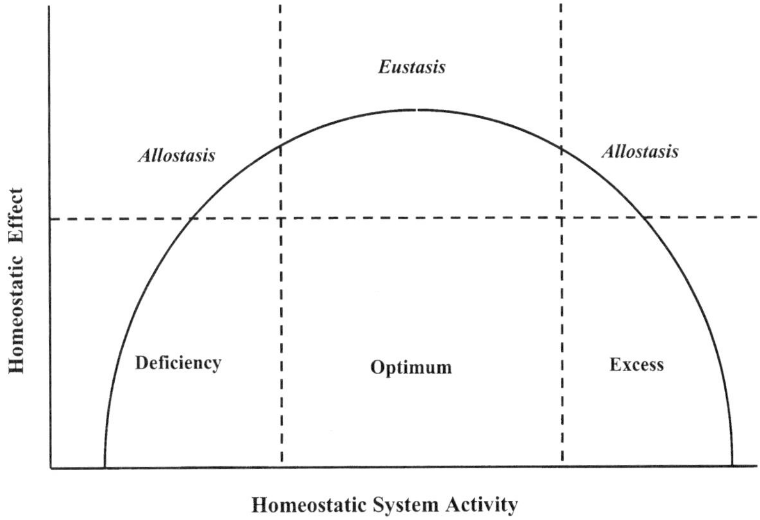

FIGURE 1. Homeostatic regulation curves. A homeostatic system, such as that between glucocorticoids and a glucocorticoid response, has an inverse U-shaped dose-response curve relation, in which there is an optimum effect range in the middle and suboptimal actions in the two sides. A homeostatic system is in *eustasis* when it operates optimally and in *allostasis* when it operates suboptimally. Allostasis signifies failure of attaining homeostasis and survival at the expense of damage to the organism.

Chronic activation of the HPA axis leads to profound *metabolic* derangements, not only because glucocorticoids inhibit the secretion of GH and sex steroids, but also because they antagonize the actions of these hormones on fat tissue catabolism (lipolysis) and muscle and bone anabolism. Increased visceral adiposity, decreased lean body (bone and muscle) mass, and osteopenia/osteoporosis are manifestations of Cushing's syndrome, melancholic depression and pseudo-Cushing's syndrome (obesity, depression, and/or alcoholism associated with hypercortisolism).[16,20–23]

Finally, increased HPA axis activity results in inhibition of the *immune/inflammatory* response, given that almost all components of the immune response are inhibited by glucocorticoids. Glucocorticoids modulate genes involved in the regulation of the innate immune response, while their actions on the adaptive immune response are to suppress cellular [T helper (Th)1-directed] immunity and to promote humoral (Th2-directed) immunity.Although glucocorticoids increase susceptibility to opportunistic infections, they are also beneficial in the presence of serious systemic inflammation, such as that observed in the acute respiratory distress syndrome and/or septic shock, when administered in a sustained fashion throughout the course of the disease. These effects are exerted both at the resting, basal state and during inflammatory stress. A circadian pattern of several immune factors exists in reverse-phase synchrony with that of plasma glucocorticoid concentrations, and these factors are suppressed during stress when glucocorticoids are elevated.[14,16,20–25]

Decreased activity of the HPA axis, on the other hand, is characterized by chronically reduced secretion of CRH and may result in hypoarousal states (TABLE 1). Pa-

TABLE 1. States associated with altered hypothalamic-pituitary-adrenal (HPA) axis activity and altered regulation or dysregulation of behavioral and/or peripheral adaptation

Increased HPA axis activity	Decreased HPA axis activity
Chronic stress	Adrenal insufficiency
Melancholic depression	Atypical/seasonal depression
Anorexia nervosa	Chronic fatigue syndrome
Malnutrition	Fibromyalgia
Obsessive-compulsive disorder	Hypothyroidism
Panic disorder	Nicotine withdrawal
Excessive exercise (obligate athleticism)	Post discontinuation of glucocorticoid therapy
Chronic active alcoholism	Following cure of Cushing's syndrome
Alcohol and narcotic withdrawal	Premenstrual tension syndrome
Diabetes mellitus	Postpartum period
Truncal obesity (metabolic syndrome X)	Following chronic stress
Childhood sexual abuse	Rheumatoid arthritis
Psychosocial short stature	Perimenopause
Attachment disorder of infancy	
"Functional" gastrointestinal disease	
Hyperthyroidism	
Cushing syndrome	
Pregnancy (last trimester)	

Adapted from Chrousos and Gold.[20]

tients with atypical or seasonal depression and the chronic fatigue syndrome demonstrate chronic hypoactivity of the HPA axis in the depressive state of the former and in the period of fatigue of the latter.[14,16,20–22] Similarly, patients with fibromyalgia often complain about fatigue and have decreased 24-h urinary cortisol excretion.[14,16,20–22] Hypothyroid patients have evidence of CRH hyposecretion and they often present with depression of the atypical type. In Cushing's syndrome, the clinical manifestations of atypical depression, hyperphagia, weight gain, fatigue, and anergia are consistent with the suppression of CRH by the elevated cortisol concentrations. The postpartum period and periods following cessation of chronic stress are also associated with transient suppression of CRH secretion and decreased HPA axis activity. Hypoactivity of the HPA axis and a defective response to inflammatory stimuli may lead to relative resistance to infections and neoplastic diseases, but increased susceptibility to autoimmune inflammatory conditions.[14,16,20–25]

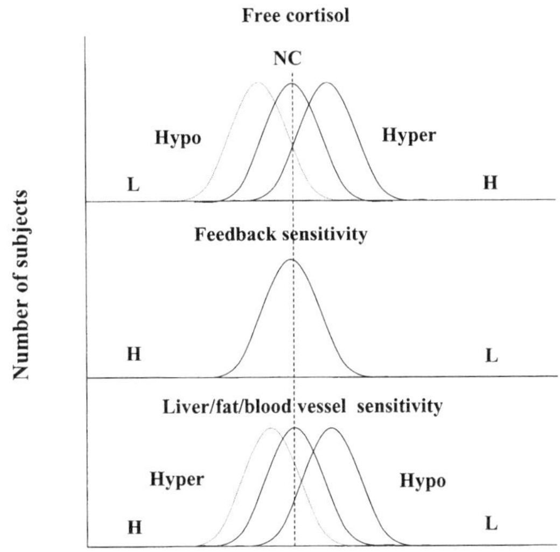

FIGURE 2. The concept of physiologic discrepancies between ubiquitous homeostatic systems, such as those of glucocorticoids, producing increased or decreased risk for certain pathologic manifestations. **Upper panel:** Cortisol production represented by urinary free cortisol (UFC) excretion; **middle panel:** negative feedback effect of cortisol at the suprahypothalamic, hypothalamic, and pituitary levels; **lower panel:** sensitivity of peripheral tissues to glucocorticoids. H, high; L, low.

TABLE 2. Expected clinical manifestations in tissue hypersensitivity or resistance to glucocorticoids

Affected Tissue	Glucocorticoid Hypersensitivity	Glucocorticoid Resistance
Central nervous system	Insomnia, anxiety, depression, impaired cognition	Fatigue, malaise, somnolence, impaired cognition
Liver	Increased gluconeogenesis, liposynthesis	Hypoglycemia
Fat	Accumulation of visceral fat	Weight loss
Blood vessels	Hypertension	Hypotension
Bone	Growth deceleration, osteoporosis	
Inflammation/Immunity	Immune suppression, anti-inflammation, susceptibility to certain infections/tumors	Increased inflammation, autoimmunity

Adapted from Chrousos *et al.*[31]

In addition to alterations in the activity of the HPA axis, changes in the sensitivity of peripheral tissues to glucocorticoids may lead to disease. Generalized glucocorticoid resistance and hypersensitivity states represent diametrically opposite examples of this process. Generalized *glucocorticoid resistance* is characterized by partial target tissue insensitivity to glucocorticoids. Compensatory elevations in circulating ACTH concentrations lead to increased production of adrenal steroids with mineralocorticoid and/or androgenic activity, as well as increased urinary free cortisol excretion in the absence of features suggestive of hypercortisolism.[26–30] On the other hand, generalized *glucocorticoid hypersensitivity* manifests with symptoms and signs suggestive of Cushing's syndrome, while biochemical and endocrine evaluation reveal hypocortisolism and compensatory decreases in HPA axis activity.[27,28,30]

Discordance between the activity of the HPA axis, the negative glucocorticoid feedback at the suprahypothalamic, hypothalamic, and pituitary levels, and the sensitivity of peripheral target tissues to glucocorticoids may be present in normal subjects (FIG. 2). Within a normal population, a gaussian curve describes daily cortisol production, represented by urinary free cortisol (UFC) excretion (FIG. 2, upper panel), the negative feedback effect (FIG. 2, middle panel), and the sensitivity of peripheral tissues to glucocorticoids (FIG. 2, lower panel). These three functions may be discrepant from each other because of genetic, developmental, or environmental causes. Decreased or increased cortisol production against a normal negative feedback effect and normal target tissue sensitivity will produce "allostatic" effects in the organism compatible with hypo- or hypercortisolism, respectively. On the other hand, decreased or increased sensitivity of target tissues to glucocorticoids in the face of normal cortisol production as a result of an appropriately functioning negative feedback system will also produce manifestations of hypo- or hypercortisolism from that target tissue. A highly increased risk for metabolic syndrome manifestations and arterial hypertension would be expected with increased activity of the HPA axis, in association with increased sensitivity of peripheral tissues to cortisol, a combination that is adverse but biologically possible. TABLE 2 summarizes the expected clinical manifestations in tissue hypersensitivity and resistance to glucocorticoids.[31]

Recent advances have allowed us to gain further insight into the mechanisms of glucocorticoid action and to embrace new concepts that form the basis of a clearer understanding of homeostasis and its disturbances throughout the human lifespan. While the study of functional defects of the glucocorticoid receptor sheds light on the molecular mechanisms of glucocorticoid action, the interplay of HPA axis activity and peripheral tissue sensitivity to glucocorticoids in defining long-term physical and/or psychological morbidity highlights the importance of integrated cellular and molecular signaling mechanisms for maintaining homeostasis and preserving normal physiology.

REFERENCES

1. SIMPSON, E.R. & M.R. WATERMAN. 1995. Steroid biosynthesis in the adrenal cortex and its regulation by adrenocorticotropin. *In* Endocrinology. L.J. DeGroot, M. Besser, H.G. Burger, *et al.*, Eds.: 1630–1641. W.B. Saunders. Philadelphia.
2. LAZAR, M.A. 2003. Mechanism of action of hormones that act on nuclear receptors. *In* Williams' Textbook of Endocrinology. P.R. Larsen, H.M. Kronenberg, S. Melmed & K.S. Polonsky, Eds.: 35–44. W.B. Saunders. Philadelphia.
3. CHROUSOS, G.P. 2001. Glucocorticoid therapy. *In* Endocrinology and Metabolism. 4th edit. P. Felig & L. Frohman, Eds.: 609–632. McGraw-Hill. New York.

4. ENCIO, I.J. & S.D. DETERA-WADLEIGH. 1991. The genomic structure of the human glucocorticoid receptor. J. Biol. Chem. **266:** 7182–7188.
5. CARSON-JURICA, M.A., W.T. SCHRADER & B.W. O'MALLEY. 1990. Steroid receptor family: structure and functions. Endocr. Rev. **11:** 201–220.
6. HOLLENBERG, S.M. & R.M. EVANS. 1988. Multiple and cooperative trans-activation domains of the human glucocorticoid receptor. Cell **55:** 899–906.
7. OAKLEY, R.H., M. SAR & J.A. CIDLOWSKI. 1996. The human glucocorticoid receptor beta isoform. Expression, biochemical properties, and putative function. J. Biol. Chem. **271:** 9550–9559.
8. OAKLEY, R.H., C.M. JEWELL, M.R. YUDT, *et al.* 1999. The dominant negative activity of the human glucocorticoid receptor beta isoform. Specificity and mechanisms of action. J. Biol. Chem. **274:** 27857–27866.
9. BAMBERGER, C.M., H.M. SCHULTE & G.P. CHROUSOS. 1996. Molecular determinants of glucocorticoid receptor function and tissue sensitivity to glucocorticoids. Endocr. Rev. **17:** 245 –261.
10. JONAT, C., H.J. RAHMSDORF, K.K. PARK, *et al.* 1990. Antitumor promotion and antiinflammation: down-modulation of AP-1 (Fos/Jun) activity by glucocorticoid hormone. Cell **62:** 1189–1204.
11. SCHEINMAN, R.I., A. GUALBERTO, C.M. JEWELL, *et al.* 1995. Characterization of mechanisms involved in transrepression of NF-kappa B by activated glucocorticoid receptors. Mol. Cell. Biol. **15:** 943–953.
12. MCKENNA, N.J., J. XU, Z. NAWAZ, *et al.* 1999. Nuclear receptor coactivators: multiple enzymes, multiple complexes, multiple functions. J. Steroid Biochem. Mol. Biol. **69:** 3–12.
13. MCKENNA, N.J. & B.W. O'MALLEY. 2002. Combinatorial control of gene expression by nuclear receptors and coregulators. Cell **108:** 465–474.
14. CHROUSOS, G.P. 1998. Stressors, stress and neuroendocrine integration of the adaptive response: The 1997 Hans Selye Memorial Lecture. Ann. N.Y. Acad. Sci. **851:** 311–335.
15. CHROUSOS, G.P. 2000. The stress response and immune function: clinical implications. The 1999 Novera H. Spector Lecture. Ann. N.Y. Acad. Sci. **917:** 38–67.
16. CHARMANDARI, E., T. KINO, E. SOUVATZOGLOU & G.P. CHROUSOS. 2003. Pediatric stress: hormonal mediators and human development. Horm. Res. **59:** 161–179.
17. CHROUSOS, G.P. 2000. The role of stress and the hypothalamic-pituitary-adrenal axis in the pathogenesis of the metabolic syndrome: neuro-endocrine and target tissue-related causes. Int. J. Obesity **24:** S50–S55.
18. GOLD, P.W. & G.P. CHROUSOS. 2002. Organization of the stress system and its dysregulation in melancholic and atypical depression: high vs low CRH/NE states. Mol. Psychiatry **7:** 254–275.
19. MCEWEN, B.S. 1998. Protective and damaging effects of stress mediators. N. Engl. J. Med. **338:** 171–179.
20. CHROUSOS, G.P. & P.W. GOLD. 1992. The concepts of stress and stress system disorders. Overview of physical and behavioral homeostasis. JAMA **267:** 1244–1252. [Erratum in JAMA **268:** 200.]
21. CHROUSOS, G.P. 1996. Organization and integration of the endocrine system. *In* Pediatric Endocrinology. M. Sperling, Ed.: 1–14. W.B. Saunders. Philadelphia.
22. HABIB, K.E., P.W. GOLD & G.P. CHROUSOS. 2001. Neuroendocrinology of stress. Endocrinol. Metab. Clin. North Am. **30:** 695–728; vii–viii.
23. ELENKOV, I.J., E.L. WEBSTER, D.J. TORPY & G.P. CHROUSOS. 1999. Stress, corticotropin-releasing hormone, glucocorticoids, and the immune/inflammatory response: acute and chronic effects. Ann. N.Y. Acad. Sci. **876:** 1–11; discussion: 11–13.
24. CHROUSOS, G.P. 2000. Stress, chronic inflammation, and emotional and physical well-being: concurrent effects and chronic sequelae. J. Allergy Clin. Immunol. **106:** S275–291.
25. FRANCHIMONT, D., T. KINO, J. GALON, *et al.* 2003. Glucocorticoids and inflammation revisited. The state-of-the-art. NIH Clinical Staff Conference. NeuroImmunoModulation **10:** 247–260.
26. CHROUSOS, G.P., S.D. DETERA-WADLEIGH & M. KARL. 1993. Syndromes of glucocorticoid resistance. Ann. Intern. Med. **119:** 1113–1124.

27. CHROUSOS, G.P., M. CASTRO, D.Y. LEUNG, *et al.* 1996. Molecular mechanisms of glucocorticoid resistance/hypersensitivity. Potential clinical implications. Am. J. Respir. Crit. Care Med. **154:** S39–43; discussion: S43–44.
28. KINO, T. & G.P. CHROUSOS. 2001. Glucocorticoid and mineralocorticoid resistance/hypersensitivity syndromes. J. Endocrinol. **169:** 437–745.
29. KINO, T., A. VOTTERO, E. CHARMANDARI & G.P. CHROUSOS. 2002. Familial/sporadic glucocorticoid resistance syndrome and hypertension. Ann. N.Y. Acad. Sci. **970:** 101–111.
30. KINO, T., M.U. DE MARTINO, E. CHARMANDARI, *et al.* 2003. Tissue glucocorticoid resistance/hypersensitivity syndromes. J. Steroid Biochem. Mol. Biol. **85:** 457–467.
31. CHROUSOS, G.P., E. CHARMANDARI & T. KINO. 2004. Glucocorticoid action networks—an introduction to systems biology. J. Clin. Endocrinol. Metab. **89:** 563–564.

Novel Repression of the Glucocorticoid Receptor by Anthrax Lethal Toxin

JEANETTE I. WEBSTER,[a] MAHTAB MOAYERI,[b] AND ESTHER M. STERNBERG[a]

[a]Section on Neuroendocrine Immunology and Behavior, National Institute of Mental Health, and [b]National Institute of Autoimmune and Infectious Diseases, National Institutes of Health, Department of Health and Human Services, Bethesda, Maryland 20892, USA

ABSTRACT: Death from anthrax has been reported to occur from systemic shock. The lethal toxin (LeTx) is the major effector of anthrax mortality. Although the mechanism of entry of this toxin into cells is well understood, its actions once inside the cell are not as well understood. LeTx is known to cleave and inactivate MAPKKs. We have recently shown that LeTx represses the glucocorticoid receptor (GR) both *in vitro* and *in vivo*. This repression is partial and specific, repressing the glucocorticoid, progesterone, and estrogen receptor α, but not the mineralocorticoid or estrogen receptor β. This toxin does not affect GR ligand or DNA binding, and we have suggested that it may function by removing/inactivating one or more of the many cofactors involved in nuclear hormone receptor signaling. Although the precise involvement of this nuclear hormone receptor repression in LeTx toxicity is unknown, examples of blunted HPA axis and glucocorticoid signaling in numerous autoimmune/inflammatory diseases suggest that such repression of critically important receptors could have deleterious effects on health.

KEYWORDS: GR; anthrax lethal toxin; HPA axis

ANTHRAX LETHAL TOXIN

Bacillus anthracis is a spore-forming gram-positive bacterium that contains two plasmids (for a review on the bacterium, please refer to the recent review by Mock and Fouet[1]). One of these plasmids, pXO1, encodes the three proteins that comprise two toxins. The proteins form a variation of the classical A-B toxins, with protective antigen (PA) and edema factor (EF) constituting the edema toxin, and PA and lethal factor (LF) the lethal toxin (LeTx).[2–6] *B. anthracis* strains that lack the pXO1 plasmid are avirulent, and the LeTX alone has been shown to manifest many of the symptoms of *B. anthracis* infection.[7–9] Thus, much research has focused on LeTx and its mechanism of action, and this toxin will also be the focus of this review.

The mechanism of LeTx entry into cells is now better understood[10] and is summarized in FIGURE 1. Macrophages have been suggested to be the target of LeTx ac-

Address for correspondence: Esther M. Sternberg, M.D., Section on Neuroendocrine Immunology and Behavior, NIMH, NIH, Bldg. 36, Rm. 1A23, 36 Convent Drive, MSC 4020, Bethesda, MD 20892-4020.Voice: +1 301-402-2773; fax: +1 301-496-6095.
sternbee@mail.nih.gov

Ann. N.Y. Acad. Sci. 1024: 9–23 (2004).
doi: 10.1196/annals.1321.003

tion as they are the only cell type that has been shown to be sensitive to rapid lysis by LeTx.[18] Macrophages from different mouse strains and various cell lines have been shown to have differential sensitivity to LeTx.[19–21] Sublytic concentrations of LeTx produce apoptosis of macrophages.[22,23] Human peripheral blood mononuclear cells (PBMC) are not lysed by LeTx but show similar pro-apoptotic behavior as murine cells.[24] Although recent data has suggested that macrophage lysis is not essential for LeTx toxicity, it appears to potentially contribute to exacerbation of the toxin effects in mice.[25] Other recent discoveries showing that resistant C57BL/6J macrophages or human macrophages can be made sensitive to LeTx by treatment with poly-D-glutamic acid (the major component of the *B. anthracis* capsule), peptidoglycan (a component of gram-positive bacterial cell walls), lipopolysaccharide (LPS; a component of gram-negative bacterial cell walls), or tumor necrosis factor-α (TNFα),[24,26] suggest a more complicated involvement of macrophage sensitivity in the pathogenesis of anthrax.

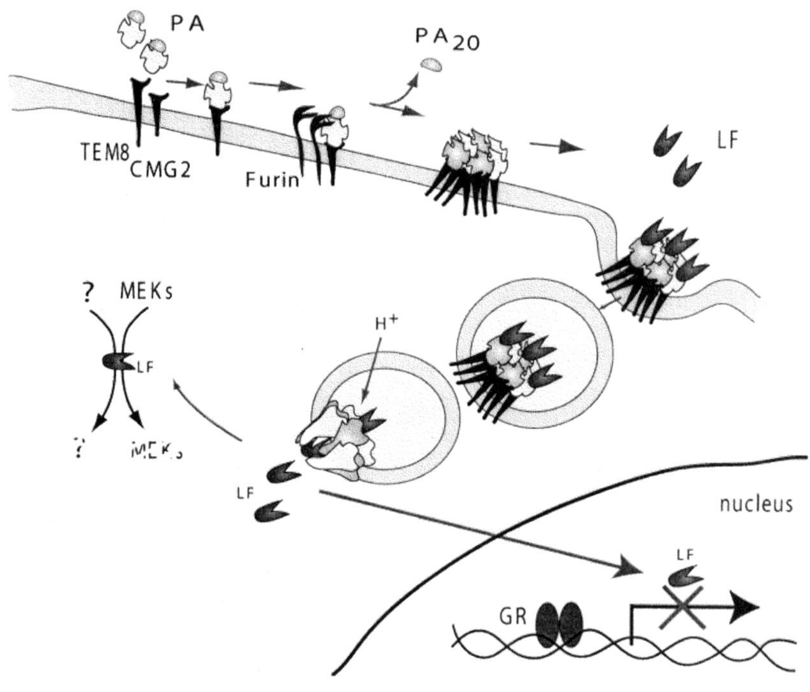

FIGURE 1. Mechanism of action of anthrax lethal toxin. PA binds to two different receptors, tumor endothelial marker 8 (TEM8) and capillary morphogenesis protein 2 (CMG2),[11,12] which seem to be ubiquitously expressed. PA is then cleaved by the enzyme furin,[13] heptermizes, and then binds EF and/or LF.[14] This complex is then internalized by clathrin-dependent raft-mediated endocytosis,[10] and the LF/EF are translocated across the endosomal membrane and into the cytosol via a pH- and voltage-dependent mechanism.[15–17] Once inside the cell, the mechanism of action of LF is less well understood. It is known to cleave and inactivate members of the MAPKK family, and we now show that it can inactivate the glucocorticoid receptor.

LF is a zinc metalloprotease that is known to cleave and inactivate some members of the mitogen activation protein (MAP) kinase kinases (MAPKK/MEK) family.[23,27–32] However, LF internalization[20,21,33] and MAPKK cleavage in sensitive and resistant macrophages and cell lines[29,30,34] are the same. Inhibition of the proteolytic LF function prevents LeTx toxicity in sensitive cells[27,31–33,35] suggesting that cleavage of MAPKKs or other potentially unidentified substrates are needed for LeTx toxicity. MAPKK cleavage alone may be sufficient for induction of the cascade of events leading to macrophage lysis; the response to cleavage, however, in different cells may define differential sensitivity to LeTx. One factor involved in this response is likely the kinesin Kif1C which has been identified as the macrophage sensitivity locus for LeTx.[21,34,36] The function of this kinesin is unknown, although it has been suggested to be involved in endoplasmic reticulum (ER) transport.[37] Recently, another group has shown that LeTx macrophage sensitivity is linked to two additional loci, Ltxs2 and Ltxs3.[38] Inhibition of the proteosome,[39] intracellular calcium release,[40,41] phospholipase C, and protein kinase C,[42] or reactive oxygen intermediates[43] have also all been shown to protect sensitive macrophages against LeTx toxicity, suggesting that these pathways may be involved in LeTx toxicity.

Death from anthrax LeTx has been reported to occur from systemic shock[44] and has been suggested to resemble cytokine-mediated LPS-induced shock.[45–47] However, the role of cytokines in this has been a matter of some debate. LeTx inhibits LPS/IFNγ stimulation of TNFα and NO in both sensitive and resistant macrophage cell lines probably through its inactivation of the MAPKK pathways.[29,30,48] LeTx (1 pg/mL for 16 h) alone in sensitive macrophages did not induce TNFα, IL-6, IL-1α, or IL-1β.[48] However, others have shown that sublytic LeTx (1 pg/mL for 6 h) concentrations alone were able to induce TNFα and IL-1β in a LeTx-sensitive macrophage cell line[47] and induce TNFα in macrophages from ICR mice.[49] A recent analysis of over 40 cytokines and inflammatory mediators in BALB/cJ and C57BL/6J mice showed an early transitory increase in numerous factors in BALB/cJ, but not C57BL/6J, mice. However, this did not result in an inflammatory cascade and no TNFα induction was seen.[25] LeTx also repressed LPS-induced cytokine (TNFα, IL-1α, IL-6, and IL-12) production in dendritic cells[50] and anthrax cell wall–induced cytokine release in PBMCs.[24] Additionally, TNF or IL-1 receptor or iNOS knockout mice did not differ in susceptibility to anthrax infection compared to control mice.[51] Although the data regarding cytokines is complicated and different groups have shown different results, there does seem to be a consensus that LeTx can inhibit cytokine release probably through inhibition of MAPKK and that it is likely that inflammatory cytokine release is not critical for LeTx lethality.

Rodent models have been used in the study of *B. anthracis* and LeTx. Different strains show differential susceptibility to both the *B. anthracis* bacterium[52] and the toxin.[21,25] BALB/c mice are sensitive[21,47] to LeTx whereas C57/6J are relatively resistant.[21,25] It has also been long known that Fischer (F344/N) rats are highly sensitive to LeTx,[53,54] with death occurring within 40 min from a lethal dose[55] that does not affect mice over 72 h.[25] F344/N rats have also been used for many years for their inflammatory resistance in comparison with their histocompatible counterparts, Lewis (LEW/N) rats, which are highly susceptible to inflammatory disease. This inflammatory sensitivity/resistance has been shown to be due to the differential hypothalamic-pituitary-adrenal (HPA) axis is these animals.

HYPOTHALAMIC-PITUITARY-ADRENAL AXIS AND GLUCOCORTICOID RESPONSES

The HPA axis and resultant glucocorticoid responses are required to maintain a balance within the body between the brain and immune systems.[56] Inflammatory or other stimuli activate the expression of corticotrophin-releasing hormone (CRH) in the hypothalamic region of the brain. This, in turn, stimulates the release of adrenocorticotropin hormone (ACTH) from the anterior pituitary gland into the blood stream. At the adrenals, ACTH stimulates the synthesis and release of glucocorticoids. Glucocorticoids feed back at the level of the hypothalamus and pituitary to downregulate the HPA axis. In man, the endogenous glucocorticoid is cortisol whereas in rodents it is corticosterone. Glucocorticoids are required not only to regulate the immune system, but are also essential for the regulation of several homeostatic mechanisms in the body, including the central nervous system, cardiovascular system, and metabolic homeostasis. Glucocorticoid regulation of the immune system will not be discussed here in detail but has been the subject of another recent review.[56]

Glucocorticoids exert their many functions through a cytosolic receptor, the glucocorticoid receptor (GR). This receptor is a member of the nuclear hormone receptor superfamily that also includes the thyroid-hormone, mineralocorticoid, estrogen, and progesterone receptors.[57] In the case of GR, in the absence of ligand, the receptor is located in the cytoplasm in a complex of proteins including Hsp90 and Hsp70. Upon ligand binding, the receptor is released, dimerizes, and translocates to the nucleus where it binds to specific DNA sequences—glucocorticoid response elements (GRE)—in order to regulate gene expression.[58] Through direct DNA binding, GR is able to upregulate gene expression, such as for the gluconeogenic enzyme tyrosine animotransferase (TAT),[59] but it can also bind to DNA sequences called negative GREs (nGRE) to repress gene activation, such as the POMC gene.[60] However, GR can also function by interfering with the action of other signaling pathways such as NFκB and AP-1, and it is through this action that glucocorticoids exert many of their anti-inflammatory actions.[61,62] Lack of GR is incompatible with life, and GR knockout mice die shortly after birth due to a defect in lung maturation.[63] However, dimerization knockout mice ($GR^{dim/dim}$), in which GRE-mediated gene activation is removed but GR interactions with NFκB and AP-1 are still possible, are viable suggesting that it is these anti-inflammatory actions of GR through protein–protein interactions rather than direct DNA binding interactions that are essential for life.[64]

F334/N rats have a hyper-HPA axis with hypersecretion of CRH, ACTH, and corticosterone in response to stimuli, thereby allowing corticosterone to suppress the immune system and resulting in their inflammatory resistant state. Conversely, LEW/N rats have a blunted HPA axis with minimal production of CRH, ACTH, and corticosterone in response to stimuli. If those stimuli are a pro-inflammatory or antigenic molecules, then the immune system will not be downregulated due to the lack of corticosterone; therefore these animals are susceptible to development of a variety of autoimmune/inflammatory diseases.[65–68] Interestingly, BALB/c mice have been shown to also have a hyper-HPA axis and C57/BJ a relative blunted HPA axis response similar to the F334/N and LEW/N rats.[69]

ANTHRAX LETHAL TOXIN REPRESSION OF GR

The observations that normally inflammatory resistant F334/N rats exhibit a severe inflammatory response resulting in death if they are simultaneously treated with the GR antagonist RU486 and streptococcal cell walls (SCW)[66] and that they are highly susceptible to death from anthrax LeTx[55] led to the hypothesis that LeTx may function as a GR antagonist in a similar manner to RU486. We have recently shown that LeTx does indeed block transactivation of dexamethasone-induced GR in both a transient transfection system, where the ability of GR to activate a reporter gene was repressed by 50%, and in cells endogenously expressing GR, where the ability of dexamethasone to induce the tyrosine aminotransferase gene was also repressed 50%. Further, we showed that in BALB/c mice the ability of dexamethasone to induce liver tyrosine aminotransferase *in vivo* was also repressed 50% by LeTx (FIG. 1). This LeTx-mediated repression of GR gene activation does not resemble a true GR antagonist, such as RU486, because it does not prevent ligand binding to the receptor.[70]

The mechanism by which LeTx is able to repress GR is novel and is still under investigation. We have shown that LeTx does not interfere with the interaction of GR and the GRE DNA sequence, at least in an *in vitro* electrophoretic mobility shift assay (EMSA). The ability of LeTx to repress nuclear hormone receptors is not specific to GR as the progesterone receptor B (PR-B) and estrogen receptor α (ERα) can also be repressed to different extents by LeTx. However, this repression of nuclear hormone receptors is also not ubiquitous because the mineralocorticoid receptor (MR) and ERβ are not repressed by LeTx. Therefore, LeTx shows some specificity in the repression of nuclear hormone receptors, and the features that determine LeTx repression are a subject of current research. The observation that, unlike RU486, LeTx does not fully repress these receptors has led us to suggest that LeTx may be removing/inactivating one or more of the many cofactors involved in the interaction of nuclear hormone receptors and the transcriptional machinery.[70] One might envisage that removal of one or more of these multiple pathways may result in partial repression as some activity could still be afforded through the remaining intact pathways.

P38 MAPK AND GR

Since LeTx has been shown to cleave and inactivate members of the MAPKK family,[23,27–32] we tested other inhibitors of MAPKKs on dexamethasone induction of GR transactivation. We have shown that inhibitors of the p38 MAPKK pathway, but not inhibitors of the other MAPKK pathways, were able to specifically repress dexamethasone-induced GR transactivation.[70] Whether LeTx repression of GR is a direct consequence of p38 MAPKK inactivation or is purely a correlative effect remains to be determined. However, if LeTx repression of GR is mediated through its inactivation of p38 MAPKK then this is a novel mechanism of action because published literature shows that activation of MAPKK pathways inactivates GR, not repression of basal MAPKK.[71–78] However, a few studies are emerging that suggest an interaction between p38 and GR. In one study, the p38 inhibitor SB203580 prevented dexamethasone induction of Hsp27 in osteoclasts, suggesting that p38 is involved in the dexamethasone induction of Hsp27.[79] In another system, dexamethasone was shown to rapidly activate p38 in a GR-independent mechanism.[80] In the absence of the coactivator PGC-1, the p38 inhibitor had no effect of GR-mediated transactivation in HeLa cells, suggesting that p38 does not act directly on GR. However, in the presence of PGC-1, coexpression of constitutively active MKK6 enhanced GR transactivation and this enhanced GR transactivation could be repressed by SB203580.[81] The extent of the involvement of p38 in LeTx repression of GR is currently under investigation.

PROTECTIVE FEATURES OF AN INTACT HPA AXIS AND GLUCOCORTICOID RESPONSE

The importance of an intact HPA axis and resultant glucocorticoid responses in maintaining body homeostasis and protecting against insults of a variety of sources has been shown in several systems; it has been shown that lack of GR is incompatible with life.[63] GR knockout mice die shortly after birth due to defects in lung maturation, and to date no patient lacking GR has ever been described. However, an animal can be adrenalectomized, thereby removing the endogenous production of glucocorticoids, and can survive if the appropriate hormones are replaced exogenously.

Removal of endogenous glucocorticoids either by adrenalecomy or hypophysectomy or removal of functional GR by the antagonist RU486 has been shown to exacerbate disease course even to the extent of death in response to numerous bacterial or viral infections. Conversely, corticosterone replacement promotes survival and disease remittance. This was shown first in mice where adrenalectomy significantly reduced the lethal amount (LD_{50}) of *Escherichia coli* serotype O111:B4 endotoxin.[82] Treatment with RU486 and SCW resulted in higher mortality rates in F334/N rats than did SCW or RU486 alone.[66] Similarly, adrenalectomy exacerbated myelin basic protein (MBP)-induced experimental allergic encephalomyelitis (EAE) in Lewis rats and replacement of corticosterone either returned to disease status to control animals or even alleviated symptoms depending on dose used.[83] Furthermore, intervention of the HPA axis by hypophysectomy resulted in increased mortality rates from Salmonella,[84] and adrenaectomy resulted in increased mortality rates from MCMV virus, which was reversed by dexamethasone treatment.[85]

More recently, RU486 and adrenalectomy enhanced *Clostridium difficile* toxin A–induced fluid secretion and inflammation[86,87] and enhanced mortality and disease symptoms from Shiga toxin in BALB/c mice.[88] Dexamethasone treatment increased survival rates of adrenalectomized animals from Shiga toxin[88,89] and dexamethasone administration or corticosterone replacement reversed the *Clostridium difficile* toxin A–enhanced inflammatory responses in adrenalectomized animals.[86,87] In fact, replacement with a physiological corticosterone dose resulted in an inflammatory response equivalent to sham-operated animals while replacement of a high corticosterone dose resulted in a reduction of the inflammatory response.[87]

DIMINISHED HPA AXIS AND GR RESPONSES IN DISEASE

A blunted HPA axis response—that is, blunted glucocorticoid secretion in response to stimuli—has been associated with numerous autoimmune/inflammatory diseases in both animal models and humans. In animals, a blunted HPA axis has been associated with autoimmune thyroiditis in chickens,[90] lupus in mice,[91,92] and numerous autoimmune/inflammatory diseases in rats.[66–68] In humans, a blunted HPA axis response has been associated with rheumatoid arthritis,[93–98] systemic lupus erythematosus (SLE),[99,100] Sjögren's syndrome,[100,101] allergic asthma and atopic skin disease,[102–106] chronic fatigue syndrome,[107–110] fibromyalgia,[107,109,111] and multiple sclerosis.[112,113]

Glucocorticoid resistance and/or diminished function of GR have been associated with numerous diseases. Familial glucocorticoid resistance is a hereditary disease usually involving mutations in the GR. To date, three point mutations and a deletion in the ligand-binding domain of GR and a point mutation in the hinge region have been identified in families with familiar glucocorticoid resistance.[114] Polymorphisms in GR could also cause changes in glucocorticoid sensitivity, as has been shown for a polymorphism in codon 363.[115] However, no relationship between this and other polymorphisms and glucocorticoid resistance was seen in a normal population.[116] For a recent update on all known GR mutations and polymorphisms in both patients and cell lines, see the recent review by Bray and Cotton.[117] A splice variant of the glucocorticoid receptor, GRβ has been suggested to function as a dominant negative repressor of GR,[118–120] although there is some evidence that this is not the case.[121–123] However, an increased expression of GRβ relative to GRα has been shown in several autoimmune/inflammatory diseases including glucocorticoid resistant asthma,[124–127] ulcerative colitis,[128,129] chronic lymphocytic leukemia,[130] nasal polyposis disease,[131] and rheumatoid arthritis.[132,133] Decreased GR numbers have been associated with various diseases, including Crohn's disease[134] and rheumatoid arthritis.[135]

Glucocorticoid insensitivity need not result from a mutation in the glucocorticoid receptor itself as there are multiple steps in glucocorticoid signaling at which a problem could arise resulting in glucocorticoid insensitivity. These include transport of the hormone in the blood, availability of the hormone, entry of the hormone into the cell, dissociation from the heat shock protein complex, dimerization, and translocation to the nucleus. Increased expression of cortisol binding globulin (CBG) would decrease the bioavailability of cortisol. This has been suggested to cause the glucocorticoid resistance seen in patients with long-standing Crohn's disease.[136] Changes in the enzyme 11β-dehydroxysteroid dehydrogenase (11β-HSD) would cause changes in the ratio of glucocorticoids in the active versus inactive state.[137] Furthermore, impairment of this enzyme has been shown in obese men,[138] and decreased 11β-HSD mRNA is seen in ulcerative colitis.[139] Multidrug resistance proteins (MDR) are members of the ABC family of transporters that have been shown to be capable of exporting glucocorticoids from cells and thereby regulating the intracellular hormonal concentration.[140–142] Overexpression of MDR1 has been shown in glucocorticoid resistant inflammatory bowel disease,[143] rheumatoid arthritis,[144] and systemic lupus erythematosus.[145] In addition, defects in the cofactors involved in the interaction between the glucocorticoid receptor and the transcriptional machinery may cause changes in glucocorticoid resistance. The HIV protein, virion-associated protein (Vpr), functions as a cofactor to enhance glucocorticoid responses resulting in the HIV-associated glucocorticoid hypersensitive state.[146] Conversely, a defect in a cofactor has been proposed to be the cause of resistance to multiple steroids in two sisters.[147,148]

THERAPEUTIC IMPLICATIONS

Reduced glucocorticoid production as a result of a blunted HPA axis response or diminished glucocorticoid sensitivity has been associated with numerous autoimmune/inflammatory diseases, as described above. Thus, although we have yet to show the extent of the involvement of LeTx repression of nuclear hormone receptors in LeTx toxicity, it is reasonable to predict that there may be potential clinical therapeutic implications of LeTx repression of GR. Thus, identification of the precise mechanism by which LeTx represses GR and other nuclear hormone receptors could provide novel targets for therapy of anthrax and possibly other bacterial toxins.

REFERENCES

1. MOCK, M. & A. FOUET. 2001. Anthrax. Annu. Rev. Microbiol. **55:** 647–671.
2. BROSSIER, F. & M. MOCK. 2001. Toxins of *Bacillus anthracis*. Toxicon **39:** 1747–1755.
3. BHATNAGAR, R. & S. BATRA. 2001. Anthrax toxin. Crit. Rev. Microbiol. **27:** 167–200.
4. LEPPLA, S.H. 2000. Anthrax toxin. *In* Bacterial Protein Toxins. K. Aktories & I. Just, Eds.: 445–472. Springer. Berlin.
5. LEPPLA, S.H. 1999. The bifactorial *Bacillus anthracis* lethal and edema toxins. *In* Comprehensive Sourcebook of Bacterial Protein Toxins. J.E. Alouf & J.H. Freer, Eds: 243–263. Academic Press. London.
6. COLLIER, R.J. & J.A. YOUNG. 2003. Anthrax toxin. Annu. Rev. Cell Dev. Biol. **19:** 45–70.
7. PEZARD, C., P. BERCHE & M. MOCK. 1991. Contribution of individual toxin components to virulence of *Bacillus anthracis*. Infect. Immun. **59:** 3472–3477.
8. FISH, D.C., F. KLEIN, R.E. LINCOLN, *et al.* 1968. Pathophysiological changes in the rat associated with anthrax toxin. J. Infect. Dis. **118:** 114–124.
9. KLEIN, F., D.R. HODGES, B.G. MAHLANDT, *et al.* 1962. Anthrax toxin: causative agent in the death of rhesus monkeys. Science **138:** 1331–1333.
10. ABRAMI, L., S. LIU, P. COSSON, *et al.* 2003. Anthrax toxin triggers endocytosis of its receptor via a lipid raft-mediated clathrin-dependent process. J. Cell. Biol. **160:** 321–328.
11. BRADLEY, K.A., J. MOGRIDGE, M. MOUREZ, *et al.* 2001. Identification of the cellular receptor for anthrax toxin. Nature **414:** 225–229.
12. SCOBIE, H.M., G.J. RAINEY, K.A. BRADLEY, *et al.* 2003. Human capillary morphogenesis protein 2 functions as an anthrax toxin receptor. Proc. Natl. Acad. Sci. USA **100:** 5170–5174.
13. KLIMPEL, K.R., S.S. MOLLOY, G. THOMAS, *et al.* 1992. Anthrax toxin protective antigen is activated by a cell surface protease with the sequence specificity and catalytic properties of furin. Proc. Natl. Acad. Sci. USA **89:** 10277–10281.
14. SINGH, Y., K.R. KLIMPEL, S. GOEL, *et al.* 1999. Oligomerization of anthrax toxin protective antigen and binding of lethal factor during endocytic uptake into mammalian cells. Infect. Immun. **67:** 1853–1859.
15. BLAUSTEIN, R.O., T.M. KOEHLER, R.J. COLLIER, *et al.* 1989. Anthrax toxin: channel-forming activity of protective antigen in planar phospholipid bilayers. Proc. Natl. Acad. Sci. USA **86:** 2209–2213.
16. WESCHE, J., J.L. ELLIOTT, P.O. FALNES, *et al.* 1998. Characterization of membrane translocation by anthrax protective antigen. Biochemistry **37:** 15737–15746.
17. ZHAO, J., J.C. MILNE & R.J. COLLIER. 1995. Effect of anthrax toxin's lethal factor on ion channels formed by the protective antigen. J. Biol. Chem. **270:** 18626–18630.
18. FRIEDLANDER, A.M. 1986. Macrophages are sensitive to anthrax lethal toxin through an acid-dependent process. J. Biol. Chem. **261:** 7123–7126.
19. FRIEDLANDER, A.M., R. BHATNAGAR, S.H. LEPPLA, *et al.* 1993. Characterization of macrophage sensitivity and resistance to anthrax lethal toxin. Infect. Immun. **61:** 245–252.

20. SINGH, Y., S.H. LEPPLA, R. BHATNAGAR, *et al.* 1989. Internalization and processing of Bacillus anthracis lethal toxin by toxin-sensitive and -resistant cells. J. Biol. Chem. **264:** 11099–11102.
21. ROBERTS, J.E., J.W. WATTERS, J.D. BALLARD, *et al.* 1998. Ltx1, a mouse locus that influences the susceptibility of macrophages to cytolysis caused by intoxication with *Bacillus anthracis* lethal factor, maps to chromosome 11. Mol. Microbiol. **29:** 581–591.
22. POPOV, S.G., R. VILLASMIL, J. BERNARDI, *et al.* 2002. Lethal toxin of *Bacillus anthracis* causes apoptosis of macrophages. Biochem. Biophys. Res. Commun. **293:** 349–355.
23. PARK, J.M., F.R. GRETEN, Z.W. LI, *et al.* 2002. Macrophage apoptosis by anthrax lethal factor through p38 MAP kinase inhibition. Science **297:** 2048–2051.
24. POPOV, S.G., R. VILLASMIL, J. BERNARDI, *et al.* 2002. Effect of *Bacillus anthracis* lethal toxin on human peripheral blood mononuclear cells. FEBS Lett. **527:** 211–215.
25. MOAYERI, M., D. HAINES, H.A. YOUNG, *et al.* 2003. *Bacillus anthracis* lethal toxin induces TNF-alpha-independent hypoxia-mediated toxicity in mice. J. Clin. Invest. **112:** 670–682.
26. KIM, S.O., Q. JING, K. HOEBE, *et al.* 2003. Sensitizing anthrax lethal toxin-resistant macrophages to lethal toxin-induced killing by tumor necrosis factor-alpha. J. Biol. Chem. **278:** 7413–7421.
27. VITALE, G., R. PELLIZZARI, C. RECCHI, *et al.* 1998. Anthrax lethal factor cleaves the N-terminus of MAPKKs and induces tyrosine/threonine phosphorylation of MAPKs in cultured macrophages. Biochem. Biophys. Res. Commun. **248:** 706–711.
28. VITALE, G., L. BERNARDI, G. NAPOLITANI, *et al.* 2000. Susceptibility of mitogen-activated protein kinase kinase family members to proteolysis by anthrax lethal factor. Biochem. J. **352:** 739–745.
29. PELLIZZARI, R., C. GUIDI-RONTANI, G. VITALE, *et al.* 1999. Anthrax lethal factor cleaves MKK3 in macrophages and inhibits the LPS/IFNγ-induced release of NO and TNFα. FEBS Lett. **462:** 199–204.
30. PELLIZZARI, R., C. GUIDI-RONTANI, G. VITALE, *et al.* 2000. Lethal factor of *Bacillus anthracis* cleaves the N-terminus of MAPKKs: analysis of the intracellular consequences in macrophages. Int. J. Med. Microbiol. **290**: 421–427.
31. KLIMPEL, K.R., N. ARORA & S.H. LEPPLA. 1994. Anthrax toxin lethal factor contains a zinc metalloprotease consensus sequence which is required for lethal toxin activity. Mol. Microbiol. **13:** 1093–1100.
32. DUESBERY, N.S., C.P. WEBB, S.H. LEPPLA, *et al.* 1998. Proteolytic inactivation of MAP-kinase-kinase by anthrax lethal factor. Science **280:** 734–737.
33. MENARD, A., E. PAPINI, M. MOCK, *et al.* 1996. The cytotoxic activity of *Bacillus anthracis* lethal factor is inhibited by leukotriene A4 hydrolase and metallopeptidase inhibitors. Biochem. J. **320:** 687–691.
34. WATTERS, J.W., K. DEWAR, J. LEHOCZKY, *et al.* 2001. Kif1C, a kinesin-like motor protein, mediates mouse macrophage resistance to anthrax lethal factor. Curr. Biol. **11:** 1503–1511.
35. HAMMOND, S.E. & P.C. HANNA. 1998. Lethal factor active-site mutations affect catalytic activity in vitro. Infect. Immun. **66**: 2374–2378.
36. WATTERS, J.W. & W.F. DIETRICH. 2001. Genetic, physical, and transcript map of the Ltxs1 region of mouse chromosome 11. Genomics **73:** 223–231.
37. DORNER, C., T. CIOSSEK, S. MULLER, *et al.* 1998. Characterization of KIF1C, a new kinesin-like protein involved in vesicle transport from the Golgi apparatus to the endoplasmic reticulum. J. Biol. Chem. **273:** 20267–20275.
38. MCALLISTER, R.D., Y. SINGH, W.D. DU BOIS, *et al.* 2003. Susceptibility to anthrax lethal toxin is controlled by three linked quantitative trait loci. Am. J. Pathol. **163:** 1735–1741.
39. TANG, G. & S.H. LEPPLA. 1999. Proteosome activity is required for anthrax lethal toxin to kill macrophages. Infect. Immun. **67:** 3055–3060.
40. BHATNAGAR, R., Y. SINGH, S.H. LEPPLA, *et al.* 1989. Calcium is required for the expression of anthrax lethal toxin activity in the macrophage like cell line J774A.1. Infect. Immun. **57:** 2107–2114.
41. SHIN, S., G.H. HUR, Y.B. KIM, *et al.* 2000. Intracellular calcium antagonist protects cultured peritoneal macrophages against anthrax lethal toxin-induced cytotoxicity. Cell. Biol. Toxicol. **16:** 137–144.

42. BHATNAGAR, R., N. AHUJA, R. GOILA, *et al.* 1999. Activation of phospholipase C and protein kinase C is required for expression of anthrax lethal toxin cytotoxicity in J774A.1 cells. Cell Signal. **11:** 111–116.
43. HANNA, P.C., B.A. KRUSKAL, R.A. EZEKOWITZ, *et al.* 1994. Role of macrophage oxidative burst in the action of anthrax lethal toxin. Mol. Med. **1:** 7–18.
44. SMITH, H., J. KEPPIE, J.L. STANLEY, *et al.* 1955. The chemical basis of the virulence of *Bacillus anthracis*. IV. Secondary shock as the major factor in death of guinea-pigs from anthrax. Br. J. Exp. Pathol. **36:** 323–335.
45. HANNA, P.C. & J.A. IRELAND. 1999. Understanding *Bacillus anthracis* pathogenesis. Trends Microbiol. **7:** 180–182.
46. HANNA, P. 1999. Lethal toxin actions and their consequences. J. Appl. Microbiol. **87:** 285–287.
47. HANNA, P.C., D. ACOSTA & R.J. COLLIER. 1993. On the role of macrophages in anthrax. Proc. Natl. Acad. Sci. USA **90:** 10198–10201.
48. ERWIN, J.L., L.M. DASILVA, S. BAVARI, *et al.* 2001. Macrophage-derived cell lines do not express proinflammatory cytokines after exposure to *Bacillus anthracis* lethal toxin. Infect. Immun. **69:** 1175–1177.
49. SHIN, S., G.H. HUR, Y.B. KIM, *et al.* 2000. Dehydroepiandrosterone and melatonin prevent *Bacillus anthracis* lethal toxin-induced TNF production in macrophages. Cell. Biol. Toxicol. **16:** 165–174.
50. AGRAWAL, A., J. LINGAPPA, S.H. LEPPLA, *et al.* 2003. Impairment of dendritic cells and adaptive immunity by anthrax lethal toxin. Nature **424:** 329–334.
51. KALNS, J., J. SCRUGGS, N. MILLENBAUGH, *et al.* 2002. TNF receptor 1, IL-1 receptor, and iNOS genetic knockout mice are not protected from anthrax infection. Biochem. Biophys. Res. Commun. **292:** 41–44.
52. WELKOS, S.L., T.J. KEENER & P.H. GIBBS. 1986. Differences in susceptibility of inbred mice to *Bacillus anthracis*. Infect. Immun. **51:** 795–800.
53. BEALL, F.A., M.J. TAYLOR & C.B. THORNE. 1962. Rapid lethal effect in rats of a third component found upon fractionating the toxin of *Bacillus anthracis*. J. Bacteriol. **83:** 1274–1280.
54. KLEIN, F., B.W. HAINES, B.G. MAHLANDT, *et al.* 1963. Dual nature of resistance mechanisms as revealed by studies of anthrax stepticemia. J. Bacteriol. **85:** 1032–1038.
55. EZZELL, J.W., B.E. IVINS & S.H. LEPPLA. 1984. Immunoelectrophoretic analysis, toxicity, and kinetics of in vitro production of the protective antigen and lethal factor components of *Bacillus anthracis* toxin. Infect. Immun. **45:** 761–767.
56. WEBSTER, J.I., L. TONELLI & E.M. STERNBERG. 2002. Neuroendocrine regulation of immunity. Annu. Rev. Immunol. **20:** 125–163.
57. EVANS, R.M. 1988. The steroid and thyroid hormone receptor superfamily. Science **240:** 889–895.
58. ARANDA, A. & A. PASCUAL. 2001. Nuclear hormone receptors and gene expression. Physiol. Rev. **81:** 1269–1304.
59. JANTZEN, H.M., U. STRAHLE, B. GLOSS, *et al.* 1987. Cooperativity of glucocorticoid response elements located far upstream of the tyrosine aminotransferase gene. Cell **49:** 29–38.
60. DROUIN, J., Y.L. SUN & M. NEMER. 1989. Glucocorticoid repression of pro-opiomelanocortin gene transcription. J. Steroid Biochem. **34:** 63–69.
61. MCKAY, L.I. & J.A. CIDLOWSKI. 1999. Molecular control of immune/inflammatory responses: interactions between nuclear factor-kappa B and steroid receptor-signaling pathways. Endocr. Rev. **20:** 435–459.
62. ADCOCK, I.M. 2000. Molecular mechanisms of glucocorticosteroid actions. Pulm. Pharmacol. Ther. **13:** 115–126.
63. COLE, T.J., J.A. BLENDY, A.P. MONAGHAN, *et al.* 1995. Targeted disruption of the glucocorticoid receptor gene blocks adrenergic chromaffin cell development and severely retards lung maturation. Genes Dev. **9:** 1608–1621.
64. REICHARDT, H.M., K.H. KAESTNER, J. TUCKERMANN, *et al.* 1998. DNA binding of the glucocorticoid receptor is not essential for survival. Cell **93:** 531–541.

65. MONCEK, F., R. KVETNANSKY, D. JEZOVA, *et al.* 2001. Differential responses to stress stimuli of Lewis and Fischer rats at the pituitary and adrenocortical level. Endocr. Regul. **35:** 35–41.
66. STERNBERG, E.M., J.M. HILL, G.P. CHROUSOS, *et al.* 1989. Inflammatory mediator-induced hypothalamic-pituitary-adrenal axis activation is defective in streptococcal cell wall arthritis-susceptible Lewis rats. Proc. Natl. Acad. Sci. USA **86:** 2374–2378.
67. STERNBERG, E.M., W.S.D. YOUNG, R. BERNARDINI, *et al.* 1989. A central nervous system defect in biosynthesis of corticotropin-releasing hormone is associated with susceptibility to streptococcal cell wall-induced arthritis in Lewis rats. Proc. Natl. Acad. Sci. USA **86:** 4771–4775.
68. WILDER, R.L., G.B. CALANDRA, A.J. GARVIN, *et al.* 1982. Strain and sex variation in the susceptibility to streptococcal cell wall-induced polyarthritis in the rat. Arthritis Rheum. **25:** 1064–1072.
69. SHANKS, N., J. GRIFFITHS, S. ZALCMAN, *et al.* 1990. Mouse strain differences in plasma corticosterone following uncontrollable footshock. Pharmacol. Biochem. Behav. **36:** 515–519.
70. WEBSTER, J.I., L.H. TONELLI, M. MOAYERI, *et al.* 2003. Anthrax lethal factor represses glucocorticoid and progesterone receptor activity. Proc. Natl. Acad. Sci. USA **100:** 5706–5711.
71. KARIN, M. & L. CHANG. 2001. AP-1-glucocorticoid receptor crosstalk taken to a higher level. J. Endocrinol. **169:** 447–451.
72. IRUSEN, E., J.G. MATTHEWS, A. TAKAHASHI, *et al.* 2002. p38 Mitogen-activated protein kinase-induced glucocorticoid receptor phosphorylation reduces its activity: role in steroid-insensitive asthma. J. Allergy Clin. Immunol. **109:** 649–657.
73. HERRLICH, P. 2001. Cross-talk between glucocorticoid receptor and AP-1. Oncogene **20:** 2465–2475.
74. LOPEZ, G.N., C.W. TURCK, F. SCHAUFELE, *et al.* 2001. Growth factors signal to steroid receptors through mitogen-activated protein kinase regulation of p160 coactivator activity. J. Biol. Chem. **276:** 22177–22182.
75. LUCIBELLO, F.C., E.P. SLATER, K.U. JOOSS, *et al.* 1990. Mutual transrepression of Fos and the glucocorticoid receptor: involvement of a functional domain in Fos which is absent in FosB. EMBO J. **9:** 2827–2834.
76. ROGATSKY, I., S.K. LOGAN & M.J. GARABEDIAN. 1998. Antagonism of glucocorticoid receptor transcriptional activation by the c-Jun N-terminal kinase. Proc. Natl. Acad. Sci. USA **95:** 2050–2055.
77. SCHULE, R., P. RANGARAJAN, S. KLIEWER, *et al.* 1990. Functional antagonism between oncoprotein c-Jun and the glucocorticoid receptor. Cell **62:** 1217–1226.
78. KRSTIC, M.D., I. ROGATSKY, K.R. YAMAMOTO, *et al.* 1997. Mitogen-activated and cyclin-dependent protein kinases selectively and differentially modulate transcriptional enhancement by the glucocorticoid receptor. Mol. Cell. Biol. **17:** 3947–3954.
79. KOZAWA, O., M. NIWA, D. HATAKEYAMA, *et al.* 2002. Specific induction of heat shock protein 27 by glucocorticoid in osteoblasts. J. Cell. Biochem. **86:** 357–364.
80. XIA, B., J. LU & G. WANG. 2003. Glucocorticoid modulation of extracellular signal-regulated protein kinase 1/2 and p38 in human ovarian cancer HO-8910 cells. Chin. Med. J. **116:** 753–756.
81. KNUTTI, D., D. KRESSLER, A. KRALLI. 2001. Regulation of the transcriptional coactivator PGC-1 via MAPK-sensitive interaction with a repressor. Proc. Natl. Acad. Sci. USA **98:** 9713–9718.
82. MCCALLUM, R.E. & R.D. STITH. 1982. Endotoxin-induced inhibition of steroid binding by mouse liver cytosol. Circ. Shock **9:** 357–367.
83. MACPHEE, I.A.M., F.A. ANTONI, D.W. MASON, *et al.* 1989. Spontaneous recovery of rats from experimental allergic encephalomyelitis is dependent on regulation of the immune system by endogenous adrenal corticosteroids. J. Exp. Med. **169:** 431–445.
84. EDWARDS, C.K.I., L.M. YUNGER, R.M. LORENCE, *et al.* 1991. The pituitary gland is required for protection against lethal effects of *Salmonella typhimurium*. Proc. Natl. Acad. Sci. USA **88:** 2274–2277.

85. RUZEK, M.C., B.D. PEARCE, A.H. MILLER, *et al.* 1999. Endogenous glucocorticoids protect against cytokine-mediated lethality during viral infection. J. Immunol. **162:** 3527–3533.
86. MYKONIATIS, A., P.M. ANTON, M. WLK, *et al.* 2003. Leptin mediates *Clostridium difficile* toxin A-induced enteritis in mice. Gastroenterology **124:** 683–691.
87. CASTAGLIUOLO, I., K. KARALIS, L. VALENICK, *et al.* 2001. Endogenous corticosteroids modulate *Clostridium difficile* toxin A-induced enteritis in rats. Am. J. Physiol. Gastrointest. Liver Physiol. **280:** G539–545.
88. GÓMEZ, F., E.R. DE KLOET & A. ARMARIO. 1998. Glucocorticoid negative feedback on the HPA axis in five inbred rat strains. Am. J. Physiol. **274:** R420–R427.
89. PALERMO, M., F. ALVES-ROSA, C. RUBEL, *et al.* 2000. Pretreatment of mice with lipopolysaccharide (LPS) or IL-1beta exerts dose-dependent opposite effects on Shiga toxin-2 lethality. Clin. Exp. Immunol. **119:** 77–83.
90. WICK, G., R. SGONC, & O. LECHNER. 1998. Neuroendocrine-immune disturbances in animal models with spontaneous autoimmune diseases. Ann. N.Y. Acad. Sci. **840:** 591–598.
91. LECHNER, O., Y. HU, M. JAFARIAN TEHRANI, *et al.* 1996. Disturbed immunoendocrine communication via the hypothalamo-pituitaty-adrenal axis in murine lupus. Brain Behav. Immun. **10:** 337–350.
92. HU, Y., H. DIETRICH, M. HEROLD, *et al.* 1993. Disturbed immuno-endocrine communication via the hypothalamo-pituitary-adrenal axis in autoimmune disease. Int. Arch. Allergy Immunol. **102:** 232–241.
93. GUTIERREZ, M.A., M.E. GARCIA, J.A. RODRIGUEZ, *et al.* 1999. Hypothalamic-pituitary-adrenal axis function in patients with active rheumatoid arthritis: a controlled study using insulin hypoglycemia stress test and prolactin stimulation. J. Rheumatol. **26:** 277–281.
94. CASH, J.M., L.J. CROFFORD, W.T. GALLUCCI, *et al.* 1992. Pituitary-adrenal axis responsiveness to ovine corticotropin releasing hormone in patients with rheumatoid arthritis treated with low dose prednisone. J. Rheumatol. **19:** 1692–1696.
95. CHIKANZA, I.C., P. PETROU, G. KINGSLEY, *et al.* 1992. Defective hypothalamic response to immune and inflammatory stimuli in patients with rheumatoid arthritis. Arthritis Rheum. **35:** 1281–1288.
96. CUTOLO, M., L. FOPPIANI, C. PERETE, *et al.* 1999. Hypothalamic-pituitary-adrenocortical axis function in premenopausal women with rheumatoid arthritis not treated with glucocorticoids. J. Rheumatol. **26:** 282–288.
97. EIJSBOUTS, A.M. & E.P. MURPHY. 1999. The role of the hypothalamic-pituitary-adrenal axis in rheumatoid arthritis. Baillieres Best Pract. Res. Clin. Rheumatol. **13:** 599–613.
98. CROFFORD, L.J., K.T. KALOGERAS, G. MASTORAKOS, *et al.* 1997. Circadian relationships between interleukin (IL)-6 and hypothalamic-pituitary-adrenal axis hormones: failure of IL-6 to cause sustained hypercortisolism in patients with early untreated rheumatoid arthritis. J. Clin. Endocrinol. Metab. **82:** 1279–1283.
99. GUTIERREZ, M.A., M.E. GARCIA, J.A. RODRIGUEZ, *et al.* 1998. Hypothalamic-pituitary-adrenal axis function and prolactin secretion in systemic lupus erythematosus. Lupus **7:** 404–408.
100. CROFFORD, L.J. 2002. The hypothalamic-pituitary-adrenal axis in the pathogenesis of rheumatic diseases. Endocrinol. Metab. Clin. North Am. **31:** 1–13.
101. JOHNSON, E.O., P.G. VLACHOYIANNOPOULOS, F.N. SKOPOULI, *et al.* 1998. Hypofunction of the stress axis in Sjogren's syndrome. J. Rheumatol. **25:** 1508–1514.
102. BUSKE-KIRSCHBAUM, A., S. JOBST, D. PSYCH, *et al.* 1997. Attenuated free cortisol response to psychosocial stress in children with atopic dermatitis. Psychosom. Med. **59:** 419–426.
103. BUSKE-KIRSCHBAUM, A., S. JOBST & D.H. HELLHAMMER. 1998. Altered reactivity of the hypothalamus-pituitary-adrenal axis in patients with atopic dermatitis: pathologic factor or symptom? Ann. N.Y. Acad. Sci. **840:** 747–754.
104. BUSKE-KIRSCHBAUM, A. & D.H. HELLHAMMER. 2003. Endocrine and immune responses to stress in chronic inflammatory skin disorders. Ann. N.Y. Acad. Sci. **992:** 231–240.

105. BUSKE-KIRSCHBAUM, A., K. VON AUER, S. KRIEGER, *et al.* 2003. Blunted cortisol responses to psychosocial stress in asthmatic children: a general feature of atopic disease? Psychosom. Med. **65:** 806–810.
106. RUPPRECHT, M., O.P. HORNSTEIN, D. SCHLUTER, *et al.* 1995. Cortisol, corticotropin, and beta-endorphin responses to corticotropin-releasing hormone in patients with atopic eczema. Psychoneuroendocrinology **20:** 543–551.
107. NEECK, G. & L.J. CROFFORD. 2000. Neuroendocrine perturbations in fibromyalgia and chronic fatigue syndrome. Rheum. Dis. Clin. North Am. **26:** 989–1002.
108. DEMITRACK, M.A., J.K. DALE, S.E. STRAUS, *et al.* 1991. Evidence for impaired activation of the hypothalamic-pituitary-adrenal axis in patients with chronic fatigue syndrome. J. Clin. Endocrinol. Metab. **73:** 1224–1234.
109. DEMITRACK, M.A. & L.J. CROFFORD. 1998. Evidence for and pathophysiologic implications of hypothalamic- pituitary-adrenal axis dysregulation in fibromyalgia and chronic fatigue syndrome. Ann. N.Y. Acad. Sci. **840:** 684–697.
110. GAAB, J., D. HUSTER, R. PEISEN, *et al.* 2002. Hypothalamic-pituitary-adrenal axis reactivity in chronic fatigue syndrome and health under psychological, physiological, and pharmacological stimulation. Psychosom. Med. **64:** 951–962.
111. CROFFORD, L.J., S.R. PILLEMER, K.T. KALOGERAS, *et al.* 1994. Hypothalamic-pituitary-adrenal axis perturbations in patients with fibromyalgia. Arthritis Rheum. **37:** 1583–1592.
112. MICHELSON, D., L. STONE, E. GALLIVEN, *et al.* 1994. Multiple sclerosis is associated with alterations in hypothalamic-pituitary-adrenal axis function. J. Clin. Endocrinol. Metab. **79**: 848–853.
113. WEI, T. & S.L. LIGHTMAN. 1997. The neuroendocrine axis in patients with multiple sclerosis. Brain **120:** 1067–1076.
114. KINO, T. & G.P. CHROUSOS. 2001. Glucocorticoid and mineralocorticoid resistance/hypersensitivity syndromes. J. Endocrinol. **169:** 437–445.
115. HUIZENGA, N.A.T.M., J.W. KOPER, P. DE LANGE, *et al.* 1998. A polymorphism in the glucocorticoid receptor gene may be associated with an increased sensitivity to glucocorticoids *in vivo*. J. Clin. Endocrinol. Metab. **83:** 144–151.
116. KOPER, J.W., R.P. STOLK, P. DE LANGE, *et al.* 1997. Lack of association between five polymorphisms in the human glucocorticoid receptor gene and glucocorticoid resistance. Hum. Genet. **99:** 663–668.
117. BRAY, P.J. & R.G. COTTON. 2003. Variations of the human glucocorticoid receptor gene (NR3C1): pathological and in vitro mutations and polymorphisms. Hum. Mutat. **21:** 557–568.
118. VOTTERO, A. & G.P. CHROUSOS. 1999. Glucocorticoid receptor β: View I. Trends Endocrinol. Metab. **10:** 333–338.
119. OAKLEY, R.H., M. SAR, J.A. CIDLOWSKI, *et al.* 1996. The human glucocorticoid receptor β isoform. Expression, biochemical properties and putative function. J. Biol. Chem. **271:** 9550–9559.
120. OAKLEY, R.H., C.M. JEWELL, M.R. YUDT, *et al.* 1999. The dominant negative activity of the human glucocorticoid receptor β isoform. Specificity and mechanisms of action. J. Biol. Chem. **274:** 27857–27866.
121. BROGAN, I.J., I.A. MURRAY, G. CERILLO, *et al.* 1999. Interaction of glucocorticoid receptor isoforms with transcription factors AP-1 and NF-κB: lack of effect of glucocorticoid receptor β. Mol. Cell. Endocrinol. **157:** 95–104.
122. CARLSTEDT-DUKE, J. 1999. Glucocorticoid receptor β: View II. Trends Endocrinol. Metab. **10:** 339–342.
123. HECHT, K., J. CARLSTEDT-DUKE, P. STIERNA, *et al.* 1997. Evidence that the β-isoform of the human glucocorticoid receptor does not act as a physiologically significant repressor. J. Biol. Chem. **272:** 26659–26664.
124. LEUNG, D.Y.M., Q. HAMID, A. VOTTERO, *et al.* 1997. Association of glucocorticoid insensitivity with increased expression of glucocorticoid receptor β. J. Exp. Med. **186:** 1567–1574.
125. HAMID, Q.A., S.E. WENZEL, P.J. HAUK, *et al.* 1999. Increased glucocorticoid receptor β in airway cells of glucocorticoid-insensitive asthma. Am. J. Respir. Crit. Care Med. **159:** 1600–1604.

126. STRICKLAND, I., K. KISICH, P.J. HAUK, *et al.* 2001. High constitutive glucocorticoid receptor β in human neutrophils enables them to reduce their spontaneous rate of cell death in response to corticosteroids. J. Exp. Med. **193:** 585–593.
127. SOUSA, A.R., S.J. LANE, J.A. CIDLOWSKI, *et al.* 2000. Glucocorticoid resistance in asthma is associated with elevated in vivo expression of the glucocorticoid receptor beta-isoform. J. Allergy Clin. Immunol. **105:** 943–950.
128. HONDA, M., F. ORII, T. AYABE, *et al.* 2000. Expression of glucocorticoid receptor β in lymphocytes of patients with glucocorticoid-resistant ulcerative colitis. Gastroenterology **118:** 859–866.
129. ORII, F., T. ASHIDA, M. NOMURA, *et al.* 2002. Quantitative analysis for human glucocorticoid receptor alpha/beta mRNA in IBD. Biochem. Biophys. Res. Commun. **296:** 1286–1294.
130. SHAHIDI, H., A. VOTTERO, C. STRATAKIS, *et al.* 1999. Imbalanced expression of the glucocorticoid receptor isoforms in cultured lymphocytes from a patient with systemic glucocorticoid resistance and chronic lymphocytic leukemia. Biochem. Biophys. Res. Commun. **254:** 559–565.
131. HAMILOS, D.L., D.Y. LEUNG, S. MURO, *et al.* 2001. GRbeta expression in nasal polyp inflammatory cells and its relationship to the anti-inflammatory effects of intranasal fluticasone. J. Allergy Clin. Immunol. **108:** 59–68.
132. DERIJK, R.H., M.J. SCHAAF, G. TURNER, *et al.* 2001. A human glucocorticoid receptor gene variant that increases the stability of the glucocorticoid receptor beta-isoform mRNA is associated with rheumatoid arthritis. J. Rheumatol. **28:** 2383–2388.
133. CHIKANZA, I.C. 2002. Mechanisms of corticosteroid resistance in rheumatoid arthritis: a putative role for the corticosteroid receptor beta isoform. Ann. N.Y. Acad. Sci. **966:** 39–48.
134. HORI, T., K. WATANABE, M. MIYAOKA, *et al.* 2002. Expression of mRNA for glucocorticoid receptors in peripheral blood mononuclear cells of patients with Crohn's disease. J. Gastroenterol. Hepatol. **17:** 1070–1077.
135. SCHLAGHECKE, R., E. KORNELY, J. WOLLENHAUPT, *et al.* 1992. Glucocorticoid receptors in rheumatoid arthritis. Arthritis Rheum. **35:** 740–744.
136. MINGRONE, G., A. DEGAETANO, M. PUGEAT, *et al.* 1999. The steroid resistance of Crohn's disease. J. Invest. Med. **47:** 319–325.
137. SECKL, J.R. & B.R. WALKER. 2001. Minireview: 11β–hydroxysteroid dehydrogenase type 1-a tissue-specific amplifier of glucocorticoid action. Endocrinology **142:** 1371–1376.
138. RASK, E., T. OLSSON, S. SODERBERG, *et al.* 2001. Tissue-specific dysregulation of cortisol metabolism in human obesity. J. Clin. Endocrinol. Metab. **86:** 1418–1421.
139. TAKAHASI, K.I., K. FUKUSHIMA, H. SASANO, *et al.* 1999. Type II 11beta-hydroxysteroid dehydrogenase expression in human colonic epithelial cells of inflammatory bowel disease. Dig. Dis. Sci. **44:** 2516–2522.
140. WEBSTER, J.I. & J. CARLSTEDT-DUKE. 2002. Involvement of multidrug-resistance proteins (MDR) in the modulation of glucocorticoid response. J. Steroid Biochem. Mol. Biol. **82:** 277–288.
141. KRALLI, A. & K.R. YAMAMOTO. 1996. An FK506-sensitive transporter selectively decreases intracellular levels and potency of steroid hormones. J. Biol. Chem. **271:** 17152–17156.
142. MEDH, R.D., R.H. LAY, T.J. SCHMIDT, *et al.* 1998. Agonist-specific modulation of glucocorticoid receptor-mediated transcription by immunosuppressants. Mol. Cell. Endocrinol. **138:** 11–23.
143. FARRELL, R.J., A. MURPHY, A. LONG, *et al.* 2000. High multidrug resistance (P-glycoprotein 170) expression in inflammatory bowel disease patients who fail medical therapy. Gastroenterology **118:** 279–288.
144. LLORENTE, L., Y. RICHAUD-PATIN, A. DIAZ-BORJON, *et al.* 2000. Multidrug resistance-1 (MDR-1) in rheumatic autoimmune disorders. Part I: Increased P-glycoprotein activity in lymphocytes from rheumatoid arthritis patients might influence disease outcome. Joint Bone Spine **67:** 30–39.
145. DIAZ-BORJON, A., Y. RICHAUD-PATIN, C. ALVARADO DE LA BARRERA, *et al.* 2000. Multidrug resistance-1 (MDR-1) in rheumatic autoimmune disorders. Part II: Increased

P-glycoprotein activity in lymphocytes from systemic lupus erythematosus patients might affect steroid requirements for disease control. Joint Bone Spine **67:** 40–48.

146. KINO, T., A. GRAGEROV, J. B. KOPP, *et al.* 1999. The HIV-1 virion-associated protein Vpr is a coactivator of the human glucocorticoid receptor. J. Exp. Med. **189:** 51–61.
147. NEW, M.I., S. NIMKARN, D.D. BRANDON, *et al.* 1999. Resistance to several steroids in two sisters. J. Clin. Endocrinol. Metab. **84:** 4454–4464.
148. NEW, M.I., S. NIMKARN, D.D. BRANDON, *et al.* 2001. Resistance to multiple steroids in two sisters. J. Steroid Biochem. Mol. Biol. **76:** 161–166.

Systemic Inflammation-Associated Glucocorticoid Resistance and Outcome of ARDS

G. UMBERTO MEDURI[a] AND CHARLES R. YATES[b]

[a]*Memphis Lung Research Program, Department of Medicine—Divisions of Pulmonary and Critical Care Medicine and* [b]*Department of Pharmaceutical Sciences University of Tennessee Health Science Center, Memphis, Tennessee 38163, USA*

ABSTRACT: Dysregulated systemic inflammation with excess activation of pro-inflammatory transcription factor nuclear factor-κB (NF-κB)—activated by inflammatory signals—compared to the anti-inflammatory transcription factor glucocorticoid receptor-α (GRα)—activated by endogenous or exogenous glucocorticoids (GCs)—is an important pathogenetic mechanism for pulmonary and extrapulmonary organ dysfunction in patients with acute respiratory distress syndrome (ARDS). Activation of one transcription factor in excess of the binding (inhibitory) capacity of the other shifts cellular responses toward increased (dysregulated) or decreased (regulated) transcription of inflammatory mediators over time. Recent data indicate that failure to improve in ARDS (unresolving ARDS) is frequently associated with failure of the activated GRs to downregulate the transcription of inflammatory cytokines despite elevated levels of circulating cortisol, a condition defined as systemic inflammation-associated acquired GC resistance; it is potentially reversible with prolonged GC supplementation.

In the first part of this paper, after a brief description of inflammation in ARDS and our model of translational research, we review the two cellular signaling pathways that are central to the regulation of inflammation—the *stimulatory* NF-κB and the *inhibitory* GRα. In the second part, we review findings of recent studies indicating that excessive inflammatory activity in patients with unresolving ARDS may induce noncompensated GC resistance in target organs. In the third part, we review factors affecting cellular response to GC and potential mechanisms involved in inflammation-associated GC resistance.

KEYWORDS: GR; glucocorticoid resistance; ARDS; HPA axis; inflammation; NF-κB

INTRODUCTION

Dysregulated systemic inflammation with persistent elevation of circulating inflammatory cytokines over time is the pathogenetic mechanism for pulmonary and extrapulmonary organ dysfunction in patients with acute respiratory distress syn-

Address for correspondence: Dr. G. Umberto Meduri, University of Tennessee Health Science Center, College of Medicine, Pulmonary Division, Memphis, TN 38163. Voice: 901-448-5258; fax: 901-448-7726.

umeduri@utmem.edu

Ann. N.Y. Acad. Sci. 1024: 24–53 (2004). © 2004 New York Academy of Sciences.
doi: 10.1196/annals.1321.004

drome (ARDS). The role of glucocorticoid (GC) supplementation in patients with systemic inflammation is undergoing a cyclical reassessment stirred by new understanding of GCs' role in modulating inflammation and immunity,[1] and the positive results of recent randomized studies. It is now appreciated that at cellular level, transcription factors nuclear factor-κB (NF-κB)—activated by inflammatory signals—and glucocorticoid receptor α (GRα)—activated by endogenous or exogenous GCs—have diametrically opposed functions (stimulatory vs. inhibitory) in regulating inflammation. NF-κB is recognized as the principal driver of the inflammatory response, responsible for the transcription of greater than one hundred genes, including tumor necrosis factor (TNF)-α, IL-1β, and IL-6.[2] Once activated, NF-κB and GRα can mutually repress each other through a protein–protein interaction that prevents their DNA binding and subsequent transcriptional activity. Activation of one transcription factor in excess of the binding (inhibitory) capacity of the other shifts cellular responses toward increased (dysregulated) or decreased (regulated) transcription of inflammatory mediators over time.[3]

In this chapter, we review recent data indicating that failure to improve in ARDS (unresolving ARDS) is frequently associated with failure of the activated GRs to downregulate the transcription of inflammatory cytokines despite elevated levels of circulating cortisol, a condition defined as systemic inflammation-associated acquired GC resistance and potentially reversible with prolonged GC supplementation. We have recently shown that, in patients with unresolving ARDS, normal peripheral blood leukocytes (PBL) exposed to plasma samples collected during prolonged methylprednisolone treatment exhibited (i) a progressive increase in cytoplasmic binding of GRα to NF-κB and (ii) a concomitant reduction in NF-κB binding to DNA and transcription of TNFα and the interleukin-1β.[3]

In the first part, after a brief description of inflammation in ARDS and our model of translational research, we review the two cellular signaling pathways that are central to the regulation of inflammation, the *stimulatory* NF-κB and the *inhibitory* GRα. In the second part, we review findings of recent studies indicating that excessive inflammatory activity in patients with ARDS may induce noncompensated GC resistance in target organs. In the third part, we will review factors affecting cellular response to GC and potential mechanisms involved in inflammation-associated GC resistance.

I

ACUTE RESPIRATORY DISTRESS SYNDROME

ARDS is a term applied to a relatively specific morphologic lesion of multifactorial etiology known as diffuse alveolar damage (DAD).[4] ARDS develops rapidly, in most patients within 12–48 h of exposure to infectious or noninfectious insults that can affect the lung directly (via the alveolar compartment) or indirectly (via the vascular compartment).[5] At presentation, (early) ARDS manifests with severe, diffuse, and nonhomogenous acute host inflammatory response (HIR) of the pulmonary lobules, leading to a breakdown in the barrier and gas exchange function of the lung. Diffuse injury to the alveolocapillary membrane (ACM)—the gas transfer surface of the lung—causes edema of the air spaces and interstitium with a protein-rich

TABLE 1. Definitions of resolving and unresolving ARDS

Findings on day 7 of ARDS	Resolving	Unresolving
LIS from day 1 to day 7 of ARDS	≥ 1 point reduction	< 1 point reduction
Clinical definition	Improver	Nonimprover
Host inflammatory response	Regulated	Dysregulated
Progression of lung histology	Adaptive	Maladaptive
Observed mortality*	14%	83%

*Data obtained from Ref. 15.

neutrophilic exudate, resulting in severe gas exchange and lung compliance abnormalities.[6]

Overall mortality in ARDS is 35–60%,[7] with most nonsurvivors dying within two weeks of disease development.[6,8] While a regulated inflammatory response is critical to survival,[5] a major predictor of poor outcome in ARDS patients is persistence of pulmonary and systemic inflammation after one week of lung injury.[9,10] Failure to downregulate the production of inflammatory mediators (dysregulated HIR) is associated with maladaptive lung repair and inability to improve ACM permeability, gas exchange, and lung mechanics over time.

The lung injury score (LIS) quantifies the physiologic respiratory impairment in ARDS through the use of a four-point score based on the levels of positive end-expiratory pressure (PEEP), ratios of partial pressure of oxygen (PaO_2) to fraction of inspired oxygen (FiO_2) (PaO_2:FiO_2), the static lung compliance (CST), and the degree of infiltration present on chest radiograph.[11] Patients failing to improve the LIS or its components by day 7 of ARDS (unresolving ARDS) have a poor outcome.[12–14] We have previously reported that patients who meet predefined criteria for unresolving ARDS (LIS on day 7 of ARDS ≥ 2.5 and < one-point reduction from day 1 of ARDS) have a mortality rate in excess of 80% (TABLE 1).[15]

MODEL OF TRANSLATIONAL RESEARCH

Because there is no animal model to study the progression of ARDS, translational clinical research has an important role to play for advancing understanding in this field. Our research model has followed a holistic level of inquiry in constructing a pathophysiological model of ARDS, attempting to fit pathogenesis (biology) with morphological (pathology) and clinical (physiology) findings (FIG. 1) observed during the longitudinal course of the disease.[16] Most importantly, we have attempted to define the differences over time between patients with an adaptive (resolving ARDS) vs. maladaptive (unresolving ARDS) reparative response, and the time span of disease reversibility (prior to reaching end-stage disease) that is potentially amenable to anti-inflammatory treatment. In patients failing to improve after one week of ARDS onset (unresolving ARDS), we also have investigated the effect of prolonged anti-inflammatory (GC) treatment. In relation to the treatment investigation, we as-

TABLE 2. Comparison of old and new trials investigating methylprednisolone use in ARDS

	1980s Trials	1990s Trials
Timing of ARDS	< 2 days	7–14 days (unresolved)
Dosage	120 mg/kg/day	2 mg/kg/day
Duration	1 day	Average 30 days
Understanding of the HIR in ARDS	Massive, short-lived	Prolonged, initial intensity affects duration
Understanding of glucocorticoid treatment in ARDS	Reversibility lost early	Reversibility lost with end-stage fibrosis
Glucocorticoid treatment	Massive dose, time-limited	Lower dose, prolonged until resolution

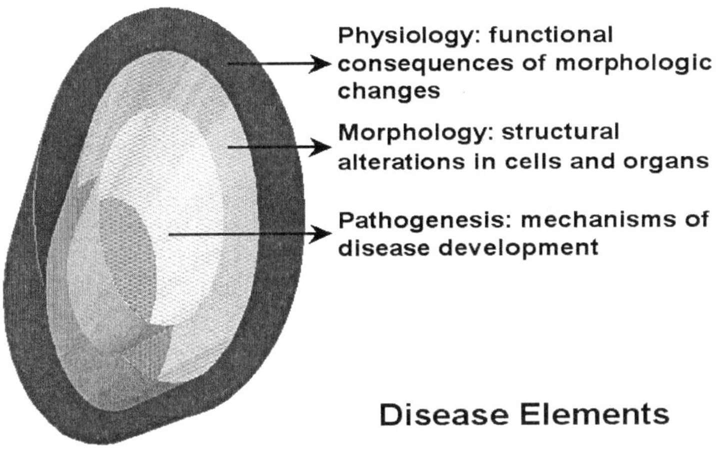

FIGURE 1. Disease elements.

sumed that the closer our treatment intervention was to the core pathogenetic mechanisms of disease development, the more likely treatment would affect all "layers" of the disease process (FIG. 1). In this context, a positive or negative biological and physiological response to treatment was used to prove or disprove the pathogenetic relevance (central vs. peripheral) of the factor or pathway affected by treatment.

Controversy continues on the use of GC treatment in ARDS[17] and sepsis.[18,19] An historical review of GC treatment in sepsis and ARDS in relation to the evolving pathophysiological understanding of systemic inflammation can be found elsewhere.[20] Similar to the original work of Ashbaugh and Maier[21] and Hooper and Kearl,[22] we investigated the use of prolonged methylprednisolone (MP) administration in patients with unresolving ARDS. The differences between GC dosage and duration of administration of older (1980s) versus newer trials are outlined in the TABLE 2.

Over the last decade our understanding of the intermediary events that occur between the reception of a biological signal by the cell membrane and the eventual conversion of that signal to a change in gene expression at the nuclear level (i.e., signal transduction) has grown immensely.[23] It is now recognized that two cellular signaling pathways are central to the regulation of inflammation, the *stimulatory* NF-κB and the *inhibitory* GR-mediated signal transduction cascades. In unstimulated cells, both NF-κB and GR are predominantly sequestered in the cytoplasm.

NUCLEAR FACTOR-κB

NF-κB is recognized as the central transcription factor that drives the inflammatory response to insults. NF-κB activation is an essential step in the experimental development of neutrophilic lung inflammation.[24–26] NF-B is found in essentially all cell types and is involved in activation of an exceptionally large number of target genes (over 100).[27] NF-κB is a heterogenous collection of dimers, composed of various combinations of the NF-κB/Rel family. The p65:p50 heterodimer was the first form of NF-κB to be identified and is the most abundant in most cell types.[27] NF-κB is maintained in an inactive form by sequestration in the cytoplasm through interaction with inhibitory proteins IκBs (the most important being IκBα) (FIG. 2).[28] Activation of NF-κB is a rapid, immediate early event that occurs within minutes after exposure to a relevant inducer, including innate immunity stimulating molecules (e.g., lipopolysaccharide and hsp60), double-stranded DNA, physical and chemical stresses, and inflammatory cytokines (e.g., TNFα and IL-1β). In response to these various stimuli, the latent NF-κB/IκB complex is activated by phosphorylation and proteolytic degradation of IκB, with exposure of the NF-κB nuclear localization sequence (NSL).[27] Proteolytic degradation of IκB is an irreversible step in the signaling pathway that constitutes a commitment to transcriptional activation.[27]

The liberated NF-κB then translocates into the nucleus and binds to promoter regions of target genes to initiate the transcription of multiple cytokines including TNFα and the interleukins (IL) IL-1β, IL-2, IL-6, chemokines such as IL-8, cell adhesion molecules (e.g., intercellular adhesion molecule-1, E-selectin), interferon, receptors involved in immune recognition such as members of the MHC, proteins involved in antigen presentation, receptors required for neutrophils adhesion and migration, and inflammation-associated enzymes (cyclooxygenase [COX], phospholipase A2 [PLA2], inducible nitric oxide [iNOS]).[28,29] Products of the genes that are stimulated by NF-κB activate this transcription factor. Thus, TNFα and IL-1β both activate and are activated by NF-κB, by forming a positive regulatory loop that amplifies and perpetuates inflammation.[30] NF-κB also operates in conjunction with other transcription factors, including activator protein-1 (AP-1).[31]

Negative regulation of NF-κB activity is very complex, and a variety of mechanisms are involved in both termination of NF-κB activation and its downregulation in response to specific signals (reviewed in Ref. 27). The critical inhibitory step, however, involves binding of newly synthesized IκB to NF-κB in the nucleus. IκBα synthesis is a process requiring NF-κB as well as activated GR (see section on Activation of the HPA-axis) transcriptional activity. IκB has NLS regions and transfer to the nucleus is an energy-dependent process.[27] After nuclear translocation, IκB is capable of terminating NF-κB activity by transporting NF-κB back to the cytoplasm.[27]

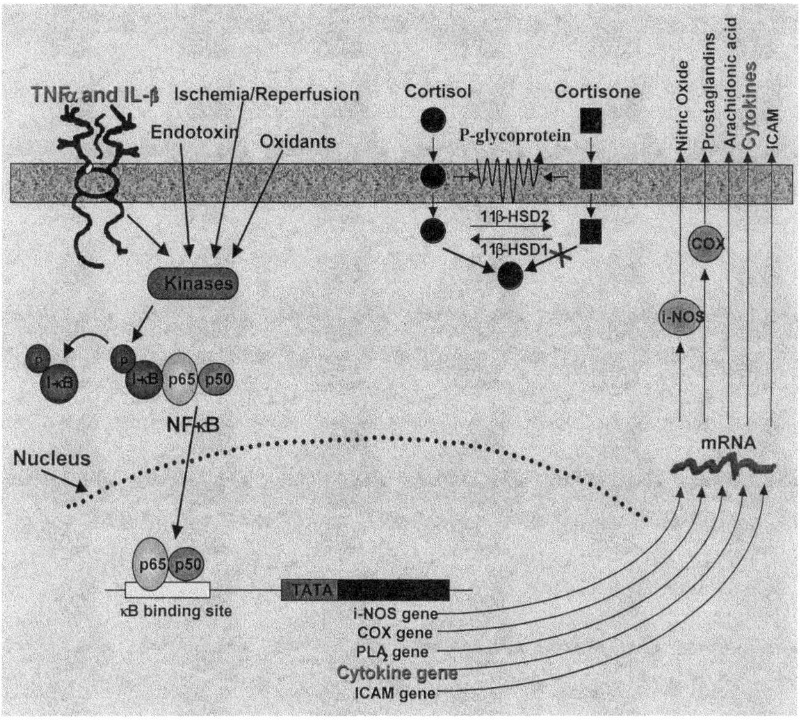

FIGURE 2. Activation of NF-κB. When cells are stimulated by inflammatory signals, specific kinases phosphorylate the inhibitory protein IκB and cause its rapid degradation. The activated form of NF-κB then moves to the nucleus initiating the transcription of mRNA of inflammatory cytokines, chemokines, cell adhesion molecules, and inflammation-associated enzymes (inducible nitric oxide [i-INOS], cyclooxygenase [.COX], phospholipase A_2 [PLA_2]). Products of the genes that are stimulated by NF-κB activate this transcription factor. Thus, TNF-α and IL-1β both activate and are activated by NF-κB by forming a positive regulatory loop that amplifies and perpetuates inflammation. See text for references.

INTEGRATED HOST DEFENSE RESPONSE PATHWAYS

Among the broad spectrum of NF-κB-induced mediators, the early response cytokines TNFα and IL-1β are uniquely important in activating most pathways of the host defense response (HDR).[32,33] Whereas the biological functions of TNFα and IL-1β are remarkably similar, the two cytokines and their receptors are biochemically unrelated. The cell most commonly associated with the initial release of these cytokines is the tissue macrophage or the blood monocyte.[32] Increasing evidence, however, suggests that nonimmune cells play crucial roles in the generation, maintenance, and resolution of both local and systemic responses.[34]

In patients with ARDS, increased NF-κB activation was shown in alveolar macrophages,[35–38] and in BAL nuclear protein extracts.[26] In experimental ARDS, initial activation of NF-κB occurs in alveolar macrophages, and the ensuing production of

TABLE 3. Pathways of the host response to infectious and noninfectious insults

Inflammation	Modulation of the immune response
Vasodilation and stasis	Fever
Increased expression of adhesion molecules	Induction of heat-shock proteins
Increased permeability of the micro-vasculature	Release of neutrophils from the bone marrow
Leucocyte extravasation[a]	Priming of phagocytic cells
Release of leukocyte products	Adaptive immune response[c]
Coagulation	Activation of the HPA-axis
Activation of coagulation	Release of ACTH with cortisol production
Inhibition of fibrinolysis	Modulation of the sympathetic nervous system[d]
Intravascular clotting	Modulation of hepatic acute-phase protein production[e]
Extravascular fibrin deposition	
Tissue repair[b]	
Angiogenesis	
Epithelial growth	
Fibroblast migration and proliferation	
Deposition of extravellular matrix/ remodeling	

[a]Initially polymorphonuclear cells and later monocytes.

[b]Tissue repair consists of replacing injured tissue by regenerating native parenchymal cells and filling defects with fibroblastic tissue.

[c]The adaptive immune response depends on two classes of specialized lymphocytes, T and B cells, with specific receptors that are somatically generated in response to antigen-presenting cells.

[d]Glucocorticoids and ACTH affect several key regulatory enzymes in catecholamine biosynthesis and influence the number and functional state of adrenergic receptor stimulation. Catecholamine–glucocorticoid interactions play an important role in maintaining vascular tone. Druing stress, epinephrine response parallels HPA axis stimulation.

[e]Elevated levels of circulating glucocorticoids synergize with IL-6 in inducing hepatic synthesis and secretion of "acute phase" reactants such as fibrinogen (coagulation and tissue repair), protease inhibitors (limit tissue damage during inflammation), complement C3 (opsonization), ceruloplasmin, haptoglobin, and C-reactive protein (inhibitor of neutrophil chemotaxis).

TNFα propagates NF-κB activation to other types of cells in the lung.[25] Once released, TNFα and IL-1β act on epithelial cells, stromal cells (fibroblasts and endothelial), the ECM, and recruited circulating cells (neutrophils, platelets, lymphocytes) to cause secondary waves of cytokine release with amplification of the HDR.[32]

The biological actions of cytokines are pleiotropic and redundant, and relevant to disease pathophysiology. TNFα and IL-1β set off a network of integrated pathways that includes inflammation,[39] coagulation (intravascular clotting and extravascular fibrin deposition),[40–43] modulation of the immune response,[34] and activation of the hypothalamic-pituitary-adrenal (HPA) axis with production of GCs.[44,45] These pathways (TABLE 3) are activated simultaneously (slight delay for activation of the HPA-axis, see later) and work in synergy to eliminate or control injurious agents and to repair any consequent tissue damage.[46] Inflammation is the most prominent of these pathways, and the term host inflammatory response (HIR) is frequently used in place of HDR.

Cellular responses in HDR are regulated by a complex interaction among cytokines with final effects on the surrounding tissue not directly induced by the initiating insult. In this regard, cytokines have concentration-dependent biologic effects.[47] Optimal levels of these cytokines are important for a successful host defense; however, at increasingly higher concentrations they mediate proportionately stronger local and systemic inflammatory responses, with predominantly destructive, rather than protective, effects on the host. If unchecked and unregulated, a protracted and exaggerated release of inflammatory mediators leads to deterioration of organ function (protracted tissue injury and maladaptive repair) rather than restoring normal anatomy and function.[15,48,49]

ACTIVATION OF THE HPA AXIS AND CONTROL OF THE HOST DEFENSE RESPONSE

An expanding body of evidence indicates that activation of the immune and inflammatory response as well as peripherally generated TNFα, IL-1β, and IL-6 activate the hypothalamic-pituitary-adrenal (HPA) axis independently at some or all of its levels (hypothalamic corticotropin-releasing hormone [CRH] neurons, pituitary corticotrophs, and the adrenal cortex); in combination their effects are synergistic.[44,45] This effect is assumed to be sufficiently delayed in relation to the initial insult (stress stimulus) to allow the appropriate defense mechanisms to become activated.[50,51] The HPA axis responds in a graded manner to greater intensities of stress with increased production of ACTH and GCs. ACTH is the predominant but not the only regulator of GCs secretion. Other factors, including angiotensin and vasopressin, also influence adrenal GCs secretion.[52] GCs are secreted directly into the circulation immediately after their synthesis. Cortisol, the major human GC, circulates bound (95%) to a corticosteroid-binding globulin (CBG) synthesized primarily by the liver, thereby providing a large reservoir that is released at sites of inflammation or tissue remodeling.[53] Due to their hormonal and lipophilic nature, GCs pass freely through the cell membrane.

GLUCOCORTICOID RECEPTOR

Glucocorticoids exert most of their effects by activating ubiquitously distributed (2,000 to 30,000 per cell) cytoplasmic heat shock protein-complexed GRs with formation of GC-GR complexes.[54] The GR belongs to the nuclear receptor superfamily with members that function as ligand-activated transcription factors. There are two splicing variants of the human hGR, isoforms hGRα and hGRβ. GRα is the classic receptor that binds to GCs and transduces their biological activities, while GRβ does not bind to GCs and exerts weak dominant negative activity on GRα.[55] After binding to the ligand, GRα dissociates from the heat shock proteins and enters the nucle-

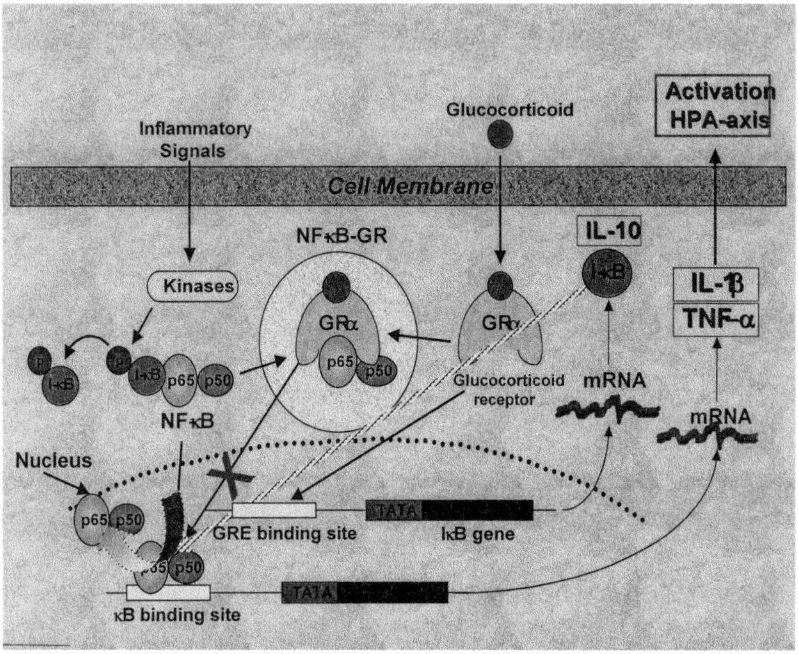

FIGURE 3. Interaction between NF-κB and the activated GR. Peripherally generated TNF-α, IL-1β, and IL-6 activate the hypothalamic-pituitary-adrenal (HPA) axis. The HPA axis responds in a graded manner to greater degree of inflammation with increased production of ACTH and cortisol. Cortisol or exogenous GCs freely cross into the cytoplasm. Once inside the cell, GC interaction with the GR is controlled by the cortisone/cortisol shuttle. 11beta-Hydroxysteroid dehydrogenase (11β–HSD) isozymes are responsible for the interconversion of active 11β-hydroxy GCs (cortisol) and inactive 11-keto GCs (cortisone). GCs bind to specific GRs (GRα) to form the activated receptor (GC-GRα). GC-GRα complexes may influence NF-κB activity in five major ways: (1) by physically interacting with the p65 subunit with formation of an inactive (GC-GRα/NF-κB) complex, (2) by inducing the synthesis of the inhibitory protein IκBα via interaction with GC-responsive element DNA in the promoter of the IκB gene, (3) by blocking degradation of IκBα via enhanced synthesis of IL-10, (4) by impairing TNF-α–induced degradation of IκBα, and (5) by competing for limited amounts of GR coactivators such as CREB-binding protein (CBP) and steroid receptor coactivator-1 (SRC-1). GC-GRα may also decrease the stability of mRNA of several proinflammatory cytokines and other molecules.

us through the nuclear pore, thanks to two nuclear localization signals (NLS1 and NLS 2). After exerting its transcriptional effects, GRα is exported to the cytoplasm via an energy-dependent transport system, also through the nuclear pore.

It is now appreciated that the GC-GR complexes modulate transcription in a *hormone-dependent manner* by binding as a dimer to GC-responsive elements (GREs) located in the promoter regions of GC responsive genes and by interfering with the activity of other transcription factors such as NF-κB on genes regulated by these factors.[56] As a dimer and/or a monomer, GR-mediated transcriptional interference is achieved by five important mechanisms (FIG. 3): (i) by physically interacting with the p65 subunit to form an inactive (GR-NF-κB) complex,[54] (ii) by inducing the transcription of the inhibitory protein *IκBα* gene[54,57,58] (iii) by blocking degradation of IκBα via enhanced synthesis of IL-10,[59–61] (iv) by impairing TNFα-induced degradation of IκBα,[62,63] and (v) by competing for limited amounts of GR coactivators such as CREB-binding protein (CBP) and steroid receptor coactivator-1 (SRC-1).[64] In addition to the transcriptional modulation described above, GCs also influence the processing of mRNA and translation of proteins probably through transactivation or transrepression of genes that regulate mRNA stability and translation.[55]

By these mechanisms, activated GRs inhibit the transcription of several cytokines,[65,66] inflammation-associated enzymes,[65,67] and adhesion molecules.[60] GCs also have an inhibitory effect on fibrogenesis[68] and act in synergy with IL-1 receptor antagonist[69] and the anti-inflammatory cytokines IL-4, IL-10, and IL-13.[70]

EXAGGERATED HOST DEFENSE RESPONSE AND ASSOCIATED GLUCOCORTICOID INADEQUACY/RESISTANCE

Glucocorticoids as end-effectors of the hypothalamic-pituitary-adrenal axis are the most important natural inhibitors of inflammation.[71] However, endogenous GCs are not always effective in suppressing life-threatening systemic inflammation, and degree of cortisolemia frequently correlates with severity of illness and mortality rate.[72–75] Unquestionably, the elevation of GC secretion in nonsurvivors is inadequate to meet the needs of the concurrent inflammatory response and its adverse systemic effects. Failure to suppress inflammation could be due to tissue resistance to GCs, inadequacy of the level and duration of endogenous GC elevation to suppress an HIR gone awry, or both.[76]

The concept of acquired GC resistance was first introduced by Kass and Finland in 1957.[77] These investigators suggested that increased blood cortisol levels in nonsurvivors of sepsis might reflect a block to steroidal activity or transport, as a consequence of the infection. In this situation, a small increase in blood levels with a low (equal or less than physiological) dose of exogenous GCs was believed to be sufficient for facilitating the passage of steroids into host cells.[77] GR-mediated resistance was originally described as a *primary* inherited familial syndrome,[66,78] and was recently recognized as an *acquired* condition. Among others, *acquired* immune tissue-specific GR resistance has been described in patients with asthma,[65,79–81] acquired immunodeficiency syndrome (AIDS),[82] and severe sepsis.[83]

Recent *in vitro* studies have shown that cytokines may induce resistance to GCs by reducing GR binding affinity to cortisol and/or GREs.[84–86] Such abnormalities of GR function were demonstrated in T cells incubated with a combination of IL-2 and IL-4,[85] IL-1β, IL-6, and interferon (IFN)-γ,[84] or IL-13.[85] GC resistance was induced in a cytokine concentration-dependent fashion and was reversed by the removal of cytokines.[85] GR receptor–mediated resistance in the presence of systemic inflammation was also studied in experimental models of sepsis and sepsis-induced ARDS.[83,87,88] In a sheep model of sepsis-induced ARDS, maximal binding capacity of GR decreased continuously after endotoxin infusion, while there was a marked elevation of cortisol levels.[87] The reduced GR binding correlated negatively ($r = -0.87$, $P < 0.01$) with phospholipase A_2 (PLA_2) activity, a gene that is stimulated by NF-κB.

In a rat model of septic shock, GR blockade by mifepristone (RU 486) exacerbated the physiologic and pathologic changes induced by endotoxemia.[88] PLA_2 activity in rats with 80% GR blockade was more marked than in those with 50% GR blockade.[88] Monocytes of patients with sepsis developed near total GC resistance *in vitro*, when cytokines, especially IL-2, were added.[83]

Several inflammatory cytokines, including TNFα, IL-1β, and IL-6 activate NF-κB.[89] It has been proposed that when cytokine-activated NF-κB forms protein–protein complexes with activated GR, the availability and activity of effective GR molecules are reduced.[54,65] This functional reduction in GR availability is associated with decreased GR-GRE DNA binding and GC-mediated anti-inflammatory activity.[54,65]

II

LONGITUDINAL STUDIES OF BIOMARKERS OF HIR IN ARDS

We have completed two longitudinal prospective studies obtaining serial quantitative measurements of biomarkers of HIR activity (TABLE 4) in the systemic circulation and in the lung in 72 patients with ARDS (24 improvers and 48 nonimprovers). In these patients, sepsis was the leading (71%) condition precipitating ARDS. The first set of studies (1992–1993) included 43 patients[15,90–92] and the second set (1994–1996) included 29 patients.[93–96] Biomarker levels were correlated longitudinally with physiological scores of pulmonary and extrapulmonary organ function over time (LIS and multiple organ dysfunction syndrome [MODS] components and scores)[11,97] and survival. In all patients, ARDS was studied at its onset (ARDS days 1–3), during its natural progression in the first 7–10 days, and in those failing to improve by day 7–10 (unresolving ARDS) in response to prolonged GC treatment.[92,98] In all of our studies, we consistently found that the biomarker levels profile significantly diverged over time between improvers and nonimprovers (TABLE 4).

Findings at the Onset of ARDS (ARDS Days 1–3)

ARDS was associated with elevation of all measured biomarkers of HIR activity in plasma and bronchoalveolar lavage (BAL). In the circulation, improvers in comparison to nonimprovers had (i) significantly lower levels of TNFα, IL-1β, IL-2, IL-4, IL-6, IL-8, sICAM-1, and IL-1ra, and (ii) similar levels of IL-10, sTNFR1, sTNFR2, PINP, and PIIINP. In the lung, improvers in comparison to nonimprovers had (i) significantly lower levels of TNFα, IL-1β, IL-6, IL-8, sICAM-1, and (ii) similar levels of IL-2, IL-4, IL-10, IL-1ra, sTNFR1, sTNFR2, PINP, PIIINP, albumin, total protein, and PMN%. Nonimprovers had significantly lower ratio of IL-10/IL-1β and IL-10/TNFα in plasma and BAL, indicating an insufficient anti-inflammatory response.[96] Moreover, nonimprovers had greater than twofold lower ratios of cortisol:ACTH, suggesting an inadequate increase in endogenous GC levels from either a blunted ACTH response and/or tissue resistance to GCs.[95] Importantly, we have consistently found in our studies[15,90,91,93–96] that plasma and BAL levels of HIR biomarkers at ARDS onset were better predictors of outcome than physiological variables and indices of severity (LIS, MODS, and APACHE III scores).[15] Among

TABLE 4. Biomarkers of HIR activity in ARDS improvers (n = 24) and nonimprovers (n = 48)

Laboratory Variables	ARDS day 1–3 Improvers vs nonimprovers	ARDS days 3–10 Improvers	ARDS days 3–10 Nonimprovers
Plasma TNFα,* IL-1β,* IL-6,* IL-8*	↓	⇓	≈
BAL TNFα, IL-1β, IL-6	↓	⇓	≈
Plasma sICAM-1*	↓	≈	⇑
BAL IL-8* and sICAM-1*	↓	⇓	≈
Plasma IL-2,* IL-4	↓	⇑	⇑
Plasma IL-10†	≈	≈	⇓
BAL IL-10†	≈	≈	⇓
Plasma IL-10/ IL-1β, IL-10/TNFα	↑	⇑	≈
BAL IL-10/ IL-1β, IL-10/TNFα	↑	⇑	≈
Plasma IL-1ra	↓	⇓	≈
Plasma sTNFR1, sTNFR2	≈	⇓	⇑
BAL IL-1ra, sTNFR1, sTNFR2	≈	⇓	≈
BAL IL-1ra/IL-1β , sTNFR1/TNFα, sTNFR2/TNFα	≈	⇑	≈
Plasma cortisol:ACTH ratio	↓	≈	≈
Plasma PINP and PIIINP	≈	≈	⇑
BAL PINP and PIIINP	≈	⇓	⇑
BAL albumin, Total protein, PMN%	≈	⇓	≈

ARDS day 1–3 improvers vs nonimprovers = comparison between the two groups at the onset of ARDS; ARDS days 3–10 improvers = changes from onset of ARDS to day 10 within the group of improvers; ARDS days 3-10 nonimprovers = changes from onset of ARDS to day 10 within the group of nonimprovers. ARDS day 1–3 (improvers vs. nonimprovers): ↑ = significantly higher; ↓ = significantly lower; ≈ = similar; ARDS day 3-10: ⇑ = significant increase over time; ⇓ = significant reduction over time; ≈ = no change over time (i.e., persistent elevation).

*Transcription regulated by nuclear factor-κB (NF-κB)

†Transcription regulated by the glucocorticoid receptor (GR) complex

Definitions: sICAM-1, soluble intercellular adhesion molecule-1; sTNFR, soluble TNF receptor; IL-1ra, IL-1 receptor antagonist; ACM, alveolo-capillary membrane; PINP and PIIINP, procollagen aminoterminal propeptide type I and III; PMN%, percentage of neutrophils.

nonimprovers/nonsurvivors, those dying early (≤ 3 days) had significantly higher plasma levels of TNFα, IL-1β IL-6, and IL-8.[12] Additionally, we found that in blood collected at the onset of ARDS TNFα plasma levels correlated strongly with TNFα mRNA levels in PBLs ($R = 0.993$, $R^2 = 0.985$; $P < 0.0001$), indicating that elevated plasma levels reflected true biological pro-inflammatory activity.[99]

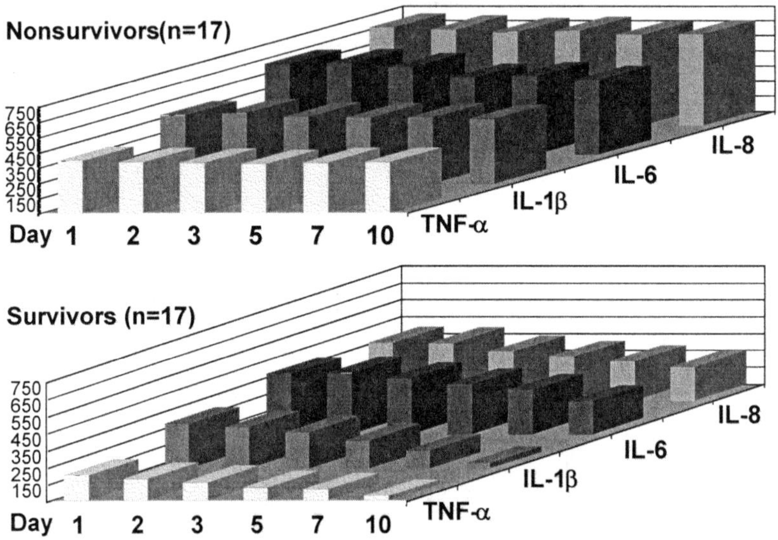

FIGURE 4. Plasma inflammatory cytokine levels over time in survivors and nonsurvivors. Plasma TNF-α, IL-1β, IL-6, and IL-8 levels from days 1 to 10 of sepsis-induced ARDS. On day 1 of ARDS, nonsurvivors ($n = 17$) had significantly higher ($P < 0.001$) TNF-α, IL-1β, IL 6, and IL-8 levels. Over time, nonsurvivors had persistent elevation, whereas survivors ($n = 17$) had a rapid decline. These findings indicate that loss of autoregulation is an early phenomenon. Data obtained from Refs. 15 and 90.

Findings during ARDS Progression (Days 3–10)

During ARDS progression, improvers had in the circulation (i) a significant reduction in levels of TNFα, IL-1β, IL-6, IL-8 (FIG. 4), IL-1ra, sTNFR1, sTNFR2, (ii) a trend toward a reduction in PINP and PIIINP, and (iii) no change in sICAM-1, IL-10, ACTH, and cortisol. In contrast, nonimprovers had in the circulation (i) persistent elevation in levels of TNFα, IL-1β, IL-6, IL-8, IL-1ra, (ii) an additional increase in IL-2, IL-4, sIACM-1, sTNFR1, sTNFR2, PINP and PIIINP, and (iii) a trend toward increased ACTH levels. BAL findings were similar to those of the plasma findings (TABLE 4). BAL IL-8 levels correlated with BAL PMN% ($R = 0.56624$, $P = 0.02$). Improvers, contrary to nonimprovers, exhibited a progressive decline in indices of vascular permeability (BAL albumin and TP levels) over time, indicating effective repair of the ACM barrier.[91]

Findings during Glucocorticoid Treatment of Unresolving ARDS

In a small, prospective, randomized, double-blind, placebo-controlled trial, we evaluate the efficacy of prolonged methylprednisolone (MP) treatment (MPT) in patients with unresolving ARDS (starting dose: 2 mg/kg/day).[98] The primary clinical outcome measures were improvement in LIS (equal or greater than one point) after 10 days of therapy and a decrease in ICU mortality. Sixteen patients received MPT and 8 received placebo. The two groups were similar at study entry. After 10 days, all patients in the MPT group improved (≥ one-point reduction in LIS) vs. 2 of 8 (25%) in the placebo group. ICU and hospital-associated mortality were significantly reduced: 0 vs. 62% ($P = 0.002$) and 12 vs. 62% ($P = 0.03$), respectively. The small number of patients may have biased the estimate of the treatment effect.

The therapeutic anti-inflammatory and anti-fibrotic efficacy of prolonged MPT was assessed with serial measurements of HIR biomarkers (TABLE 4). MPT was associated with a rapid and sustained reduction in mean plasma and BAL TNFα, IL-1β, IL-6, IL-8, sICAM-1, IL-1ra, sTNFR1, sTNFR2, PINP, and PIIINP, and with increases in IL-10 and in anti-inflammatory to pro-inflammatory cytokines ratios (IL-1ra/IL-1β, IL-10/TNFα, IL-10/IL-1β).[3,93,96] Placebo administration was not associated with reductions in HIR biomarkers. We believe that failure of older trials investigating massive doses of MP in early ARDS (TABLE 2) was attributable to the short duration of administration and not to timing of administration.

Hypothesis and Design To Test the Hypothesis

By investigating the longitudinal relationship between biomarkers of HIR activity and disease progression, we have identified the following attributes (i and iii are not reviewed in this chapter) of the HIR in ARDS: (i) nonspecific (similar qualitative response to insults of different etiology), (ii) broad (multiple cytokines elevated over time with a strong agreement among cytokines over time), (iii) autonomous (not influenced by intercurrent events, including nosocomial infections), and, most importantly, (iv) downregulation of HIR activity occurs early in improvers, but is absent in nonimprovers; it could be restored late in ARDS with prolonged GC treatment. Within the present knowledge of NF-κB and GR activities, a nonspecific and broad HIR is consistent with activation of NF-κB, whereas downregulation in association with GC treatment suggests that an exaggerated and protracted HIR may operate through pathways regulated by GC and its GC-GR complex.

We hypothesized that in the presence of an excess activation of NF-κB vs. GRα, the transcription of inflammatory mediators will increase over time, while in the presence of an excess activation of GRα vs. NF-κB, the transcription of inflammatory mediators will decrease over time (FIG. 5). Furthermore, we tested the secondary hypothesis that if endogenous GC inadequacy and/or peripheral tissue resistance are important pathophysiologic factors in a *dysregulated, protracted* HIR in ARDS, then *prolonged* GC therapy may be useful—not as an anti-inflammatory treatment *per se*, but as hormonal supplementation necessary to compensate for the host's inability to produce appropriately elevated levels of cortisol for the degree of peripheral GC resistance.[20] It follows that if *acquired* GR resistance played a role in the pathogenesis of unresolving ARDS, adequate hormonal supplementation should restore GRs anti-inflammatory function, decreasing the (1) production of inflammatory cytokines, (2) cytokine-driven HPA-axis activity, and (3) cytokine-driven organ dysfunction. Since NF-^{k}B and GRα are ubiquitous, we have investigated their relationship in normal circulating cells exposed to patients' plasma in experiments replicating systemic inflammation, and directly in lung tissue obtained from patients with unresolving ARDS.

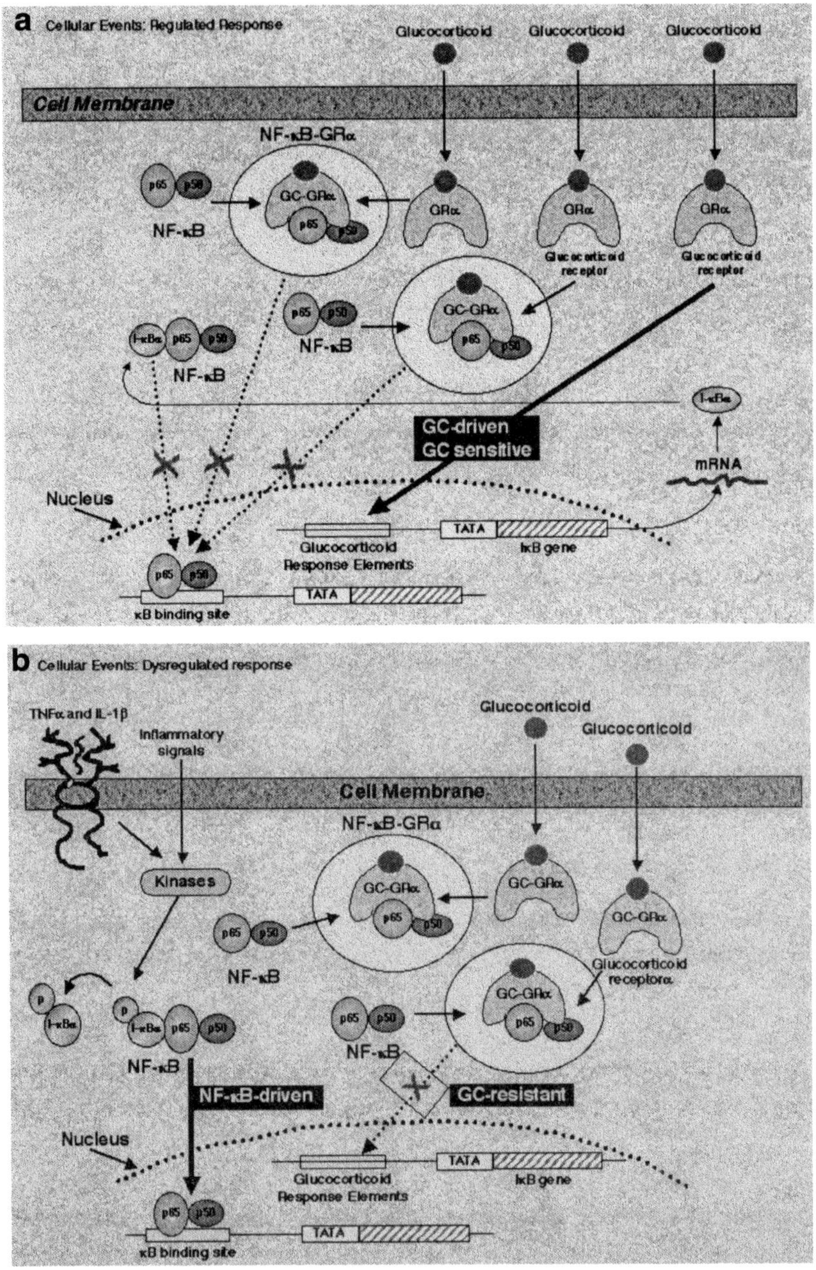

FIGURE 5. Regulated versus dysregulated response. Once activated, NF-κB and GRα can mutually repress each other through a protein–protein interaction that prevents their DNA binding and subsequent transcriptional activity. Activation of one transcription factor in excess of the binding (inhibitory) capacity of the other shifts cellular responses toward increased (dysregulated) or decreased (regulated) transcription of inflammatory mediators

Systemic Inflammation ARDS—Relationship between NF-κB and GRα?

Using an *ex vivo* model of systemic inflammation in ARDS, we investigated intracellular upstream and downstream events associated with DNA binding of NF-κB and GRα in the PBLs (naïve cells) obtained from a healthy volunteer.[95] PBLs were incubated for 3 h with plasma samples obtained longitudinally from 29 patients with ARDS, and then processed for fractionation into cytosolic and nuclear components, RNA extraction, and intracellular labeling. The primary objectives of these studies were to quantify the relationships among circulating levels of inflammatory cytokines TNFα and IL-1β and HPA-axis hormones (ACTH and cortisol), and intracellular activities mediated by NF-κB and GRα.

We found that exposure of naïve cells to ARDS patients' plasma (ARDS days 1–10) lead to NF-κB and GRα activation that reflected the patterns of systemic inflammation over time (as defined by plasma TNFα and IL-1β levels) and the physiological evolution of the disease (as defined by the LIS).[95] PBLs stimulated with plasma obtained from improvers with declining inflammatory cytokine levels over time demonstrated a progressive rise in GRα-mediated activities (GC-GRα binding to NF-κB, GC-GRα binding to GRE DNA, stimulation of inhibitory protein IκBα, and stimulation of IL-10 transcription), and a concomitant reduction in NF-κB κb-binding and transcription of TNFα and IL-1β (regulated, GRα-driven response; FIG. 5a). In contrast, PBLs stimulated with plasma from nonimprovers with exaggerated and sustained elevation in plasma inflammatory cytokine levels demonstrated only a modest increase in GC-GRα–mediated activity (despite elevated ACTH and cortisol levels), and a progressive rise in NF-κB activation over time (dysregulated, NF-κB–driven (FIG. 5b). Importantly, the response patterns observed in the PBLs following exposure to plasma from nonimprovers were most striking in nonsurvivors.

Effect of MPT on Systemic Inflammation ARDS—Relationship between NF-κB and GRα

Patients with unresolving ARDS and treated with methylprednisolone had rapid, progressive, and sustained reductions in plasma TNFα, IL-1β, IL-6, ACTH, and cortisol levels over time.[98] The progressive reduction in plasma TNFα and IL-1β levels was associated with parallel improvements in pulmonary and extrapulmonary organ dysfunction scores. Normal PBL exposed to plasma samples collected during MPT vs. placebo treatment exhibited rapid, progressive, significant increases in GR-mediated activities (GR binding to NF-κB, GR binding to GRE DNA, stimulation of inhibitory protein IκBα, and stimulation of IL-10 transcription), and significant reductions in NF-κB κb-binding and transcription of TNFα and IL-1β (FIG. 6). With MPT, the intracellular relationship between the NF-κB and GRα signaling pathways changed from an initial NF-κB–driven and GR-resistant state to a GR-driven

over time. (**a**) GC-GRα activation sufficient to maintain NF-κB levels in homeostasis (GC-GRα–regulated response) and to achieve a reduction in inflammation over time is shown. (**b**) An excess of NF-κB activation is shown, leading to protracted inflammation (NF-κB–driven dysregulated response) over time.

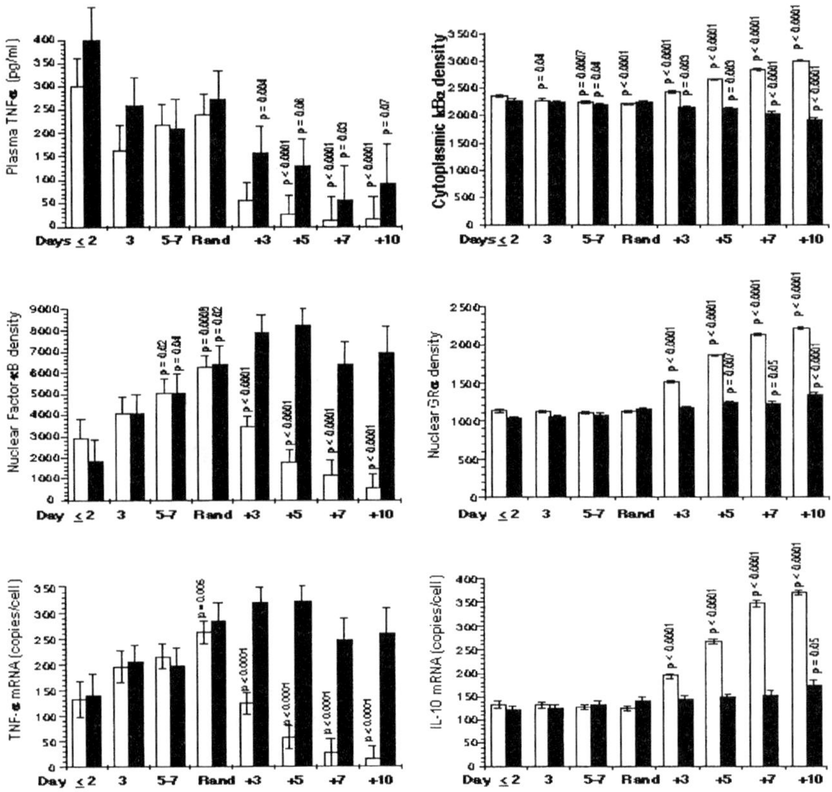

FIGURE 6. Extracellular and intracellular findings before and after randomization (Rand) in the methylprednisolone group (*open bar*) and placebo (*closed bar*). Values are expressed as mean ± standard error. The intracellular changes were observed by exposing peripheral blood leukocytes of healthy volunteers to plasma samples collected before and after randomization. (*Top left)* Plasma TNF-α levels; (*middle left*) nuclear factor-κB binding to κB response elements; (*bottom left*) messenger RNA TNF-α; (*bottom left*) messenger RNA TNF-α; (*top right*) levels of cytoplasmic IκBα; (*middle right)* GR binding to GC response elements; (*bottom right*); (*bottom right)* messenger RNA IL-10. From Meduri *et al.*[3] Reproduced with permission from the *American Journal of Respiratory and Critical Care Medicine.*

and GR-sensitive one. The responses observed during MPT support the concept of inflammation-dependent *acquired* GC resistance in patients with ARDS and underscore the central role played by activated GR in regulating NF-κB–driven inflammation.

Pulmonary Inflammation in ARDS—Relationship between NF-κB and GRα

We hypothesized that in patients with unresolving ARDS, DAD with severe pulmonary fibroproliferation (PFP) is associated with predominant nuclear NF-κB uptake in resident cells, while mild PFP is associated with predominant nuclear GRα uptake and lower nuclear NF-κB uptake. In lung tissue obtained by open lung biopsies from eight patients with unresolving ARDS, we found histologically heterogeneity with mild and severe forms of the fibroproliferation involving different lobules in the same biopsy slide.[100] Immunohistochemical analysis consistently showed NF-κB and GRα uptake in all cells including type I and type II pneumocytes, endothelial cells, fibroblasts, and alveolar macrophages. This finding is consistent with the concept that all resident cells in the lungs are activated in the HDR. The intensity and the patterns of nuclear vs. cytoplasmic distribution differed in areas with mild vs. severe PFP, and a reciprocal relationship was observed for nuclear uptake of NF-κB and GRα.[100] Regions with severe vs. mild PFP had predominant NF-κB nuclear uptake (13 ± 1.3 vs. 7 ± 2.9; $P = 0.01$), less nuclear GRα uptake (7 ± 2.9 vs. 10 ± 3.5; $P = 0.23$), and lower GRα/NF-κB nuclear uptake ratio (0.53 ± 0.18 vs. 1.52 ± 0.23; $P = 0.007$). These immunohistochemistry findings support our original hypothesis and underscore the regulatory role of GR in the adaptive vs. maladaptive histological evolution of ARDS.

III

FACTORS AFFECTING CELLULAR RESPONSE TO GLUCOCORTICOIDS

Glucocorticoid responsiveness is defined as the ability of a cell type or target gene to exhibit measurable changes in response to GCs.[54] The myriad determinants of GC responsiveness can be summarized as follows: (1) GCs must be available in sufficient concentration at the site of action (i.e., favorable concentration vs. time profile); (2) physicochemical properties of GCs must permit sustainable intracellular concentrations; (3) functional GR and requisite transcription factors must be present; and (4) the target gene of interest must be active in a given cell. If one of these factors is defective, the system does not respond properly.[76]

Factors Affecting Intracellular Accumulation of Free Hormone

The stress response results in an increase in steady-state free cortisol levels via at least three separate mechanisms: (1) enhanced synthesis, (2) decreased protein binding secondary to reduced expression of binding proteins, and (3) reduced elimination (FIG. 7).

Hormone production. Inflammatory mediators (e.g., IL-1β and IL-6) stimulate the hypothalamic paraventricular nucleus resulting in enhanced production and release of CRH into the hypophyseal portal system. CRH directs the anterior pituitary to release ACTH. ACTH engages two different ACTH receptor subtypes in zona fasciculata cells of the adrenal glands, high affinity and low affinity. ACTH stimulates synthesis of cortisol probably via induction of enzymes involved in steroidogenesis (CYP17, CYP21, and 3β-HSD). There is a poor correlation between cortisol production rates and total circulating cortisol levels. Interestingly, the correlation is greatly improved when total cortisol levels are corrected for serum cortisol binding globulin concentration.[101] Approximately 90% of circulating cortisol is nonlinearly bound to

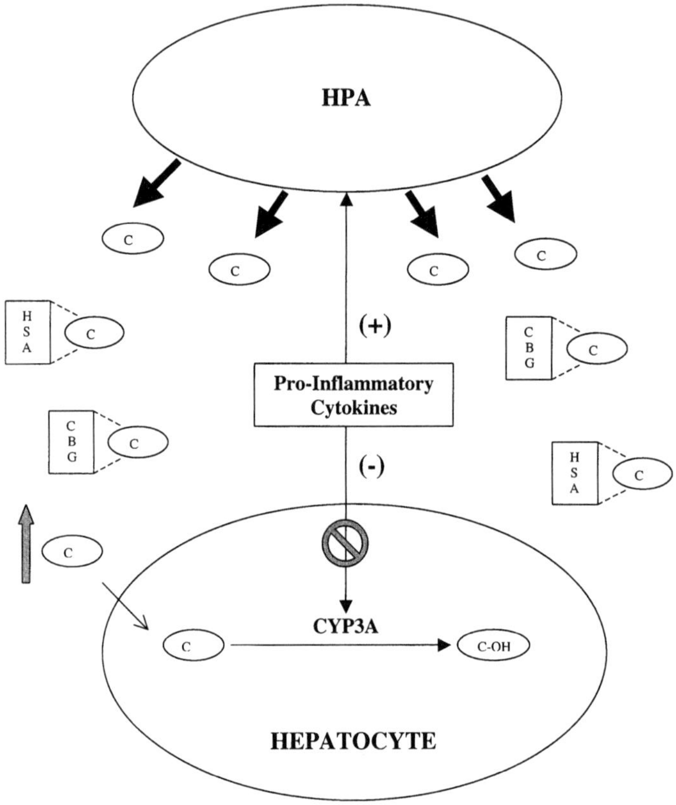

FIGURE 7. Effect of inflammation of circulating cortisol levels. Pro-inflammatory cytokines stimulate the hypothalamic-pituitary axis (HPA) to increase cortisol secretion. Cortisol circulates in the plasma bound to low-affinity, high-capacity (human serum albumin, HSA) and high-affinity low-capacity (cortisol binding globulin, CBG) proteins. Inflammation leads to (1) elevated levels of unbound (free) cortisol secondary to decreased plasma HSA and CBG levels and (2) diminished cytochrome P-450 (CYP3A) mediated conversion of cortisol to its inactive metabolite 6β-hydroxy cortisol (6-OH). Thus, the overall effect of inflammation is increased circulating free levels of cortisol.

cortisol binding globulin (CBG). Consequently, the fraction of free cortisol increases as total plasma cortisol levels increase.

Hormone elimination. Hepatic metabolism is the major elimination pathway for free cortisol with less than 1% of unchanged cortisol undergoing renal excretion. Cortisol elimination pathways include reduction of the A ring, reduction of the 20-ketone function, and by 6β-hydroxylation. In humans, conversion of cortisol to its inactive metabolite, 6β-hydroxy cortisol, is catalyzed by cytochrome P-450 3A (CYP3A), the most abundant hepatic CYP-P450 drug metabolizing enzyme.[102] Inflammation and infection result in a generalized reduction in CYP-P450 activity.[103–108] Reduced removal rates results in increased circulating cortisol levels independent of cortisol synthesis. The effects of inflammation and infection on P-450 activity are believed to be attributable to stimulation of the cellular immune response. Animal models in which inflammation and infection have been shown to reduce the clearance of hepatically metabolized drugs have been developed, which clearly establish the link between inflammation and/or infection and alterations in the *in vitro* and *in vivo* metabolism of drugs in animals.[109,110] Administration of LPS to humans resulted in a significant reduction in the hepatic metabolism of antipyrine, hexobarbital, and theophylline, drugs commonly used as probes for assessing P-450 activity.[108] Moreover, the reduction in antipyrine clearance was correlated with peak values of TNFα and IL-6, suggesting a relationship between the reduction in P-450 activity and intensity of the inflammatory response.[103,108]

Circulating cortisol levels. The majority of circulating cortisol is bound to the plasma proteins cortisol binding globulin (CBG; binds cortisol with high affinity and low capacity) and human serum albumin (HSA; binds cortisol with low affinity and high capacity). The unbound fraction of cortisol is responsible for activity. Albumin is considered a negative acute phase reactive protein whose synthesis is negatively regulated by TNFα and IL-1β.[111] Previous studies showed that infusion of LPS reduced albumin binding capacity with no change in albumin concentration, resulting in an increased unbound fraction of drug.[112,113] A reduction in the circulating levels of CBG and albumin during the acute-phase response results in an increase in circulating free cortisol levels. Additional factors contributing to hypercortisolemia in patients with systemic inflammation include decreased binding to CBG and decreased cortisol extraction from the blood.[71,114–116] Septic patients have a rapid depletion in CBG with increased free (nonbound) cortisol.[114,116,117] The mechanism for the high rate of degradation or removal of CBG is unknown, but it parallels an increase in C-reactive protein (an acute-phase protein produced by the liver in response to inflammatory cytokines) levels.[114] In survivors, in contrast to nonsurvivors, CBG and cortisol levels return to normal values during recovery and correlate positively with a reduction in PLA_2 levels ($r > 0.833$; $p < 0.0001$).[114,118] Furthermore, nonsurvivors of sepsis, in contrast to survivors, also have impaired removal of injected cortisol from the plasma.[72]

P-Glycoprotein as a determinant of intracellular cortisol accumulation. Our laboratory characterized methylprednisolone disposition in ARDS patients in an attempt to gain insight into the mechanism(s) of GC resistance.[119] Interestingly, we found that the one individual who failed to respond to methylprednisolone therapy actually experienced deteriorating clearance over the course of therapy. As a result, this individual was exposed to a greater amount of methylprednisolone, i.e., greater area under the curve (AUC), when compared to individuals who responded to methylprednisolone therapy. These data, coupled with the fact that patients in many instances develop ARDS despite elevated cortisol levels, suggest that GC resistance may be related to mechanism(s) other than a reduced amount of GC made available to the target tissue (e.g., GC intracellular penetration).

It was thought previously that GCs move freely into and out of the cell by simple diffusion only. However, recent data demonstrated that many immune response effector cells (e.g., $CD4^+$ T cells, $CD8^+$ T cells, and natural killer [NK] cells) express a member of the large ATP-binding cassette superfamily of transport proteins also called traffic ATPases, P-glycoprotein (P-gp), in the cell membrane that impedes cellular penetration of various small molecules.[120,121] This observation is particularly significant considering that GCs are P-gp substrates with individual GC transport efficiency related to log P value.[122] P-gp inhibitors exert a statistically significantly dose-dependent increase in intracellular cortisol levels in human T lymphocytes.[123]

Recent data indicate that P-gp expression is capable of reducing intracellular GC penetration resulting in reduced GC binding and transactivating potency. In J774.2 mdr1b cells that have been selected for overexpression of P-gp, total dexamethasone binding was approximately threefold lower in comparison to J774.2 parental cells with minimal constitutive P-gp expression.[124]

Inhibition of P-gp with the potent inhibitor verapamil restored total dexamethasone binding in J774.2 mdr1b cells to approximately that noted in J774.2 cells. LLC-PK, a pig kidney epithelium-derived cell line, and a variant cell line that stably expresses the product of the human P-gp gene (L-MDR1) were transiently transfected with a luciferase reporter gene driven by an NF-κB response element and treated with cortisol and methylprednisolone in order to determine the effect of P-gp expression on GC transactivating potency (defined here as EC_{50}).[124] GR-mediated transcriptional activation increased nonlinearly with increasing steroid concentrations, reaching a plateau at higher concentrations. Cortisol EC_{50} was approximately threefold lower in L-MDR1 cells (EC_{50} = 24 nM) compared to LLC-PK cells (EC_{50} = 74 nM). Similarly, expression of P-gp reduced the GR-mediated transcriptional activation of methylprednisolone by threefold in L-MDR1 cells (EC_{50} = 13 nM) compared to LLC-PK cells (EC_{50} = 4 nM).

Inflammation may affect GC cellular penetration. Primary rat hepatocytes exposed to TNFα exhibited a twofold increase in mdr1b mRNA expression and immunodetectable P-gp protein.[125] Moreover, accumulation of intracellular rhodamine 123 (a P-gp substrate) was decreased, suggesting an enhancement of functional P-gp transporter.[125]

The fact that P-gp transports cortisol and is expressed in the adrenal glands suggests that P-gp may play a role in active secretion of cortisol. However, Cufer *et al.* found that administration of valspodar resulted in no statistically significant changes in plasma cortisol levels when compared to controls.[126] Interestingly, the authors noted a significant reduction in plasma cortisol levels when anticancer agents (mitotane or doxorubicin) were concomitantly administered with valspodar.[126] One possible interpretation of these results is that P-gp plays an insignificant role in basal cortisol secretion. However, P-gp may contribute to enhanced cortisol secretion in response to inflammatory stimuli. The role of P-gp in the host inflammatory response has yet to be studied in mice nullizygous for P-gps (*mdr1a* and *mdr1b*).

Cortisone/cortisol shuttle. Once inside the cell, GC interaction with the GR is controlled by the cortisone/cortisol shuttle (FIG. 2). 11beta-Hydroxysteroid dehydrogenase (11β-HSD) isozymes are responsible for the interconversion of active 11β-hydroxy GCs (cortisol) and inactive 11-keto GCs (cortisone). 11β-HSD1 is believed to possess dehydrogenase (cortisol to cortisone) and reductase (cortisone to cortisol) activity. However, *in vitro* and *in vivo* experiments suggest that 11β-HSD1 reductase activity predominates.[127–130] 11β-HSD2 possesses dehydrogenase activity only. 11β-HSD1 is expressed in liver, adipose tissue, vascular smooth muscle, brain, lung, anterior pituitary, and gonads. 11β-HSD1 is believed to amplify GC action due to the fact that 11β-HSD1 activates GCs in tissues with low mineralocorticoid receptor expression, but abundant, relatively low-affinity GRs. Reactivation of inert 11-keto GCs (e.g., cortisone) by 11β-HSD1 is believed to be important during periods of inadequate circulating cortisol levels (e.g., during diurnal trough). 11β-HSD2 is found primarily in mineralocorticoid target tissues, such as kidney, sweat glands, salivary glands, and colonic mucosa.[131] The mineralocorticoid receptor has a similar affinity for cortisol and aldosterone. By converting cortisol to cortisone, 11β-HSD2 provides selective activation of mineralocorticoid receptors by aldosterone in the kidney. Transgenic knockout of the 11β-HSD2 gene leads to GC-dependent mineralocorticoid excess and hypertension.[132] 11β-HSD2 synthesis and activity appear to be constitutive (i.e., not highly regulated). Conversely, 11β-HSD1 synthesis and activity are affected by various stimuli including GCs, stress, sex steroids, growth hormone, cytokines, and peroxisome proliferator activated receptor agonists.[131,133] For example, IL-1β and TNFα exhibit a dose-dependent increase in the reductase, but not the dehydrogenase, activity of 11β-HSD1.[134]

Factors Affecting GC-GR Signaling Cascade

Altered GR binding capacity and/or affinity. GR binding affinity and capacity are reduced in response to inflammatory stimuli. Mice administered endotoxin exhibited a dose and time-dependent decrease in the hepatic GR binding capacity (Bmax), which resulted in abrogation of GC-induced increase in phosphoenolpyruvate carboxykinase activity.[135] In two separate canine shock models, endotoxin administration or intestinal ischemia, there was a significant reduction in leukocyte GR specific binding.[136,137] Similar findings of reduced GR binding capacity have been reported in patients with septic shock.[136,138] Treatment of mouse L929 fibroblasts with nitric oxide (NO) donors (e.g., S-nitroso-acetyl-DL-penicillamine) resulted in a time- and dose-dependent decrease in GR Bmax and Kd. Significantly elevated NO levels, secondary to increased inducible nitric oxide synthetase activity, are found in septic patients compared to healthy controls. Thus, the authors postulated that decreased GR binding capacity resulting from S-nitrosylation of critical GR sulfhydryl groups may explain GC resistance observed in some patients with septic shock.[139] In a sheep model of acute lung injury, high-dose endotoxin infusion led to a continuous time-dependent decrease in Bmax of GR in lung cytoplasm extracts compared to untreated controls (66 ± 2 vs. 113 ± 3 fmol/mg protein at 2 h; 52 ± 3 vs. 105 ± 6 fmol/mg protein at 4 h; and 37 ± 2 vs. 105 ± 5 fmol/mg protein at 6 h). Interestingly, there was a negative correlation between Bmax and phospholipase A2 activity.[140] Bronchial epithelial-derived cell lines, BEAS S6 and 2B, exhibited significant reductions in dexamethasone binding affinity, i.e., increased Kd, when exposed to either lipopolysaccharide or IL-1β.[141]

Numerous *in vitro* and *in vivo* experiments have demonstrated that acute exposure to GCs results in a significant reduction in GR binding capacity using whole-cell binding assays.[142–147] Diminished GR density has been shown to be a direct result of GC-induced GR mRNA and protein downregulation in both human and animal cell lines.[148–152]

Perturbations in GR crosstalk. GCs exert potent anti-inflammatory and immunosuppressive activity primarily as a result of their ability to transrepress several key transcription factors (e.g., NF-κB and AP-1), a process known as crosstalk.[153,154] In support of this notion, LPS challenge experiments in a GR–mutant mouse model (GR^{dim}), which is deficient in dimerization and DNA binding, demonstrated that the anti-inflammatory and immunosuppressive activity of GCs is largely independent of transactivation (i.e., DNA binding).[155] Thus, an overabundance of pro-inflammatory transcription factors in relation to ligand-activated GRs can be hypothesized to contribute to GC resistance in inflamed tissues. Work presented in earlier in this chapter supports the presence of inflammation-induced GC resistance in patients with unresolving ARDS.

Two distinct GC-resistant asthmatic phenotypes with regard to GR–binding capacity have been described. The reversible type 1 phenotype is associated with a reduced Kd and normal Bmax whereas the irreversible type 2 phenotype is associated with a reduced Bmax and a normal Kd.[80] Both phenotypes are associated with a blunted GC response in PBMCs.[156] The exact mechanism underlying GC resistance in asthma is unknown. However, there is a demonstrated increase in DNA binding of AP-1 in PBMCs from GC-resistant asthmatics compared to GC-sensitive asthmatics, suggesting that titration of ligand-activated GRs by excess AP-1 may alleviate the anti-inflammatory activity of GCs.[157] Excess basal c-fos expression, the inducible form of AP-1, accounts for the enhanced AP-1 DNA binding activity found in PBMCs obtained from GC-resistant asthmatics as compared to PBMCs obtained from GC-sensitive asthmatics.[158]

The human GR gene encodes two protein isoforms: a cytoplasmic form (GRα), which binds hormone, translocates to the nucleus, and regulates gene transcription, and a nuclear localized β isoform GRβ), which does not bind known ligands and attenuates GRα action. Webster *et al.*[159] identified a NF-κB binding site in the GR promoter, indicating that the cytokine-mediated increase in both GRα and GRβ mRNA is mediated by NF-κB–induced increase in transcription of the GR gene. In HeLaS3 and human CEMC7 lymphoid cells, activation of NF-κB by TNFα lead to a dose- and time-dependent disproportionate increase in the steady-state levels of the GRβ (3.5–4.8-fold) protein isoform over GRα(1.5–2.0-fold), making GRβ the predominant endogenous receptor isoform.[159] The increase in GRβ protein expression correlated with the development of GC resistance.[159]

CONCLUSIONS AND AREA OF FUTURE INVESTIGATION

A dysregulated inflammatory response lies at the heart of the pathogenesis of ARDS. This dysregulation manifests as an excess activation of pro-inflammatory transcription factors, e.g., NF-κB and AP-1, compared to the anti-inflammatory transcription factor GR. The imbalance of pro- and anti-inflammatory mediators is hypothesized to result in noncompensated GC resistance in target organs. Discovery of next-generation anti-inflammatory drugs designed to circumvent GC resistance hinges on our ability to gain a more in depth understanding of the GR signaling cascade. Experimental evidence obtained from a GR mutant mouse model (GR^{dim}), which is deficient in dimerization and DNA binding, demonstrates that the anti-inflammatory and immunosuppressive activity of GCs is associated with the transrepressive activity of GCs. These observations led to the discovery of dissociating GCs, GR ligands devoid of DNA binding that are capable of eliciting a selective interaction between the ligand-activated GR and either AP-1 or NF-κB. Theoretically, dissociating GCs make it possible to delineate genes whose transactivation and/or transrepression are directly related to the interaction of the activated GR with AP-1, NF-κB, or the GRE. Dissection of GR signaling pathways will be particularly useful, considering cDNA microarray results from ARDS, and sepsis patients will likely reveal novel genes related to disease-susceptibility and/or GC response. Thus, it is important to gain a greater understanding of GR signaling pathways in order to determine how these novel genes are regulated, and to facilitate their evaluation as potential drug targets.

ACKNOWLEDGMENT

This work was supported in part by a grant the St. Francis of Assisi Foundation of Memphis.

REFERENCES

1. FRANCHIMONT, D. *et al.* 2003. Glucocorticoids and inflammation revisited: the state of the art. NIH Clinical Staff Conference. Neuroimmunomodulation **10:** 247–260.
2. BAEUERLE, P.A. & D. BALTIMORE. 1988. Activation of DNA-binding activity in an apparently cytoplasmic precursor of the NF-kappa B transcription factor. Cell **53:** 211–217.
3. MEDURI, G.U. *et al.* 2002. Prolonged methylprednisolone treatment suppresses systemic inflammation in patients with unresolving acute respiratory distress syndrome. Evidence for inadequate endogenous glucocorticoid secretion and inflammation-induced immune cell resistance to glucocorticoids. Am. J. Respir. Crit. Care Med. **165:** 983–991.
4. KATZENSTEIN, A.L., C.M. BLOOR & A.A. LEIBOW. 1976. Diffuse alveolar damage—the role of oxygen, shock, and related factors. A review. Am. J. Pathol. **85:** 209–228.
5. HUDSON, L.D. *et al.* 1995. Clinical risks for development of the acute respiratory distress syndrome. Am. J. Respir. Crit. Care Med. **151:** 293–301.
6. STEINBERG, K.P. & L.D. HUDSON. 2000. Acute lung injury and acute respiratory distress syndrome. The clinical syndrome. Clin. Chest Med. **21:** 401–2417, vii.
7. KRAFFT, P. *et al.* 1996. The acute respiratory distress syndrome: definitions, severity and clinical outcome. An analysis of 101 clinical investigations. Intensive Care Med. **22**: 519–529.
8. SLOANE, P.J. *et al.* 1992. A multicenter registry of patients with acute respiratory distress syndrome. Physiology and outcome. Am. Rev. Respir. Dis. **146**: 419–426.
9. MEDURI, G.U. *et al.* 1994. Corticosteroid rescue treatment of progressive fibroproliferation in late ARDS. Patterns of response and predictors of outcome. Chest **105:** 1516–1527.
10. MEDURI, G.U., M. ELTORKY & H.T. WINER-MURAM. 1995. The fibroproliferative phase of late adult respiratory distress syndrome. Semin. Respir. Infect. **10:** 154–175.
11. MURRAY, J.F. *et al.* 1988. An expanded definition of the adult respiratory distress syndrome. Am. Rev. Respir. Dis. **138:** 720–723.
12. BONE, R.C. *et al.* 1989. An early test of survival in patients with the adult respiratory distress syndrome. The PaO2/FIo2 ratio and its differential response to conventional therapy. Prostaglandin E1 Study Group. Chest **96:** 849–851.
13. BERNARD, G.R. *et al.* 1987. High-dose corticosteroids in patients with the adult respiratory distress syndrome. N. Engl. J. Med. **317:** 1565–1570.
14. MEDURI, G.U. 1997. Host defense response and outcome in ARDS. Chest **112:** 1154–1158.
15. HEADLEY, A.S., E. TOLLEY & G.U. MEDURI. Infections and the inflammatory response in acute respiratory distress syndrome. Chest **111:** 1306–1321.
16. COTRAN, R.S., V. KUMAR & S.L. ROBBINS. 1994. Cellular injury and cellular death. *In* Pathologic Basis of Disease. R.S. Cotran, V. Kumar & S.L. Robbins, Eds.: 1–34. W.B. Saunders. Philadelphia.
17. BROWER, R.G. *et al.* 2001. Treatment of ARDS. Chest **120:** 1347–1367.

18. LEFERING, R. & E.A. NEUGEBAUER. 1995. Steroid controversy in sepsis and septic shock: a meta-analysis. Crit. Care Med. **23:** 1294–1303.
19. CRONIN, L. *et al.* 1995.Corticosteroid treatment for sepsis: a critical appraisal and meta-analysis of the literature. Crit. Care Med. **23:** 1430–1439.
20. MEDURI, G.U. 1999. An historical review of glucocorticoid treatment in Sepsis. Disease pathophysiology and the design of treatment investigation. Sepsis **3:** 21–38.
21. ASHBAUGH, D.G. & R.V. MAIER. 1985. Idiopathic pulmonary fibrosis in adult respiratory distress syndrome. Diagnosis and treatment. Arch. Surg. **120:** 530–535.
22. HOOPER, R.G. & R.A. KEARL. 1990. Established ARDS treated with a sustained course of adrenocortical steroids. Chest **97:** 138–143.
23. SHANLEY, T.P. & H.R. WONG. 2001. Signal transduction pathways in acute lung injury: NF-κB and AP-1. *In* Molecular Biology of Acute Lung Injury. H.R. Wong & T.P. Shanley, Eds.: 1–16. Kluwer. Boston.
24. ROSS, S.D. *et al.* 2000. Attenuation of lung reperfusion injury after transplantation using an inhibitor of nuclear factor-kappaB. Am. J. Physiol. Lung Cell Mol. Physiol. **279:** L528–536.
25. LENTSCH, A.B. *et al.* 1999. Essential role of alveolar macrophages in intrapulmonary activation of NF-kappaB. Am. J. Respir. Cell Mol. Biol. **20:** 692–698.
26. CHRISTMAN, J.W., R.T. SADIKOT & T.S. BLACKWELL. 2000. The role of nuclear factor-kappa B in pulmonary diseases. Chest **117:** 1482–1487.
27. KARIN, M. & Y. BEN-NERIAH. 2000. Phosphorylation meets ubiquitination: the control of NF-[kappa]B activity. Annu. Rev. Immunol. **18:** 621–663.
28. BAEUERLE, P.A. & D. BALTIMORE. 1996. NF-kappa B: ten years after. Cell **87:** 13–20.
29. YAMAMOTO, Y. & R.B. GAYNOR. 2001. Therapeutic potential of inhibition of the NF-kappaB pathway in the treatment of inflammation and cancer. J. Clin. Invest. **107:** 135–142.
30. BARNES, P.J. & M. KARIN. 1997. Nuclear factor-kappa B: a pivotal transcription factor in chronic inflammatory diseases. N. Engl. J. Med. **336:** 1066–1071.
31. STEIN, B. *et al.* 1993. Cross-coupling of the NF-kappa B p65 and Fos/Jun transcription factors produces potentiated biological function. EMBO J. **12:** 3879–3891.
32. BAUMANN, H. & J. GAULDIE. 1994. The acute phase response. Immunol. Today **15:** 74–80.
33. HEUMANN, D. & M.P. GLAUSER. 1994. Pathogenesis of sepsis. Sci. Med. **1:** 28–37.
34. STRIETER, R.M. *et al.* 2001. Innate immune mechanisms triggering lung injury. *In* Molecular Biology of Acute Lung Injury. H.R. Wong & T.P. Shanley, Eds.: 17–33. Kluwer. Boston.
35. SCHWARTZ, M.D. *et al.* 1996. Nuclear factor-kappa B is activated in alveolar macrophages from patients with acute respiratory distress syndrome. Crit. Care Med. **24:** 1285–1292.
36. MAUS, U. *et al.* 1998. Increased proinflammatory cytokine gene expression in alveolar macrophages is associated with changes in NF-kB DNA binding activity in septic ARDS. AJRCCM **157:** A459.
37. MOINE, P. *et al.* 2000. NF-kappaB regulatory mechanisms in alveolar macrophages from patients with acute respiratory distress syndrome. Shock **13:** 85–91.
38. CARTER, A.B., M.M. MONICK & G.W. HUNNINGHAKE. 1998. Lipopolysaccharide-induced NF-kappaB activation and cytokine release in human alveolar macrophages is PKC-independent and TK- and PC-PLC- dependent. Am. J. Respir. Cell. Mol. Biol. **18:** 384–391.
39. STRIETER, R.M. *et al.* 1993. Cytokines. 2. Cytokines and lung inflammation: mechanisms of neutrophil recruitment to the lung. Thorax **48:** 765–769.
40. VAN DER POLL, T. *et al.* 1990. Activation of coagulation after administration of tumor necrosis factor to normal subjects. N. Engl. J. Med. **322:** 1622–1627.
41. ELIAS, J.A. *et al.* 1990. Cytokine networks in the regulation of inflammation and fibrosis in the lung. Chest **97:** 1439–1445.
42. KING, R.J., M.B. JONES & P. MINOO. 1989. Regulation of lung cell proliferation by polypeptide growth factors. Am. J. Physiol. **257:** L23–38.
43. POSTLETHWAITE, A.E. & J.M. SEYER. 1990. Stimulation of fibroblast chemotaxis by human recombinant tumor necrosis factor alpha (TNF-alpha) and a synthetic TNF-alpha 31–68 peptide. J. Exp. Med. **172:** 1749–1756.

44. PERLSTEIN, R.S. *et al.* 1993. Synergistic roles of interleukin-6, interleukin-1, and tumor necrosis factor in the adrenocorticotropin response to bacterial lipopolysaccharide *in vivo*. Endocrinology **132:** 946–952.
45. HERMUS, A.R. & C.G. SWEEP. 1990. Cytokines and the hypothalamic-pituitary-adrenal axis. J. Steroid Biochem. Mol. Biol. **37:** 867–871.
46. GALLIN, J.K., I.M. GOLDSTEIN & R. SNYDERMAN. 1988. Overview. *In* Inflammation: Basic Principles and Clinical Correlates., J.I. Gallin, I.M. Goldstein & R. Snyderman, Eds.: 1–3. Raven Press. New York.
47. CERAMI, A. 1992. Inflammatory cytokines. Clin. Immunol. Immunopathol. **62**): S3–10.
48. CHRISTOU, N.V. 1996. Host defense mechanisms of surgical patients. Friend or foe? Arch. Surg. **131:** 1136–1140.
49. GUIRAO, X. & S.F. LOWRY. 1996. Biologic control of injury and inflammation: much more than too little or too late. World J. Surg. **20:** 437–446.
50. MUNCK, A., P.M. GUYRE & N.J. HOLBROOK. 1984. Physiological functions of glucocorticoids in stress and their relation to pharmacological actions. Endocr. Rev. **5:** 25–44.
51. ZUCKERMAN, S.H., J. SHELLHAAS & L.D. BUTLER. 1989. Differential regulation of lipopolysaccharide-induced interleukin 1 and tumor necrosis factor synthesis: effects of endogenous and exogenous glucocorticoids and the role of the pituitary-adrenal axis. Eur. J. Immunol. **19:** 301–305.
52. VERMES, I. *et al.* 1995. Dissociation of plasma adrenocorticotropin and cortisol levels in critically ill patients: possible role of endothelin and atrial natriuretic hormone. J. Clin. Endocrinol. Metab. **80:** 1238–1242.
53. HAMMOND, G.L. 1995. Potential functions of plasma steroid-binding proteins. Trends Endocrinol. Metab. **6:** 298–304.
54. BAMBERGER, C.M., H.M. SCHULTE & G.P. CHROUSOS. 1996. Molecular determinants of glucocorticoid receptor function and tissue sensitivity to glucocorticoids. Endocr. Rev. **17:** 245–261.
55. BAMBERGER, C.M. *et al.* 1995. Glucocorticoid receptor beta, a potential endogenous inhibitor of glucocorticoid action in humans. J. Clin. Invest. **95:** 2435–2441.
56. DIDONATO, J.A., F. SAATCIOGLU, AND M. KARIN. 1996. Molecular mechanisms of immunosuppression and anti-inflammatory activities by glucocorticoids. Am. J. Respir. Crit. Care Med. **154:** S11–15.
57. SCHEINMAN, R.I. *et al.* 1995. Role of transcriptional activation of I kappa B alpha in mediation of immunosuppression by glucocorticoids. Science **270:** 283–286.
58. WISSINK, S. *et al.* 1998. A dual mechanism mediates repression of NF-kappa B activity by glucocorticoids. Mol. Endocrinol. **12:** 355–363.
59. WANG, P. *et al.* 1995. Interleukin (IL)-10 inhibits nuclear factor kappa B (NF kappa B) activation in human monocytes. IL-10 and IL-4 suppress cytokine synthesis by different mechanisms. J. Biol. Chem. **270:** 9558–9563.
60. LENTSCH, A.B. *et al.* 1997. *In vivo* suppression of NF-kappa B and preservation of I kappa B alpha by interleukin-10 and interleukin-13. J. Clin. Invest. **100:** 2443–2448.
61. SHAMES, B.D. *et al.* 1998. Interleukin-10 stabilizes inhibitory kappaB-alpha in human monocytes. Shock **10:** 389–394.
62. HOFFMAN, S.L. *et al.* 1984. Reduction of mortality in chloramphenicol-treated severe typhoid fever by high-dose dexamethasone. N. Engl. J. Med. **310:** 82–88.
63. POPPERS, D.M., P. SCHWENGER & J. VILCEK. 2000. Persistent tumor necrosis factor signaling in normal human fibroblasts prevents the complete resynthesis of I kappa B-alpha. J. Biol. Chem. **275:** 29587–29593.
64. SHEPPARD, K.A. *et al.* 1998. Nuclear integration of glucocorticoid receptor and nuclear factor- kappaB signaling by CREB-binding protein and steroid receptor coactivator-1. J. Biol. Chem. **273:** 29291–29294.
65. BARNES, P.J., A.P. GREENING & G.K. CROMPTON. 1995. Glucocorticoid resistance in asthma. Am. J. Respir. Crit. Care Med. **152:** S125–140.
66. CHROUSOS, G.P., DETERA-WADLEIGH, S.D. & M. KARL. 1993. Syndromes of glucocorticoid resistance. Ann. Intern. Med. **119:** 1113–1124.
67. PRUZANSKI, W. & P. VADAS. 1991. Phospholipase A2—a mediator between proximal and distal effectors of inflammation. Immunol .Today **12:** 143–146.

68. MEDURI, G.U. *et al.* 1991. Fibroproliferative phase of ARDS. Clinical findings and effects of corticosteroids. Chest **100:** 943–952.
69. SANTOS, A.A. *et al.* 1993. Elaboration of interleukin 1-receptor antagonist is not attenuated by glucocorticoids after endotoxemia. Arch. Surg. **128:** 138–143; discussion: 143–144.
70. HART, P.H. *et al.* 1990. Augmentation of glucocorticoid action on human monocytes by interleukin- 4. Lymphokine Res. 1990. **9:** 147–153.
71. CHROUSOS, G.P. 1995. The hypothalamic-pituitary-adrenal axis and immune-mediated inflammation. N. Engl. J. Med. 1995. **332:** 1351–1362.
72. MELBY, J.C. & W.W. SPINK. 1958. Comparative studies on adrenalcortical function and cortisol metabolism in healthy adults and in patients with shock due to infection. J. Clin. Invest. **37:** 1791–1798.
73. REINCKE, M. *et al.* 1993. The hypothalamic-pituitary-adrenal axis in critical illness: response to dexamethasone and corticotropin-releasing hormone. J. Clin. Endocrinol. Metab. **77:** 151–156.
74. BRIEGEL, J. *et al.* 1991. Contribution of cortisol deficiency to septic shock. Lancet **338:** 507–508.
75. ANNANE, D. *et al.* 2000. A 3-level prognostic classification in septic shock based on cortisol levels and cortisol response to corticotropin. JAMA **283:** 1038–1045.
76. MEDURI, G.U. & G.P. CHROUSOS. 1998. Duration of glucocorticoid treatment and outcome in sepsis: is the right drug used the wrong way? Chest **114:** 355–360.
77. KASS, E.H. & M. FINLAND. 1957. Adrenocortical hormones and the management of infection. Annu. Rev. Med. **8:** 1–18.
78. LAMBERTS, S.W. *et al.* 1992. Cortisol receptor resistance: the variability of its clinical presentation and response to treatment. J. Clin. Endocrinol. Metab. **74:** 313–321.
79. LANE, S.J. & T.H. LEE. 1991. Glucocorticoid receptor characteristics in monocytes of patients with corticosteroid-resistant bronchial asthma. Am. Rev. Respir. Dis. **143:** 1020–1024.
80. SHER, E.R. *et al.* 1994. Steroid-resistant asthma. Cellular mechanisms contributing to inadequate response to glucocorticoid therapy. J. Clin. Invest. **93:** 33–39.
81. ADCOCK, I.M. *et al.* 1995. Differences in binding of glucocorticoid receptor to DNA in steroid-resistant asthma. J. Immunol. **154:** 3500–3505.
82. NORBIATO, G. *et al.* 1992. Cortisol resistance in acquired immunodeficiency syndrome. J. Clin. Endocrinol. Metab. **74:** 608–613.
83. MOLIJN, G.J. *et al.* 1995. Differential adaptation of glucocorticoid sensitivity of peripheral blood mononuclear leukocytes in patients with sepsis or septic shock. J. Clin. Endocrinol. Metab. **80:** 1799–1803.
84. ALMAWI, W.Y. *et al.* 1991. Abrogation of glucocorticoid-mediated inhibition of T cell proliferation by the synergistic action of IL-1, IL-6, and IFN-gamma. J. Immunol. **146:** 3523–3527.
85. KAM, J.C. *et al.* 1993. Combination IL-2 and IL-4 reduces glucocorticoid receptor-binding affinity and T cell response to glucocorticoids. J. Immunol. **151:** 3460–3466.
86. SPAHN, J.D. *et al.* 1996. A novel action of IL-13: induction of diminished monocyte glucocorticoid receptor-binding affinity. J. Immunol. **157:** 2654–2659.
87. LIU, L.Y. *et al.* 1993. Changes of pulmonary glucocorticoid receptor and phospholipase A2 in sheep with acute lung injury after high dose endotoxin infusion. Am. Rev. Respir. Dis. **148:** 878–881.
88. FAN, J. *et al.* 1994. Effect of glucocorticoid receptor (GR) blockade on endotoxemia in rats. Circ. Shock **42:** 76–82.
89. BAEUERLE, P.A. & V.R. BAICHWAL. 1997. NF-kappa B as a frequent target for immunosuppressive and anti- inflammatory molecules. Adv. Immunol. **65:** 111–137.
90. MEDURI, G.U. *et al.* 1995. Persistent elevation of inflammatory cytokines predicts a poor outcome in ARDS. Plasma IL-1 beta and IL-6 levels are consistent and efficient predictors of outcome over time. Chest **107:** 1062–1073.
91. MEDURI, G.U. *et al.* 1995. Inflammatory cytokines in the BAL of patients with ARDS. Persistent elevation over time predicts poor outcome. Chest **108:** 1303–1314.
92. MEDURI, G.U. *et al.* 1995. Plasma and BAL cytokine response to corticosteroid rescue treatment in late ARDS. Chest **108:** 1315–1325.

93. MEDURI, G.U. *et al.* 1998. Procollagen types I and III aminoterminal propeptide levels during acute respiratory distress syndrome and in response to methylprednisolone treatment. Am. J. Respir. Crit. Care Med. **158:** 1432–1441.
94. GOLDEN, E. *et al.* 2000. Interleukin-8 and Soluble Intercellular adhesion molecule-1 during acute respiratory distress syndrome and in response to prolonged methylprednisolone treatment (Abstr.). Shock **13:** 42S.
95. STENTZ, F. *et al.* 2001. Mechanisms of NF-κB and glucocorticoid receptor (GRα) in activation and regulation of systemic inflammation (SI) in ARDS (Abstr.). Am. J. Respir. Crit. Care Med. **163:** A450.
96. HEADLEY, A.S. *et al.* 2000. Infections, SIRS, and CARS during ARDS and in response to prolonged glucocorticoid treatment (Abstr.). Am. J. Respir. Crit. Care Med. **161:** A378.
97. BERNARD, G.R. *et al.* 1995. Quantification of organ failure for clinical trials. Am. J. Respir. Crit. Care Med. **151:** A323.
98. MEDURI, G.U. *et al.* 1998. Prolonged methylprednisolone treatment improves lung function and outcome of unresolving ARDS. A randomized, double-blind, placebo-controlled trial. JAMA **280:** 159–165.
99. CARRATU, P. *et al.* 2002. TNF-α and LT-α gene polymorphism in acute respiratory distress syndrome (ARDS) (Abstr.). Am. J. Respir. Crit. Care Med. **165:** A474.
100. PUGAZHENTHI, M. *et al.* 2001. Nuclear uptake of nuclear factor-κB (NF-κB) and glucocorticoid receptor (GR) in lung tissue of patients with unresolving ARDS (U-ARDS) (Abstr.). Am. J. Respir. Crit. Care Med. **163:** A452.
101. BRIGHT, G.M. 1995. Corticosteroid-binding globulin influences kinetic parameters of plasma cortisol transport and clearance. J. Clin. Endocrinol. Metab. **80:** 770–775.
102. GED, C. *et al.* 1989. The increase in urinary excretion of 6 beta-hydroxycortisol as a marker of human hepatic cytochrome P450IIIA induction. Br. J. Clin. Pharmacol. **28:** 373–387.
103. STANLEY, L.A. *et al.* 1988. Potentiation and suppression of mouse liver cytochrome P-450 isozymes during the acute-phase response induced by bacterial endotoxin. Eur. J. Biochem. **174:** 31–36.
104. SEWER, M.B., D.R. KOOP & E.T. MORGAN. 1997. Differential inductive and suppressive effects of endotoxin and particulate irritants on hepatic and renal cytochrome P-450 expression. J. Pharmacol. Exp. Ther. **280:** 1445–1454.
105. CHEN, Y.L. *et al.* 1992. Effects of interleukin-6 on cytochrome P450-dependent mixed-function oxidases in the rat. Biochem. Pharmacol. **44:** 137–148.
106. CHEN, J.Q. *et al.* 1995. Suppression of the constitutive expression of cytochrome P-450 2C11 by cytokines and interferons in primary cultures of rat hepatocytes: comparison with induction of acute-phase genes and demonstration that CYP2C11 promoter sequences are involved in the suppressive response to interleukins 1 and 6. Mol. Pharmacol. **47:** 940–947.
107. MORGAN, E.T. *et al.* 1994. Selective suppression of cytochrome P-450 gene expression by interleukins 1 and 6 in rat liver. Biochim. Biophys. Acta **1219:** 475–483.
108. SHEDLOFSKY, S.I. *et al.* 1994. Endotoxin administration to humans inhibits hepatic cytochrome P450-mediated drug metabolism. J. Clin. Invest. **94:** 2209–2214.
109. PARENT, C. *et al.* 1992. Effect of inflammation on the rabbit hepatic cytochrome P-450 isoenzymes: alterations in the kinetics and dynamics of tolbutamide. J. Pharmacol. Exp. Ther. **261:** 780–787.
110. MONSHOUWER, M. *et al.* 1995. Selective effects of a bacterial infection (Actinobacillus pleuropneumoniae) on the hepatic clearances of caffeine, antipyrine, paracetamol, and indocyanine green in the pig. Xenobiotica **25:** 491–499.
111. PERLMUTTER, D.H. *et al.* 1986. Cachectin/tumor necrosis factor regulates hepatic acute-phase gene expression. J. Clin. Invest. **78:** 1349–1354.
112. NADAI, M. *et al.* 1993. Alterations in pharmacokinetics and protein binding behavior of cefazolin in endotoxemic rats. Antimicrob. Agents Chemother. **37:** 1781–1785.
113. WANG, L. *et al.* 1993. The effect of lipopolysaccharide on the disposition of xanthines in rats. J. Pharm. Pharmacol. **45:** 34–38.
114. PUGEAT, M. *et al.* 1989. Decreased immunoreactivity and binding activity of corticosteroid-binding globulin in serum in septic shock. Clin. Chem. **35:** 1675–1679.

115. MELBY, J.C., R.H. EGDAHL & W.W. SPINK. 1960. Secretion and metabolism of cortisol after injection of endotoxin. J. Lab. Clin. Med. **56:** 50–62.
116. PERROT, D. *et al.* 1993. Hypercortisolism in septic shock is not suppressible by dexamethasone infusion. Crit. Care Med. **21:** 396–401.
117. SAVU, L. *et al.* 1981. Serum depletion of corticosteroid binding activities, an early marker of human septic shock. Biochem. Biophys. Res. Commun. **102:** 411–419.
118. VADAS, P. *et al.* 1988. Concordance of endogenous cortisol and phospholipase A2 levels in gram-negative septic shock: a prospective study. J. Lab. Clin. Med. **111**(5): 584–590.
119. YATES, C.R. *et al.* 2001. Time-variant increase in methylprednisolone clearance in patients with acute respiratory distress syndrome: a population pharmocokinetic study. J. Clin. Pharmacol. **41:** 1–10.
120. CHAUDHARY, P.M., E.B. MECHETNER & I.B. RONINSON. 1992. Expression and activity of the multidrug resistance P-glycoprotein in human peripheral blood lymphocytes. Blood **80:** 2735–2739.
121. DRACH, D. *et al.* 1992. Subpopulations of normal peripheral blood and bone marrow cells express a functional multidrug resistant phenotype. Blood **80:** 2729–27 34.
122. VAN KALKEN, C.K. *et al.* 1993. Cortisol is transported by the multidrug resistance gene product P-glycoprotein. Br. J. Cancer **67:** 284–289.
123. FARRELL, R.J. *et al.* 2002. P-glycoprotein-170 inhibition significantly reduces cortisol and ciclosporin efflux from human intestinal epithelial cells and T lymphocytes. Aliment. Pharmacol. Ther. **16:** 1021–1031.
124. YATES, C.R. *et al.* 2003. Structural determinants of P-glycoprotein-mediated transport of glucocorticoids. Pharm. Res. **20:** 1794–1803.
125. HIRSCH-ERNST, K.I. *et al.* 1998. Induction of mdr1b mRNA and P-glycoprotein expression by tumor necrosis factor alpha in primary rat hepatocyte cultures. J. Cell. Physiol. **176:** 506–515.
126. CUFER, T. *et al.* 2000. Decreased cortisol secretion by adrenal glands perfused with the P-glycoprotein inhibitor valspodar and mitotane or doxorubicin. Anticancer Drugs **11:** 303–309.
127. RICKETTS, M.L. *et al.* 1998. Regulation of 11 beta-hydroxysteroid dehydrogenase type 1 in primary cultures of rat and human hepatocytes. J. Endocrinol. **156:** 159–168.
128. JAMIESON, P.M. *et al.* 1995. 11 beta-Hydroxysteroid dehydrogenase is an exclusive 11 beta-reductase in primary cultures of rat hepatocytes: effect of physicochemical and hormonal manipulations. Endocrinology **136:** 4754–4761.
129. JAMIESON, P.M. *et al.* 2000. 11 beta-Hydroxysteroid dehydrogenase type 1 is a predominant 11 beta-reductase in the intact perfused rat liver. J. Endocrinol. **165:** 685–692.
130. GE, R.S. & M.P. HARDY. 2000. Initial predominance of the oxidative activity of type I 11beta-hydroxysteroid dehydrogenase in primary rat Leydig cells and transfected cell lines. J. Androl. **21:** 303–310.
131. STEWART, P.M. & Z.S. KROZOWSKI. 1999. 11 beta-Hydroxysteroid dehydrogenase. Vitam. Horm. **57:** 249–324.
132. KOTELEVTSEV, Y. *et al.* 1999. Hypertension in mice lacking 11beta-hydroxysteroid dehydrogenase type 2. J. Clin. Invest. **103:** 683–689.
133. SECKL, J.R. & B.R. WALKER. 2001. Minireview: 11beta-hydroxysteroid dehydrogenase type 1—a tissue-specific amplifier of glucocorticoid action. Endocrinology **142:** 1371–1376.
134. ESCHER, G. *et al.* 1997. Tumor necrosis factor alpha and interleukin 1beta enhance the cortisone/cortisol shuttle. J. Exp. Med. **186:** 189–198.
135. STITH, R.D. & R.E. MCCALLUM. 1983. Down regulation of hepatic glucocorticoid receptors after endotoxin treatment. Infect. Immun. **40:** 613–621.
136. HUANG, Z.H., H. GAO & R.B. XU. 1987. Study on glucocorticoid receptors during intestinal ischemia shock and septic shock. Circ. Shock **23:** 27–36.
137. LI, F. & R.B. XU. 1988. Changes in canine leukocyte glucocorticoid receptors during endotoxin shock. Circ. Shock **26:** 99–105.

138. MOLIJN, G.J. *et al.* 1995. Temperature-induced down-regulation of the glucocorticoid receptor in peripheral blood mononuclear leucocyte in patients with sepsis or septic shock. Clin. Endocrinol. (Oxf.) **43:** 197–203.
139. GALIGNIANA, M.D., G. PIWIEN-PILIPUK & J. ASSREUY. 1999. Inhibition of glucocorticoid receptor binding by nitric oxide. Mol. Pharmacol. **55:** 317–323.
140. LIU, L.Y. *et al.* 1993. Changes of pulmonary glucocorticoid receptor and phospholipase A2 in sheep with acute lung injury after high dose endotoxin infusion. Am. Rev. Respir. Dis. **148:** 878–881.
141. VERHEGGEN, M.M. *et al.* 1996. Modulation of glucocorticoid receptor expression in human bronchial epithelial cell lines by IL-1 beta, TNF-alpha and LPS. Eur. Respir. J. **9:** 2036–2043.
142. SCHLECHTE, J.A., B.H. GINSBERGN & B.M. SHERMAN. 1982. Regulation of the glucocorticoid receptor in human lymphocytes. J. Steroid Biochem. **16:** 69–74.
143. TORNELLO, S. *et al.* 1982. Regulation of glucocorticoid receptors in brain by corticosterone treatment of adrenalectomized rats. Neuroendocrinology **35:** 411–417.
144. SAPOLSKY, R.M., L.C. KREY & B.S. MCEWEN. 1984. Stress down-regulates corticosterone receptors in a site-specific manner in the brain. Endocrinology **114:** 287–292.
145. LACROIX, A., G.D. BONNARD & M.E. LIPPMAN. 1984. Modulation of glucocorticoid receptors by mitogenic stimuli, glucocorticoids and retinoids in normal human cultured T cells. J. Steroid Biochem. **21:** 73–80.
146. CIDLOWSKI, J.A. & N.B. CIDLOWSKI. 1981. Regulation of glucocorticoid receptors by glucocorticoids in cultured HeLa S3 cells. Endocrinology **109:** 1975–1982.
147. SVEC, F. & M. RUDIS. 1981. Glucocorticoids regulate the glucocorticoid receptor in the AtT-20 cell. J. Biol. Chem. **256:** 5984–5987.
148. BURNSTEIN, K.L., C.M. JEWELL & J.A. CIDLOWSKI. 1990. Human glucocorticoid receptor cDNA contains sequences sufficient for receptor down-regulation. J. Biol. Chem. **265:** 7284–7291.
149. DONG, Y. *et al.* 1988. Regulation of glucocorticoid receptor expression: evidence for transcriptional and posttranslational mechanisms. Mol. Endocrinol. **2:** 1256–1264.
150. HOECK, W., S. RUSCONI & B. GRONER. 1989. Down-regulation and phosphorylation of glucocorticoid receptors in cultured cells. Investigations with a monospecific antiserum against a bacterially expressed receptor fragment. J. Biol. Chem. **264:** 14396–14402.
151. ROSEWICZ, S. *et al.* 1988. Mechanism of glucocorticoid receptor down-regulation by glucocorticoids. J. Biol. Chem. **263:** 2581–2584.
152. VEDECKIS, W.V., M. ALI & H.R. ALLEN. 1989. Regulation of glucocorticoid receptor protein and mRNA levels. Cancer Res. **49:** 2295s–2302s.
153. BEATO, M., P. HERRLICH & G. SCHUTZ. 1995. Steroid hormone receptors: many actors in search of a plot. Cell **83:** 851–857.
154. KARIN, M. 1998. New twists in gene regulation by glucocorticoid receptor: is DNA binding dispensable? Cell **93:** 487–490.
155. REICHARDT, H.M. *et al.* 2001. Repression of inflammatory responses in the absence of DNA binding by the glucocorticoid receptor. EMBO J. **20:** 7168–7173.
156. LANE, S.J. & T.H. LEE. 1996. Mononuclear cells in corticosteroid-resistant asthma. Am. J. Respir. Crit. Care Med. **154:** S49–51; discussion: S52.
157. ADCOCK, I.M. *et al.* 1995. Abnormal glucocorticoid receptor-activator protein 1 interaction in steroid-resistant asthma. J. Exp. Med. **182:** 1951–1958.
158. LANE, S.J. *et al.* 1998. Corticosteroid-resistant bronchial asthma is associated with increased c-fos expression in monocytes and T lymphocytes. J. Clin. Invest. **102:** 2156–2164.
159. WEBSTER, J.C. *et al.* 2001. Proinflammatory cytokines regulate human glucocorticoid receptor gene expression and lead to the accumulation of the dominant negative beta isoform: a mechanism for the generation of glucocorticoid resistance. Proc. Natl. Acad. Sci. USA **98:** 6865–6870.

Physiological and Pathological Consequences of the Interactions of the p53 Tumor Suppressor with the Glucocorticoid, Androgen, and Estrogen Receptors

SAGAR SENGUPTA[a] AND BOHDAN WASYLYK[b]

[a]*Laboratory of Human Carcinogenesis, National Cancer Institute, National Institutes of Health, Bethesda, Maryland 20892, USA*

[b]*Institut de Génétique et de Biologie Moléculaire et Cellulaire, CNRS/INSERM/ULP, 1 Rue Laurent Fries, 67404 Illkirch cedex, France*

ABSTRACT: The p53 tumor suppressor plays a key role in protection from the effects of different physiological stresses (DNA damage, hypoxia, transcriptional defects, etc.), and loss of its activity has dire consequences, such as cancer. Its activity is finely tuned through interactions with other important regulatory circuits in the cell. Recently, striking evidence has emerged for crosstalk with another class of important regulators, the steroid hormone receptors, and in particular the glucocorticoid (GR), androgen (AR), and estrogen (ER) receptors. These receptors are important in maintaining homeostasis in response to internal and external stresses (GR) and in the development, growth, and maintenance of the male and female reproductive systems (AR and ER, respectively). We review how p53 interacts closely with these receptors, to the extent that they share the same E3 ubiquitin ligase, the MDM2 oncoprotein. We discuss the different physiological contexts in which such interactions occur, and also how these interactions have been undermined in various pathological situations. We will describe future areas for research, with special emphasis on GR, and how certain common features, such as cytoplasmic anchoring of p53 by the receptors, may become targets for the development of therapeutic interventions. Given the importance of GR in inflammation, erythropoiesis, and autoimmune diseases, and the importance of AR and ER in prostate and breast cancer (respectively), the studies on p53 interactions with the steroid receptors will be an important domain in the near future.

KEYWORDS: hematological malignancies; stress erythropoiesis; Addison's disease; Cushing's syndrome; anemia; skin carcinogenesis; cancer progression

Address for correspondence: Bohdan Wasylyk, Institut de Génétique et de Biologie Moléculaire et Cellulaire, CNRS/INSERM/ULP, 1 Rue Laurent Fries, BP 10142, 67404 Illkirch cedex, France. Voice: 33 3 88 65 34 11; fax: 33 3 88 65 32 01.
boh@igbmc.u-strasbg.fr

Ann. N.Y. Acad. Sci. 1024: 54–71 (2004). © 2004 New York Academy of Sciences.
doi: 10.1196/annals.1321.005

INTRODUCTION

Glucocorticoids (GCs) in humans maintain homeostasis in response to internal or external stresses via the regulation of the function of the glucocorticoid receptor (GR) (NR3C1[1]). GRs remain in the cytoplasm in large macromolecular complexes bound to chaperones such as hsp90. GCs bind to the ligand-binding domain (LBD) of GR, leading to a transformation in the GR-chaperone complex, followed by the discharge of the chaperones and redistribution of GR to the nucleus where it modulates different sets of genes, by activation or repression, depending on the particular physiological condition.[2–4]

The tumor suppressor p53 is frequently inactivated in various types of human cancer.[5] Different physiological stresses, including DNA damage and hypoxia, stabilize and activate p53. The tumor suppressor carries out its functions by acting as a sequence-dependent transcription factor that activates (and sometimes represses) diverse sets of genes under different stress conditions, ultimately leading to growth arrest, apoptosis, or senescence. p53's function as a tumor suppressor also depends on how it interacts physically and functionally with many cellular and viral proteins. These interactions modulate the activity of p53 and integrate its functions into other cellular circuits.[6–9] p53 crosstalks with different steroid-hormone receptors, including GR. In this review we (1) recapitulate how p53 and GR are known to interact under different physiological and pathological conditions; (2) enumerate the different molecular mechanisms of such interactions; (3) propose how p53 and GR may regulate each other's function in physiological conditions; (4) compare GR–p53 interactions with those for the androgen (AR) and estrogen (ERα) receptors (NR3C4 and NR3A1, respectively[1]); and (5) discuss further directions for the study of p53–GR interactions.

DIVERSE EFFECTS OF p53–GR INTERACTIONS

The functional interactions between p53 and GR in different physiological conditions can be broadly classified under two categories (1) antagonism between p53 and GR (FIG. 1) and (2) complementation between p53 and GR (FIG. 2).

NEGATIVE REGULATION OF p53 AND GR

Links between p53 and GR

Since both p53 and GR modulate diverse signals during different types of physiological stress conditions, researchers have looked for the link between p53 activation and GC-induced apoptosis. The premise was that GC upregulation would directly lead to p53 activation, which would in turn lead to activation of the apoptotic cascade by both the intrinsic and extrinsic pathways.[7] Indeed many of the genes that are transcriptional targets of p53 may also be upregulated during the GC-mediated apoptotic program. For example, an early report demonstrated that transgenic mice overexpressing the pro-apoptotic gene product Bax showed accelerated apoptosis in response to dexamethasone (Dex).[10]

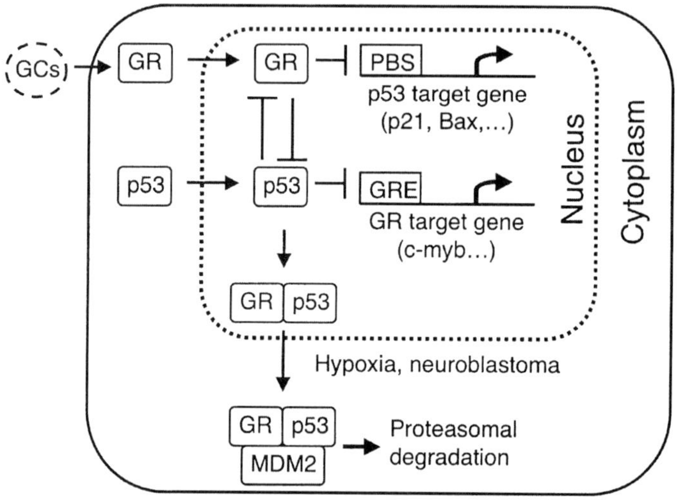

FIGURE 1. Negative crosstalk between p53 and GR. The figure illustrates some of the mechanisms described in the text. GCs stimulate nuclear import of GR. Induction of p53 can lead to mutual functional inhibition leading to decreased expression of genes containing PBS (p53 binding sites) and GRE (glucocorticoid response elements). Complex formation between GR and p53 leads to nuclear export and MDM2-mediated cytoplasmic degradation through the proteasome pathway. Degradation does not occur in neuroblastoma cells leading to cytoplasmic sequestration and complex formation of both p53 and GR.

However, other reports suggested that the GC-induced apoptotic program was independent of the activation of p53. For example, unlike the response to ionizing radiation (IR), GC-induced apoptosis remained unaffected in thymocytes of p53 knockout mice.[11,12] In contrast, adrenalectomy-induced p53 expression and granule cell degeneration were prevented by daily administration of corticosterone,[13] thereby suggesting that instead of aiding GC-induced apoptosis, p53 and GC-induced functions may in fact counteract each other, at least in certain physiological conditions.

These results signify that p53 activation causes an enhancement of the resistance to GC-induced cell death. Indeed, it has been reported that modulation of GC-induced apoptosis was linked to p53 gene dosage. GC-induced apoptosis was significantly enhanced in mouse thymocytes lacking one or both functional *p53* alleles.[14] Reciprocally, it was demonstrated that GCs protect against apoptosis induced by p53, at least in certain cellular contexts, via the modulation of Bcl-2 protein levels.[15]

The above studies indicate that p53 and GR may repress each other's functions. Regulation between different interacting proteins can take place by multiple and mutually non-exclusive mechanisms. For example, p53 is negatively regulated by its own transcription target MDM2, both by transcriptional inactivation and proteosomal degradation (see Refs. 16 and 17 for reviews of p53 and MDM2). It is quite possible that p53 and GR also have different levels of regulation, possibly depending on the physiological and cellular context. We discuss below how p53 and GR may mutually inhibit each other's functions by two mutually non-exclusive mechanisms.

Physical Interaction and Cytoplasmic Sequestration

Physical interactions between p53 and GR can lead to mutual inhibition. Using *in vitro* translated GR and GST-p53, Yu and colleagues showed that wild-type p53 physically interacts with GR.[18] Subsequently, we confirmed that the two proteins physically interact by *in vivo* reciprocal immunoprecipitations with either exogenous or endogenous proteins in human normal and cancer cells.[19,20] Using deletion mutants, we have demonstrated that the core domain and the nuclear localization signal (NLS) of p53 interact with GR.[20]

Using transient transfection systems, others and we have demonstrated that p53 and GR inhibit each other's transactivation properties.[19,21] The physical interaction of p53 and GR in the presence of ligand can result in cytoplasmic sequestration. In neuroblastoma (NB), which is one of the most common forms of childhood cancer, p53 is wild-type and cytoplasmic.[22] We found that in NB cells p53 and GR are sequestered in the cytoplasm. The p53-GR complex can be dissociated by GR antagonists, resulting in the accumulation of p53 in the nucleus, activation of p53-responsive genes, growth arrest, and apoptosis.[19]

This observation raises the possibility that sequestration by GR could account for the cytoplasmic localization of p53 in pathological and natural situations. This could be the case in various situations. Wild-type p53 is cytoplasmic in inflammatory breast carcinoma,[23] undifferentiated neuroblastomas,[24] colon adenocarcinomas,[25] differentiated oligodendrocytes, neurons, and PC12 pheochromocytoma cells.[26] It is conceivable that p53 is complexed with GR in the cytoplasm in at least some of the above pathological and physiological conditions. In such situations, GR can be envisaged to be a cytoplasmic anchor for p53. In fact, apart from GR, a number of proteins have also been proposed to serve as cytoplasmic anchors for p53, thereby blocking its normal nuclear localization. These include hsp70,[27] vimentin,[28] tubulin,[29] F-actin,[30] and a recently identified Parkin-like ubiquitin ligase, Parc.[31] These studies suggest that GR could be a component of a Parc complex and also raises the possibility that GR can act either alone or in concert with the other anchor proteins to sequester p53 in the cytoplasm, thereby regulating its function as a transcription factor.

The next aim was to determine whether cytoplasmic sequestration–mediated crosstalk between p53 and GR occurs in physiological conditions. It was already known that p53 is cytoplasmic in Balb/c 3T3 cells in certain stages of the cell cycle,[32] and in normal mouse embryonic stem cells,[33] raising the possibility of the presence of cytoplasmic anchor(s) under physiological conditions. Moreover, GCs and p53 can have antagonistic effects under various physiological conditions. For example, GCs increase glucose metabolism in response to physiological stress, such as hypoxia. Furthermore, p53 activation by hypoxia leads to decreased glucose metabolism as well as increased apoptosis.[34] Dex treatment (leading to GR activation) of normal human umbilical vein endothelial cells (HUVEC) under hypoxic conditions resulted in decreased protein levels of both p53 and GR and consequently inhibition of the transactivation of both p53 and GR target genes. Dex enhanced the proteosomal degradation of p53 and GR by stimulating the formation of a triple complex containing p53, GR, and the E3-ubiquitin ligase MDM2 in the cytoplasm, thereby inhibiting hypoxia-induced p53 tumor–suppressive functions, including apoptosis. In the absence of the GR agonist, Dex, p53 can go to the nucleus and carry out its function as a transcription factor, while GR remains sequestered in the cytoplasm.[20] In neuroblastoma cells, both p53-GR[19] and p53-MDM2[35] complexes are formed in the cytoplasm, suggesting that the cytoplasmic degradation function of MDM2 may be defective in these cells.

The identification of the crosstalk between p53 and GR during hypoxia also led to the discovery that GR is a new substrate for MDM2, thereby extending the range of targets for this E3 ubiquitin ligase.[36] Disruption of the p53-MDM2 interaction prevented Dex-induced ubiquitylation of GR and p53. The ubiquitylation of GR requires p53, the interaction of p53 with MDM2, and the E3 ligase activity of MDM2.[20] It has been demonstrated recently that crosstalk exists between GR and ER. The ER agonist–dependent decrease in GR levels is coupled with an increase in MDM2 protein,[37] confirming our earlier results that MDM2 targets GR by the ubiquitin proteasome pathway for degradation.

Transcriptional Repression

Apart from cytoplasmic sequestration, p53 and GR can also regulate each other's expression at the transcriptional level. The first indication of an interaction between p53 and GR (and GCs) at the transcriptional level was the finding that p53 stimulates the promoter activity of *sgk* (a member of the serine/ threonine protein kinase family that is transcriptionally regulated by GCs). Four of the p53 binding sites in the *sgk* promoter are specifically recognized by the p53 protein, and at least one of these sites can confer p53 transactivation to a heterologous promoter in mammary epithelial cells.[38] Subsequently, the same group also demonstrated that both p53 and GRs coordinately regulate *sgk* promoter activity. Using gel shift analysis, it was demonstrated that p53 inhibits the binding of GR to the *sgk* or a consensus GRE, thereby indicating a transcriptional level of regulation.[21] Using transient transfection assays it was also shown that p53 mutants have no or lesser effect on GR-mediated transactivation—the effect being dependent on the type of p53 mutant.[18]

p53 and GR can also interact through transcriptional modulation of one or more target genes that govern specific physiological processes. For example, we found that the negative regulation between p53 and GR during "stress erythropoiesis" is modulated by the transcriptional regulation of the GR target gene *c-myb*. Various conditions, including hypoxia encountered at high altitude, blood loss and erythroleukemia, instigate stress erythropoiesis, which involves GC-induced proliferation of erythroid progenitors (ebls). Mouse fetal liver ebls that lack p53 proliferate better than wild-type cells in the presence of Dex, thereby indicating that p53 inhibits stress erythropoiesis.[39] GR regulates the expression of genes required for ebl proliferation, like *c-myb*.[40] Using quantitative RT-PCR, we have shown that in fetal liver cultures lacking p53, Dex induced the expression of *c-myb* to significantly higher levels compared to wild-type cultures, thereby favoring GC-induced proliferation rather than differentiation.[41] These results showed that the transcriptional regulation of *c-myb*, a key regulator of ebl proliferation, may also be an important intermediary in a p53 GR–mediated decision between proliferation and differentiation.

SYNERGY BETWEEN p53 AND GR FUNCTIONS

Positive Interactions between p53 and GR in Neuronal Cells

Increased p53 gene expression has been observed during cell death in adrenalectomy[13] and in different types of neural cells.[42] GR promotes neuronal cell death, in contrast to the mineralocorticoid receptor (MR).[43] The functions of GR and MR in turn depend on the relative ratios of the pro- and anti-apoptotic members of the bcl-2 family. Bax is essential for GR-mediated cell death in neuronal cells. Under similar conditions, activation of p53 and its downstream target, Bax, are also essential for GR-mediated death-signaling cascade—indicating that p53 makes a positive contribution to GR-mediated functions.[44] Subsequently, it was demonstrated that GR activation by Dex resulted in translocation of p53 to the nucleus and enhanced transcription of different endogenous p53-responsive genes (such as *p21* and *GADD45*), ultimately leading to cell-cycle arrest.[45] These observations support the hypothesis that activation of GR enhances the basal activity of p53, as neither its protein or mRNA levels are altered.

p53 and GR Together Regulate the Expression of p21

It has been established that p53-mediated growth arrest, induced by different types of DNA damage, involves the cyclin-dependent kinase inhibitor, p21.[46] The first evidence that p53 and GR may synergize under certain physiological situations

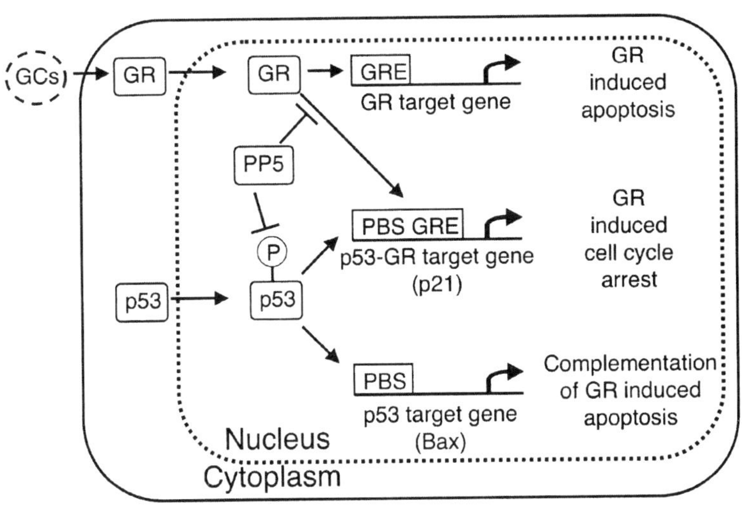

FIGURE 2. Positive crosstalk between p53 and GR. In certain physiological situations, GR and p53 positively regulate genes involved in apoptosis, leading to complementary effects on a common pathway. The p21 gene, that encodes a cell-cycle inhibitor, contains both PBS (p53 binding sites) and GRE (glucocorticoid response elements) in its promoter region, and is coordinately regulated by GR and p53. The phosphatase PP5 negatively regulates GR and p53, the latter through dephosphorylation of serine-15 on the activated tumor suppressor.

was obtained when it was demonstrated that GCs also stimulate p21 gene expression by targeting multiple transcriptional elements within a steroid response element of the p21 promoter in rat hepatoma cells.[47] In addition, Dex has been found to stimulate p21 promoter activity independent of the p53 DNA binding element, possibly via an atypical response involving a CCAAT enhancer binding protein.[48]

The above results raise the possibility that under certain physiological contexts both p53 and GR can act in concert, thereby amplifying the signal. This hypothesis was strengthened when it was found that the Ser/Thr protein phosphatase PP5 promotes cellular proliferation by inhibiting both GC and p53-mediated signaling pathways that normally lead to p21-mediated growth arrest.[49] These data indicate that PP5 suppresses a single pathway, in which p53 is a functional participant in GR-induced expression of p21 and subsequent cell-cycle arrest. Experiments with antisense oligonucleotides against p53 revealed that the suppression of p53 expression was associated with: (1) an increase in the rate of cellular proliferation; (2) a decrease in the basal expression of p21; and (3) a block of the Dex-induced p21 expression and G1 arrest. Suppression of PP5 expression results in Dex-mediated phosphorylation of p53 at Ser-15. Hence it was concluded that PP5 may suppress a GR-induced signaling cascade that influences p53 phosphorylation at Ser-15 and expression of p21.[50]

THE DIFFERENT ROLES OF GR DURING CELL PROLIFERATION AND APOPTOSIS

The contradictory readouts of p53 and GR interactions raise an interesting question: What determines whether these two transcription factors act in positive or in negative regulatory loops? To understand these apparent divergent functions it is in turn necessary to understand the diverse functions of GR and GCs. A well-known function of GR is as an inducer of apoptosis. GCs induce apoptosis in most nucleated cells of vascular origin, such as thymocytes, melanoma cells, and peripheral blood monocytes,[51–54] thus playing a central role in the anti-inflammatory process.

However, recently, another equally important role of GR (and GCs) has come to light. GCs repress apoptosis and increase proliferation of cancer cell lines,[55] neutrophils,[56] retinal pigment cells,[57] fibroblasts,[58] and serum-depleted T lymphocytes that express bcl-2.[59] GCs have protective effects against apoptosis in human and rat hepatocytes and in hepatoma cells.[60–62] Administration of Dex to rats reduces the rate of apoptosis in rat livers.[60–62] Interestingly, under these conditions, GCs increase the intracellular levels of the anti-apoptotic proteins Bcl-2 and Bcl-xL, and suppress the levels of the pro-apoptotic protein Bcl-xS.[15] Hence, there is a role for GCs in protecting cells of tissues and organs in which inflammation takes place.

LINKING THE DIVERGENT FUNCTIONS OF GR WITH p53

The divergent functions of GC and GR in different physiological contexts can be attributed to their interactions with different sets of pro- and anti-proliferation factors and co-modulators that are present in particular physiological context and cellular types. Filter hybridization, expression data, and functional assays have identified multiple key apoptosis molecules, which are regulated by GCs in a pro- or anti-apoptotic manner.[63] The overall primary response to a particular physiological stress depends on the cell type,[64] which overrides and modulates the response of individual gene products. Hence the two transcription factors, p53 and GR, both of which individually can induce apoptosis under certain circumstances, can also inhibit each other's functions under alternative physiological and pathological conditions. For example, the negative regulation of p53 and GR occurs in response to different physiological stresses (hypoxia or stress erythropoiesis), resulting in an elevation of the levels of active p53. On the other hand, when p53 protein levels are not altered (i.e., remain at a basal level, as in neuronal cells), the tumor suppressor can act in a GR-induced pathway leading to growth suppression.

It is interesting to consider why p53 and GR should crosstalk under physiological conditions. Both p53 and GR mediate stress responses, but with intrinsic differences in their actions. GR, when complexed with GCs, is involved in fight-or-flight responses, and maintains homeostasis during internal or environmental changes. Hence, GR is considered to be generally involved in a survival response. On the other hand, p53 responds to extensive genomic damage and other stresses by inducing cell death. p53 can be considered to be a death response. Hence, p53 and GR can be considered to be the protagonists in the question of life or death, and should therefore have antagonistic effects on each other's function (FIG. 3). Hypoxia is only one of the physiological processes that are known to be oppositely regulated by p53 and GR. Other processes where similar negative crosstalk can occur need further investigation.

Finally, it is interesting to consider the effects of another molecule that can be compared with the pro- and anti-apoptotic functions of GCs. Nitric oxide (NO) is a small diffusible highly reactive molecule, known for its diverse activity throughout biology. NO also affects cellular decisions of life and death by turning on apoptotic pathways or by shutting them off. Long-lasting production of NO acts as a pro-apoptotic modulator by activating caspase family proteases through the release of

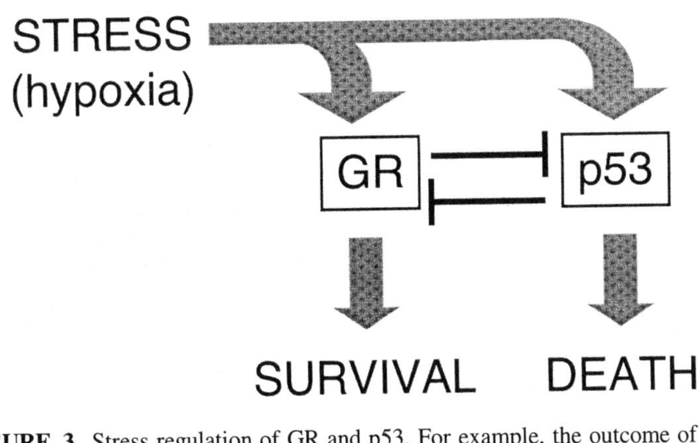

FIGURE 3. Stress regulation of GR and p53. For example, the outcome of the stress response to hypoxia can be either cell survival or cell death. The decision is made through mutual antagonism between key factors involved in the survival (GR) and death (p53).

mitochondrial cytochrome *c*, activation of JNK/SAPK, and importantly, upregulation and activation of p53. However at low or physiological concentrations NO prevents apoptosis induced by growth factor withdrawal, Fas, TNFα, and lipopolysaccharide. In both its roles, NO alters the expression of the different pro- and anti-apoptotic members of the Bcl-2 family proteins.[65,66] Hence, under different physiological contexts, NO, like GCs, can alter the fate of the cells between life and death. These complexities are cell- and stress-specific, depending on various protein–protein interactions (including p53). Most tantalizingly, GCs also inhibit the nitric oxide synthase (NOS) pathway under certain conditions[67,68]—thereby indicating an integrated response to cellular stress.

COMPARISON WITH THE ANDROGEN AND ESTROGEN RECEPTORS

The Androgen Receptor

Androgens are critical for the development, growth, and maintenance of the male reproductive system, including the prostate. Androgens bind to AR and regulate its activity as a transcription factor.[69] AR is important in pathological situations, such as human prostate cancers, where it regulates the expression of genes involved in proliferation and differentiation.[70] Prostate cancer is the second leading cause of male cancer death in the United States.[71] It can be treated by anti-androgen therapy, but this treatment is not curative and the cancers frequently evolve to hormone independence. p53 also has a role in prostate growth and hormone sensitivity. Primary prostate tumors usually have wild-type p53, whereas advanced, hormone-resistant cancers often express mutant p53.[72–74] MDM2 overexpression is associated with prostate cancer progression and aggressive behavior.[75,76] These observations indicate that the activities of AR, p53, and MDM2 are connected.

Several reports provide molecular and functional evidence for a link between p53 and AR. The genes that are regulated by AR in the ventral prostate of rats were studied following androgen treatment of androgen-depleted animals.[77] Interestingly, under these conditions, many of the known p53 target genes are upregulated (e.g., MDM2), suggesting that AR inhibits p53 protein abundance or activity *in vivo*. This possibility was studied using a prostate cancer cell line that expresses wild-type p53 (LNCaP). Agonist activation of AR led to a decrease in nuclear p53 by a posttranscription mechanism, showing that AR regulates p53 localization and stability, and, ultimately, its function.[77]

Several other studies have shown that p53 also regulates AR function. LNCaP cells with decreased levels of wild-type p53 were found to form tumors in castrated nude mice, in contrast to the parental cells.[78] Therefore, reducing the level of p53 confers a hormone-resistant phenotype on prostate cancer cells. Expression of "gain of function" p53 mutants in LNCaP cells was found to inhibit endogenous wild-type p53 activity, to confer androgen-independent growth to the cells, and to decrease AR protein levels and activity.[79] In another study, the effects of p53 expression on AR activity were investigated in p53-null AR-null PC-3 prostate cancer cells. p53 expression inhibited AR-dependent activation of the PSA promoter, whereas p53 mutants were much less active. p53 was found to inhibit specific binding of AR to DNA, probably as a result of inhibition of dimerization of the receptor.[80] In conclusion, these studies provide evidence for negative crosstalk between p53 and AR.

There is convincing evidence that MDM2, the E3 ligase for p53, is also an E3 ligase for AR. MDM2 has been found to form a complex with AR and the kinase Akt (PKB). Phosphorylation by Akt of AR and MDM2 induces MDM2-mediated ubiquitylation of AR and degradation of AR by the proteasome pathway.[81] p53's role in these interactions was not studied. It remains possible that p53 is not involved, since MDM2 has been shown to have p53-independent as well as p53-dependent roles in prostate cancer cell lines.[82] Overall, these studies provide compelling evidence for functional interactions between p53, MDM2, and AR.

The Estrogen Receptor

Estrogens are steroid hormones that regulate the proliferation and differentiation of female reproductive organs, such as the breast.[83] Most physiological functions of estrogens are transduced through the two ERs in humans, ERα and ERβ (NR2A1 and 2, respectively[1]). ERα expression is a useful predictor of prognosis and therapeutic response in breast cancer,[84] the leading cause of cancer deaths in women.[71] p53 mutations are rare in early stage ERα-positive tumors,[85] which are well-differentiated and dependent on hormone for tumor growth.[86] The frequency of p53 mutation increases with progression to ERα-negative hormone-refractory tumors.[85] MDM2 is upregulated in ERα-positive breast cancer at the mRNA and protein levels, and MDM2 expression is lower in ERα-negative tumors.[87] These correlations are reflected at the functional and physical levels.

Estrogens have been shown to increase p53 protein levels through an indirect mechanism that involves increased transcription of the *c-myc* gene followed by c-Myc activation of the p53 promoter.[88] Estrogens have also been shown to stabilize wild-type p53 protein in the absence of increased p53 gene transcription.[89] Curiously, mutant p53 protein turnover was found to be insensitive to estrogens. The stabilization of p53 by estrogens is linked to MDM2, since exogenous MDM2 expression can increase estrogen-mediated p53 accumulation.[90] ERα has been shown to form a triple complex with p53 and MDM2, and protect p53 from MDM2-mediated degradation in a ligand-independent manner.[91] Estrogens have also been shown to induce cytoplasmic accumulation and functional inactivation of wild-type p53 in MCF-7 cells,[92] which could mirror the cytoplasmic accumulation observed in breast cancer and normal lactating breast tissue.[23] In MCF-7 cells engineered to express GR, estrogen activation of endogenous ERα led to increased p53 activity, p53 and ERα recruitment to the MDM2 promoter, and increased p53 and MDM2 accumulation in the cytoplasm.[37] Interestingly, p53 was not recruited to the p21 promoter, suggesting that ERα activates p53 in a particular manner that confers promoter selectivity to p53. A common feature of these studies is that estrogen activation of ERα leads to increased expression and cytoplasmic accumulation of p53 and MDM2.

p53 overexpression has also been shown to inhibit ERα activity by preventing specific interactions with its cognate motif.[93] In contrast, MDM2 enhances ERα activity. MDM2 has been shown to be preferentially recruited to unligated ERα bound to the endogenous *pS2* promoter, where it is thought to induce ubiquitylation and turnover of the inactive receptor, in order for ERα to be able to continue to respond

TABLE 1. Effects of the interactions of p53 and MDM2 with GR, AR, and ER

	GR	AR	ER
Receptor stability	Down	Down	Up
p53 stability	Down	Down	Up
MDM2 stability	?	?	Up
MDM2 gene transcription	?	?	Up
Cytoplasmic accumulation of receptor	Yes	Yes	Yes
Receptor is a target for MDM2 E3 ligase	Yes	Yes	Yes
Triple complex: receptor-p53-MDM2	Yes	?	Yes
Inhibition of receptor dimerization	?	Yes	No
Inhibition of receptor DNA binding	Yes	Yes	Yes
p53 inactivation by the receptor	Yes	Yes	Yes
p53 reactivation by downregulation of the receptor	Yes	?	Yes
Pathological consequence of receptor–p53 interaction	Yes	?	?
Physiological consequences of receptor–p53 interaction	Yes	?	?

to hormone.[94] In another report, MDM2 was shown to interact with the receptor and stimulate transcription, possibly by acting as a co-activator.[95] The latter two studies agree that MDM2 enhances ERα activity, but they provide different explanations for why this happens.

The overall effects of p53-ER crosstalk are negative, leading to the inactivation of p53 as well as ER. Globally they are similar to the interactions of AR with p53, suggesting that the common features of breast and prostate cancer[96] extend to the interactions between p53 and these steroid hormone receptors. However, there are differences in the overall mechanisms of interaction of p53 and MDM2 with the three steroid hormone receptors (TABLE 1), which may reflect true differences, or they could be due to the heterogeneous systems and relatively crude techniques that have been used to analyze the interactions.

FUTURE DIRECTIONS IN THE STUDY OF p53–GR INTERACTIONS

Studies of the interaction between p53 and GR are very recent, yet in this short period substantial progress has been achieved. However much remains to be done to understand the molecular biology and the functional consequences of the interaction between these two distinct transcription factors. The future directions in this field can be broadly divided into two parts: (1) biology and applications of p53–GR interactions and (2) physiological conditions and pathological consequences of p53–GR regulation (FIG. 4).

Biology and Applications of p53–GR Interactions

p53 and GR are both regulated by phosphorylation.[97,98] The mechanism that regulates the ligand-dependent interaction between p53 and GR could also be mediated

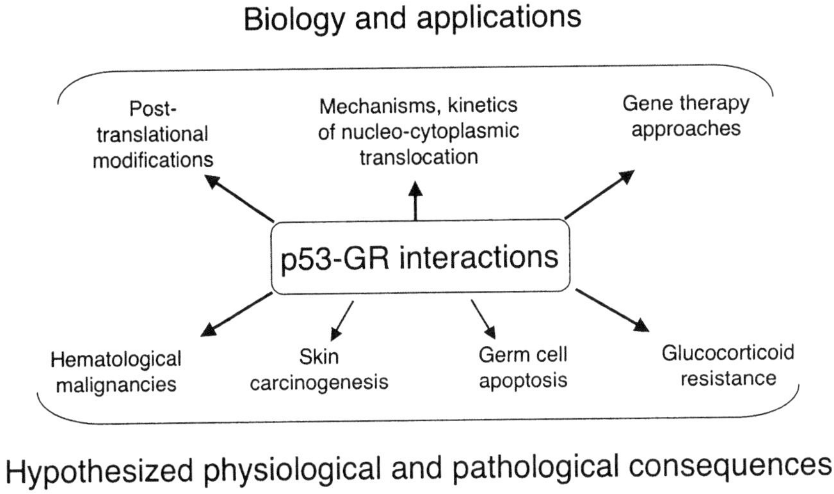

FIGURE 4. Future directions. The scientific advancements that have been made concerning p53–GR interactions raise a number of important questions that can be subdivided into "biology and applications" and "hypothesized physiological and pathological consequences." The individual themes are discussed in the text.

by a phosphorylation event. Interestingly, similar to GR antagonists, a protein kinase inhibitor has been shown to be able to induce apoptosis in neuroblastoma cells.[99] Hence, it would be interesting to determine whether phosphorylation of GR is a prerequisite for its interaction with p53. Similar considerations could be extended to other posttranslational modifications of p53 and GR.

Ligand-dependent translocation of GR from the cytoplasm to the nucleus is well characterized.[100] The precise mechanism for the nuclear-cytoplasmic translocation of p53 and GR merits further investigation. It would be interesting to know more about the mechanics, dynamics, and kinetics of GR transport in and out of the nucleus especially with respect to p53. We have already demonstrated that the nuclear localization signal (NLS) of p53 physically interacts with GR,[20] and conversely, it would be of interest to know which region(s) of GR interact(s) with p53. Recently, it has been reported that treatment of mice with the GR antagonist, RU486, causes a decrease in apoptosis and also a decrease in the level of p53 in hepatic cells.[101] In light of the diverse pleiotrophic effects of RU486, it will be worthwhile to identify more specific inhibitors of GR–p53 interactions. The determination of the minimum amino acid residues of both p53 and GR involved in complex formation would provide information needed to generate small molecules capable of disrupting the p53–GR interaction. These molecules can be expected to be a vast improvement on the use of GR antagonists like RU486, since they will have been specifically designed to inhibit the interaction between p53 and GR. These small molecules can be tested in cell lines and, if successful, can be investigated in mice and possibly primate models before extending to human clinical trials. Experiments can also be carried out to test whether p53 is complexed with GR in other cancer types where p53 has reported to be cytoplasmic. Cells representing different types of cancer could be used to establish cell-based screening procedures to isolate molecules capable of reactivating p53. Molecules isolated in these screens could become useful products for the treatment of certain forms of cancer.

Possible Physiological Conditions and Pathological Consequences of p53–GR Regulation

Hematological Malignancies

GCs and GR play essential roles in maintaining basal and stress-related homeostasis, and in particular, in stimulating erythropoiesis.[102–106] Pathological changes of GC levels affect erythropoiesis. Insufficient corticosteroid production, in Addison's disease, is associated with anemia, whereas elevated glucocorticoid levels, in Cushing's syndrome patients, is connected with increased red blood cell (RBC) count, hemoglobin, and hematocrit values. Recently, we reported that p53 antagonizes the proliferative effects of GR and inhibits the effect of c-Myb, thereby favoring differentiation.[39] The tumor suppressor p53 is frequently inactivated in human cancers, including leukemia and lymphoma.[107] Amongst its different functions, p53 may be involved in hematopoietic re-population. Numerous *in vitro* studies indicate that p53 is involved in proliferation, differentiation, and apoptosis of hematopoietic cells.[108–110]

Diamond-Blackfan anemia (DBA) is a congenital hypoplastic anemia. The majority of patients respond to treatment with GCs, and erythropoiesis can often be maintained with low doses of hormone. Erythroid precursors isolated from DBA patients are arrested at an early stage of differentiation. Molecular studies have identified mutations in the gene encoding the ribosomal protein RPS19 in 25% of patients with DBA.[111] Interestingly, in several cases RPS19 is mutated in the nucleolar localization signal, resulting in a failure to localize in the nucleolus and a decrease in protein expression.[112] MDM2 is known to bind to ribosomal proteins and RNA and can localize to the nucleolus.[36] Given, in addition, the connections between p53, MDM2, GR and erythroid progenitors, it would be pertinent to study whether p53 is involved in DBA, at the very least in the cases linked to RPS19 mutation.

As a natural extension of our recent work, where we had demonstrated negative crosstalk between p53 and GR during erythropoiesis,[39] it would be interesting to explore possibilities of how p53 and GR might crosstalk during hematological malignancies in humans. For example, p53 and GR may form a complex in Cushing's syndrome patient cells and tissues, since cytoplasmic sequestration of p53 has been reported in some of these patients.[113] It is possible that GR functions are modulated by both wild-type and mutant p53s in different hematological malignant conditions, and vice versa. The limited microarray analysis that is available does not extend to the cytokine-dependent target genes that are modulated by p53 and GR.[114] Hence, extensive molecular profiling of p53 and GR target genes in different multiple myeloma and leukemic cells could help to correlate GR status and function with that of p53. Such a database could have far-reaching implications for the treatment of patients with anti-cancer drugs.

There is a large amount of evidence that GCs suppress apoptosis in various cell types (including neutrophils).[63] Hence, a re-evaluation of the usage of GCs in conjunction with different chemotherapeutic drugs that induce apoptosis is necessary.[64]

The p53 status and the timing of GC administration should be taken into consideration to optimize the efficiency of anticancer drug therapy.

Skin Carcinogenesis

p53 is the key UV-responsive gene in skin whose mutation is thought to initiate carcinogenesis.[115] p53, expressed at low levels in unexposed skin, is induced by UV in basal layer cells,[116,117] resulting in cell-cycle arrest and apoptosis, which protects from cancer induction.[118] UV damage to the p53 gene inhibits the normal protective response, thereby predisposing the cells to cancer.[119,120] GCs are known to be potent inhibitors of tumor promotion in skin, when applied at early stages of the induction of carcinogenesis.[121] However, established skin papillomas are resistant to growth inhibition by GCs.[122] It has been suggested that alteration of both the expression and the function of GR may be an important mechanism of tumor promotion in skin,[123] due to a striking increase in the transcriptional activity of GR.[124] Hence the scope exists for the study of the role of GCs and GR during UV-induced skin carcinogenesis, especially in relation to p53 status and function. It is quite possible that p53 and GR modulate each other's activity and function in different stages of UV-induced skin carcinogenesis.

Germ Cell Apoptosis

In testes, neutrophils play an important role in regulating inflammatory immune responses. The adhesion of neutrophils to endothelial cells is inhibited by GCs.[125] GCs also inhibit apoptosis and increase proliferation of neutrophils.[56] High levels of p53 protein are detected in germ cells, with the greatest expression in primary spermatocytes.[126] The abdominal heat induced by experimental cryptorchidism results in germ line apoptosis by both p53-dependent and -independent pathways. A p53-dependent pathway is responsible for the initial phase of germ cell apoptosis, whereas other biochemical triggers of apoptosis are involved in a later phase of cell death.[127] It has been recently reported that Dex can suppress apoptosis of testicular germ cells induced by testicular ischemia.[128] Reactive oxygen species arising from neutrophils induce apoptosis of testicular germ cells. Reactive oxygen species are downstream mediators of p53 dependent apoptosis.[129] Hence it can be argued that during stress conditions (such as testicular ischemia) p53-dependent apoptosis in germ cells may be inhibited by GCs. This hypothesis can be addressed by utilizing the knockout mouse models of both p53 and GR. This study should complement the recent report that demonstrated that GR-mediated modulation of a p53-dependent processes is not related to the defect in mammary gland development as recently reported for GR^{dim} mice (i.e., mice in which the transactivation function of GR is lost).[130]

Pathological Conditions Exhibiting Glucocorticoid Resistance

The positive or negative regulation of GR by p53 can be conceptualized to be involved in either GC sensitivity or GC resistance. A number of pathological states have been associated with changes in tissue sensitivity to GCs.[131] GC resistance has been associated not only with GR mutations but also in several autoimmune/inflammatory states, such as rheumatoid arthritis (RA), Crohn's disease, ulcerative colitis (UC), and asthma. The above pathological conditions are often associated with the resistance of inflamed tissues to GCs.[132–134] In many of the above autoimmune/inflammatory states, p53 is upregulated. For example, p53 is found to be upregulated in the inflamed and regenerated mucosa in UC and Crohn's disease.[135] Most importantly, p53 is transcriptionally active and positively correlated with nitric oxide synthase.[136] Again p53 protein is expressed in RA fibroblast-like synoviocytes (FLSs), and its overexpression is a characteristic feature of RA.[137] It is quite possible that the GC resistance seen in the above autoimmune/inflammatory pathological conditions is because of the negative crosstalk between p53 and GR. This hypothesis deserves to be investigated.

ACKNOWLEDGMENTS

We thank BioAvenir (Aventis, Rhone-Poulenc), the Centre National de la Recherche Scientifique, the Institut National de la Santé et de la Recherche Médicale, the Hôpital Universitaire de Strasbourg, the Association pour la Recherche sur le Cancer, the Ligue Nationale Française contre le Cancer (Equipe labellisée), the Ligue Régionale (Haut-Rhin) contre le Cancer, the Ligue Régionale (Bas-Rhin) contre le Cancer, and the EU (FP5 project QLK6-2000-00159) for financial assistance.

REFERENCES

1. NUCLEAR, RECEPTOR, AND COMMITTEE. 1999. Cell **97:** 161–163.
2. TRONCHE, F., C. KELLENDONK, H.M. REICHARDT & G. SCHUTZ. 1998. Curr. Opin. Genet. Dev. **8:** 532–538.
3. REICHARDT, H.M. & G. SCHUTZ. 1998. Mol. Cell. Endocrinol. **146:** 1–6.
4. KELLENDONK, C., F. TRONCHE, H.M. REICHARDT & G. SCHUTZ. 1999. J. Steroid Biochem. Mol. Biol. **69:** 253–259.
5. HOLLSTEIN, M., D. SIDRANSKY, B. VOGELSTEIN & C.C. HARRIS. 1991. Science **253:** 49–53.
6. APPELLA, E. 2001. Eur. J. Biochem. **268:** 2763
7. HAUPT, S., M. BERGER, Z. GOLDBERG & Y. HAUPT. 2003. J. Cell. Sci. **116:** 4077–4085.
8. OREN, M., A. DAMALAS, T. GOTTLIEB, *et al.* 2002. Ann. N.Y. Acad. Sci. **973:** 374–383.
9. WOODS, D. B. & K.H. VOUSDEN. 2001. Exp. Cell Res. **264:** 56–66.
10. BRADY, H.J., G.S. SALOMONS, R.C. BOBELDIJK & A.J. BERNS. 1996. EMBO J. **15:** 1221–1230.
11. CLARKE, A.R., C.A. PURDIE, D.J. HARRISON, *et al.* 1993. Nature **362:** 849–852.
12. LOWE, S.W., E.M. SCHMITT, S.W. SMITH, *et al.* 1993. Nature **362:** 847–849.
13. SCHREIBER, S.S., S. SAKHI, M.M. DUGICH-DJORDJEVIC & N.R. NICHOLS. 1994. Exp. Neurol. **130:** 368–376.
14. MORI, N., J. YAMATE, A.P. STASSEN, *et al.* 1999. Oncogene **18:** 4282–4285.
15. SASSON, R., K. TAJIMA & A. AMSTERDAM. 2001. Endocrinology **142:** 802–811.
16. ALARCON-VARGAS, D. & Z. RONAI. 2002. Carcinogenesis **23:** 541–547.
17. MOMAND, J., H.H. WU & G. DASGUPTA. 2000. Gene **242:** 15–29.
18. YU, C., N. YAP, D. CHEN & S. CHENG. 1997. Cancer Lett. **116:** 191–196.
19. SENGUPTA, S., J.L.VONESCH, C. WALTZINGER, *et al.* 2000. EMBO J. **19:** 6051–6064.
20. SENGUPTA, S. & B. WASYLYK. 2001. Genes Dev. **15:** 2367–2380.
21. MAIYAR, A.C., P.T. PHU, A.J. HUANG & G.L. FIRESTONE. 1997. Mol. Endocrinol. **11:** 312–329.
22. VOGAN, K., M. BERNSTEIN, J.M. LECLERC, *et al.* 1993. Cancer Res. **53:** 5269–5273.
23. MOLL, U.M., G. RIOU & A.J. LEVINE. 1992. Proc. Natl. Acad. Sci. USA **89:** 7262–7266.
24. MOLL, U.M., M. LAQUAGLIA, J. BENARD & G. RIOU. 1995. Proc. Natl. Acad. Sci. USA **92:** 4407–4411.

25. BOSARI, S., G. VIALE, M. RONCALL, *et al.* 1995. Am. J. Pathol. **147:** 790–798.
26. EIZENBERG, O., A. FABER-ELMAN, E.O. GOTTLIEB, *et al.* 1996. Mol. Cell. Biol. **16:** 5178–5185.
27. GANNON, J.V. & D.P. LANE. 199. Nature **349:** 802–806.
28. KLOTZSCHE, O., D. ETZRODT, H. HOHENBERG, *et al.* 1998. Oncogene **16:** 3423–3434.
29. GIANNAKAKOU, P., D.L. SACKETT, Y. WARD, *et al.* 2000. Nat. Cell. Biol. **2:** 709–717.
30. METCALFE, S., A. WEEDS, A.L. OKOROKOV, *et al.* 1999. Oncogene **18:** 2351–2355.
31. NIKOLAEV, A., M. LI, N. PUSKAS, *et al.* 2003. Cell **112:** 29–40.
32. SHAULSKY, G., A. BEN-ZE'EV & V. ROTTER. 1990. Oncogene **5:** 1707–1711.
33. ALADJEM, M.I., B.T. SPIKE, L.W. RODEWALD, *et al.* 1998. Curr. Biol. **8:** 145–155.
34. RIVA, C., C. CHAUVIN, C. PISON & X. LEVERVE. 1998. Anticancer Res. **18:** 4729–4736.
35. ZAIKA, A., N. MARCHENKO & U.M. MOLL. 1999. J. Biol. Chem. **274:** 27474–27480.
36. GANGULI, G. & B. WASYLYK. 2003. Mol. Cancer Res.**1:** 1027–1035.
37. KINYAMU, H.K. & T.K. ARCHER. 2003. Mol. Cell. Biol. **23:** 5867–5881
38. MAIYAR, A.C., A.J. HUANG, P.T. PHU, *et al.* 1996. J. Biol. Chem. **271:** 12414–12422
39. GANGULI, G., J. BACK, S. SENGUPTA & B. WASYLYK. 2002. EMBO Rep. **3:** 569–574.
40. VON LINDERN, M., W. ZAUNER, G. MELLITZER, *et al.* 1999. Blood **94:** 550–559.
41. VALERIE, K. 1999. *In* Biopharmaceutical Drug Design and Development. S. Wu-Pong & Y. Rojanasakul, Eds.: 69–105. Humana Press, Inc. Totowa, NJ.
42. MILLER, F.D., C.D. POZNIAK & G.S. WALSH. 2000. Cell Death Differ. **7:** 880–888.
43. SOUSA, N. & O.F. ALMEIDA. 2002. Rev. Neurosci. **13:** 59–84.
44. ALMEIDA, O.F., G.L. CONDE, C. CROCHEMORE, *et al.* 2000. FASEB J. **14:** 779–790.
45. CROCHEMORE, C., T.M. MICHAELIDIS, D. FISCHER, *et al.* 2002. FASEB J. **16:** 761–770.
46. WALDMAN, T., K.W. KINZLER & B. VOGELSTEIN. 1995. Cancer Res. **55:** 5187–5190.
47. CHA, H.H., E.J. CRAM, E.C. WANG, *et al.* 1998. J. Biol. Chem. **273:** 1998–2007.
48. CRAM, E.J., R.A. RAMOS, E.C. WANG, *et al.* 1998. J. Biol. Chem. **273:** 2008–2014.
49. ZUO, Z., G. URBAN, J.G. SCAMMELL, *et al.* 1999. Biochemistry **38:** 8849–8857.
50. URBAN, G., T. GOLDEN, I.V. ARAGON, *et al.* 2003. J. Biol. Chem. **278:** 9747–9753.
51. SCHMIDT, M., H.G. PAUELS, N. LUGERING, *et al.* 1999. J. Immunol. **163:** 3484–3490.
52. SIKORA, E., G.P. ROSSINI, E. GRASSILLI, *et al.* 1996. Acta Biochim. Pol. **43:** 721–731.
53. MARTIN, S.J. & D.R. GREEN. 1995. Crit. Rev. Oncol. Hematol. **18:** 137–153.
54. CHAUHAN, D., P. PANDEY, A. OGATA, *et al.* 1997. Oncogene **15:** 837–843.
55. CHANG, T.C., M.W. HUNG, S.Y. JIANG, *et al.* 1997. FEBS Lett. **415:** 11–15.
56. DAFFERN, P.J., M.A. JAGELS & T.E. HUGLI. 1999. Am. J. Respir. Cell Mol. Biol. **21:** 259–267.
57. HE, S., H.M. WANG, J. YE, *et al.* 1994. Curr. Eye Res. **13:** 257–261.
58. LI, S., M. MAWAL-DEWAN, V.J. CRISTOFALO & C. SELL. 1998. J. Cell Physiol. **177:** 396–401.
59. HUANG, S. & J.A. CIDLOWSKI. 1999. FASEB J. **13:** 467–476.
60. YAMAMOTO, M., K. FUKUDA, N. MIURA, *et al.* 1998. Hepatology **27:** 959–966.
61. BAILLY-MAITRE, B., G. DE SOUSA, K. BOULUKOS, *et al.* 2001. Cell Death Differ. **8:** 279–288.
62. EVANS-STORMS, R.B. & J.A. CIDLOWSKI. 1997. Exp. Cell Res. **230:** 121–132.
63. HERR, I., E. UCUR, K. HERZER, *et al.* 2003. Cancer Res. **63:** 3112–3120.
64. AMSTERDAM, A., K. TAJIMA & R. SASSON. 2002. Biochem. Pharmacol. **64:** 843–850.
65. BRUNE, B. 2003. Cell Death Differ. **10:** 864–869.
66. CHUNG, H.T., H.O. PAE, B.M. CHOI, *et al.* 001. Biochem. Biophys. Res. Commun. **282:** 1075–1079.
67. WHITWORTH, J., C.G. SCHYVENS, Y. ZHANG, *et al.* 2002. J. Hypertens. **20:** 1035–1043.
68. KORHONEN, R., A. LAHTI, M. HAMALAINEN, *et al.* 2002. Mol. Pharmacol. **62:** 698–704.
69. LEE, H.J. & C. CHANG. 2003. Cell Mol. Life Sci. **60:** 1613–1622.
70. CULIG, Z., H. KLOCKER, G. BARTSCH, *et al.* 2003. J. Urol. **170:** 1363–1369.
71. JEMAL, A., T. MURRAY, A. SAMUELS, *et al.* 2003. CA Cancer J. Clin. **53:** 5–26.
72. APAKAMA, I., M.C. ROBINSON, N.M. WALTER, *et al.* 1996. Br. J. Cancer **74:** 1258–1262.
73. HEIDENBERG, H.B., I.A. SESTERHENN, J.P. GADDIPATI, *et al.* 1995. J. Urol. **154:** 414–421.
74. NAVONE, N.M., P. TRONCOSO, L.L. PISTERS, *et al.* 1993. J. Natl. Cancer Inst. **85:** 1657–1669.

75. LEITE, K.R., M.F. FRANCO, M. SROUGI, *et al.* 2001. Mod. Pathol. **14:** 428–436.
76. OSMAN, I., M. DROBNJAK, M. FAZZARI, *et al.* 1999. Clin. Cancer Res. **5:** 2082–2088.
77. NANTERMET, P.V., J. XU, Y. YU, *et al.* 2004. J. Biol. Chem. **279:** 1310–1322.
78. BURCHARDT, M., T. BURCHARDT, A. SHABSIGH, *et al.* 2001. Prostate **48:** 225–230.
79. NESSLINGER, N.J., X.B. SHI & R.W. DEVERE WHITE. 2003. Cancer Res. **63:** 2228–2233.
80. SHENK, J.L., C.J. FISHER, S.Y. CHEN, *et al.* 2001. J. Biol. Chem. **276:** 38472–38479.
81. LIN, H.K., L. WANG, Y.C. HU, *et al.* 2002. EMBO J. **21:** 4037–4048.
82. ZHANG, Z., M. LI, H. WANG, *et al.* 2003. Proc. Natl. Acad. Sci. USA **100:** 11636–11641.
83. FEIGELSON, H.S. & B.E. HENDERSON. 1996. Carcinogenesis **17:** 2279–2284.
84. OSBORNE, C.K. 1998. Breast Cancer Res. Treat. **51:** 227–238.
85. PHAROAH, P.D., N.E. DAY & C. CALDAS. 1999. Br. J. Cancer **80:** 1968–1973.
86. PETERSEN, O.W., L. RONNOV-JESSEN, V.M. WEAVER & M.J. BISSELL. 1998. Adv. Cancer Res. **75:** 135–161.
87. HORI, M., J. SHIMAZAKI, S. INAGAWA & M. ITABASHI. 2002. Breast Cancer Res. Treat. **71:** 77–83.
88. HURD, C., S. DINDA, N. KHATTREE & V.K. MOUDGIL. 1999. Oncogene **18:** 1067–1072.
89. OKUMURA, N., S. SAJI, H. EGUCHI, *et al.* 2002. Jpn. J. Cancer Res. **93:** 867–873.
90. SAJI, S., S. NAKASHIMA, S. HAYASHI, *et al.* 1999. Jpn. J. Cancer Res. **90:** 210–218.
91. LIU, G., SCHWARTZ, J.A. & S.C. BROOKS. 2000. Cancer Res. **60:** 1810–1814.
92. MOLINARI, A.M., P. BONTEMPO, E.M. SCHIAVONE, *et al.* 2000. Cancer Res. **60:** 2594–2597.
93. LIU, G., J.A. SCHWARTZ & S.C. BROOKS. 1999. Biochem. Biophys. Res. Commun. **264:** 359–364.
94. REID, G., M.R. HUBNER, R. METIVIER, *et al.* 2003. Mol. Cell **11:** 695–707.
95. SAJI, S., N. OKUMURA, H. EGUCHI, *et al.* 2001. Biochem. Biophys. Res. Commun. **281:** 259–265.
96. HENDERSON, B.E. & H.S. FEIGELSON. 2000. Carcinogenesis **21:** 427–433.
97. BODWELL, J.E., J.C. WEBSTER, C.M. JEWELL, *et al.* 1998. J. Steroid Biochem. Mol. Biol. **65:** 91–99.
98. APPELLA, E. & C.W. ANDERSON. 2001. Eur. J. Biochem. **268:** 2764–2772.
99. RONCA, F., S.L. CHAN & V.C. YU. 1997. J. Biol. Chem. **272:** 4252–4260.
100. PRATT, W. B. & D.O. TOFT. 1997. Endocr. Rev. **18:** 306–360.
101. YOUSSEF, J.A. & M.Z. BADR. 2003. Mol. Cancer **2:** 3.
102. GOLDE, D.W., N. BERSCH & M.J. CLINE. 1976. J. Clin. Invest. **57:** 57–62.
103. UDUPA, K.B., H.M. CRABTREE & D.A. LIPSCHITZ. 1986. Br. J. Haematol. **62:** 705–714.
104. ZITO, G.E. & E.C. LYNCH. 1977. J. Am. Med. Assoc. **237:** 991–992.
105. LIANG, R., T.K. CHAN & D. TODD. 1994. Leuk. Lymphoma **13:** 411–415.
106. GREENSTEIN, S., K. GHIAS, N.L. KRETT & S.T. ROSEN. 2002. Clin. Cancer Res. **8:** 1681–1694.
107. HAINAUT, P. & M. HOLLSTEIN. 2000. Adv. Cancer Res. **77:** 81–137.
108. KASTAN, M.B., A.I. RADIN, S.J. KUERBITZ, *et al.* 1991. Cancer Res. **51:** 4279–4286.
109. LOTEM, J. & L. SACHS. 1993. Blood **82:** 1092–1096.
110. PROKOCIMER, M. & V. ROTTER. 1994. Blood **84:** 2391–2411.
111. DA COSTA, L., T.N. WILLIG, J. FIXLER, *et al.* 2001. Curr. Opin. Pediatr. **13:** 10–15.
112. DA COSTA, L., G. TCHERNIA, P. GASCARD, *et al.* 2003. Blood **101:** 5039–5045.
113. KONTOGEORGOS, G., N. KAPRANOS, E. THODOU, *et al.* 1999. Pituitary **1:** 207–212.
114. KOLBUS, A., M. BLAZQUEZ-DOMINGO, S. CAROTTA, *et al.* 2003. Blood **102:** 3136–3146.
115. SOEHNGE, H., A. OUHTIT & O.N. ANANTHASWAMY. 1997. Front. Biosci. **2:** D538–D551.
116. JONASON, A.S., S. KUNALA, G.J. PRICE, *et al.* 1996. Proc. Natl. Acad. Sci. USA **93:** 14025–14029.
117. HALL, P.A., P.H. MCKEE, H.D. MENAGE, *et al.* 1993. Oncogene **8:** 203–207.
118. JIANG, W., H.N. ANANTHASWAMY, H.K. MULLER & M.L. KRIPKE. 1999. Oncogene **18:** 4247–4253.

119. BERG, R.J., H.J. VAN KRANEN, H.G. REBEL, *et al.* 1996. Proc. Natl. Acad. Sci. USA **93:** 274–278.
120. ZIEGLER, A., A.S. JONASON, D.J. LEFFELL, *et al.* 1994. Nature **372:** 773–776.
121. VERMA, A.K., C.T. GARCIA, C.L. ASHENDEL & R.K. BOUTWELL. 1983. Cancer Res. **43:** 3045–3049.
122. STRAWHECKER, J.M. & J.C. PELLING. 1992. Carcinogenesis **13:** 2075–2080.
123. BUDUNOVA, I.V., S. CARBAJAL, H. KANG, *et al.* 1997. Mol. Carcinog. **18:** 177–185.
124. VIVANCO, M.D., R. JOHNSON, P.E. GALANTE, *et al.* 1995. EMBO J. **14:** 2217–2228.
125. BOCHSLER, P.N., D.O. SLAUSON & N.R. NEILSEN. 1990. J. Leukoc. Biol. **48:** 306–315.
126. SCHWARTZ, D., N. GOLDFINGER & V. ROTTER. 1993. Oncogene **8:** 1487–1494.
127. YIN, Y., W.C. DEWOLF & A. MORGENTALER. 1998. Biol. Reprod. **58:** 492–496.
128. YAZAWA, H., I. SASAGAWA, Y. SUZUKI & T. NAKADA. 2001. Fertil. Steril. **75:** 980–985.
129. JOHNSON, T.M., Z.X. YU, V.J. FERRANS, *et al.* 1996. Proc. Natl. Acad. Sci. USA **93:** 11848–11852.
130. REICHARDT, H.M., K. HORSCH, H.J. GRONE, *et al.* 2001. Eur. J. Endocrinol. **145:** 519–527.
131. KINO, T., M.U. DE MARTINO, E. CHARMANDARI, *et al.* 2003. J. Steroid Biochem. Mol. Biol. **85:** 457–467.
132. FRANCHIMONT, D., E. LOUIS, P. DUPONT, *et al.* 1999. Dig. Dis. Sci. **44:** 1208–1215.
133. BAMBERGER, C.M., H.M. SCHULTE & G.P. CHROUSOS. 1996. Endocr. Rev. **17:** 245–261.
134. KINO, T. & G.P. CHROUSOS. 2001. J. Endocrinol. **169:** 437–445.
135. KRISHNA, M., B. WODA, L. SAVAS, *et al.* 1995. Mod. Pathol. **8:** 654–657.
136. HOFSETH, L.J., S. SAITO, S. HUSSAIN, *et al.* 2003. Proc. Natl. Acad. Sci. USA **100:** 143–148.
137. SUN, Y. & H.S. CHEUNG. 2002. Semin. Arthritis Rheum. **31:** 299–310.

Interaction of the Glucocorticoid Receptor and the Chicken Ovalbumin Upstream Promoter–Transcription Factor II (COUP-TFII)

Implications for the Actions of Glucocorticoids on Glucose, Lipoprotein, and Xenobiotic Metabolism

MASSIMO U. DE MARTINO, SALVATORE ALESCI, GEORGE P. CHROUSOS, AND TOMOSHIGE KINO

Pediatric and Reproductive Endocrinology Branch, National Institute of Child Health and Human Development, National Institutes of Health, Bethesda, Maryland 20892-1583, USA

ABSTRACT: Glucocorticoids exert their extremely diverse effects on numerous biologic activities of humans via only one protein module, the glucocorticoid receptor (GR). The GR binds to the glucocorticoid response elements located in the promoter region of target genes and regulates their transcriptional activity. In addition, GR associates with other transcription factors through direct protein–protein interactions and mutually represses or stimulates each other's transcriptional activities. The latter activity of GR may be more important than the former one, granted that mice harboring a mutant GR, which is active in terms of protein–protein interactions but inactive in terms of transactivation via DNA, survive and procreate, in contrast to mice with a deletion of the entire GR gene that die immediately after birth. We recently found that GR physically interacts with the chicken ovalbumin upstream promoter–transcription factor II (COUP-TFII), which plays a critical role in the metabolism of glucose, cholesterol, and xenobiotics, as well as in the development of the central nervous system in fetus. GR stimulates COUP-TFII–induced transactivation by attracting cofactors via its activation function-1, while COUP-TFII represses the GR-governed transcriptional activity by tethering corepressors, such as the silencing mediator for retinoid and thyroid hormone receptors (SMRT) and the nuclear receptor corepressors (NCoRs) via its C-terminal domain. Their mutual interaction may play an important role in gluconeogenesis, lipoprotein metabolism, and enzymatic clearance of clinically important compounds and bioactive chemicals, by regulating their rate-limiting enzymes and molecules, including the phosphoenolpyruvate carboxykinase (PEPCK), the cytochrome P450 CYP3A and CYP7A, and several apolipoproteins. It appears that glucocorticoids exert their intermediary effects partly via physical interaction with COUP-TFII.

Address for correspondence: Tomoshige Kino, M.D., Ph.D., Pediatric and Reproductive Endocrinology Branch, National Institute of Child Health and Human Development, National Institutes of Health, Bldg. 10, Rm. 9D42, 10 Center Drive MSC 1583, Bethesda, MD 20892-1583. Voice: 301-496-6417; fax: 301-402-0884.

kinot@mail.nih.gov

Ann. N.Y. Acad. Sci. 1024: 72–85 (2004).
doi: 10.1196/annals.1321.006

KEYWORDS: protein–protein interaction; gluconeogenesis; phosphoenolpyruvate carboxykinase (PEPCK); cholesterol; apolipoproteins; P450 CYP7A; xenobiotic metabolism; P450 CYP3A

INTRODUCTION

Glucocorticoids are steroid hormones secreted by the adrenal glands, important for maintenance of basal and stress-related homeostasis.[1] They regulate a variety of biologic processes and exert a profound influence on many physiologic functions, such as the energy catabolism, by regulating the metabolic rate of glucose, fatty acids, and cholesterol, and the clearance of bioactive compounds in the liver.[2,3] Glucocorticoids are also used as potent immunosuppressive agents in the management of many inflammatory, autoimmune, and lymphoproliferative diseases, while they produce many adverse effects, such as glucose intolerance/overt diabetes mellitus, and dyslipidemia, due to their effects on the corresponding metabolic pathways.[4]

Glucocorticoids exert their diverse effects through the glucocorticoid receptor (GR).[1] GR belongs to the nuclear receptor superfamily and functions as a ligand-inducible transcription factor.[5] Ligand-activated GR binds the glucocorticoid response elements (GREs) located in the enhancer region of the glucocorticoid-responsive genes and alters their transcriptional activity.[6] More importantly, GR affects other signal transduction cascades through mutual protein–protein interactions with other transcription factors, which act downstream of these signaling events.[7] Ligand-activated GR physically interacts with such transcription factors and suppresses/enhances their transcriptional activity with several mechanisms.[8–10] This activity of GR may be more important than the GRE-mediated one, granted that mice harboring a mutant GRα, which is active in terms of protein–protein interactions but inactive in terms of transactivation via DNA, survive and procreate, in contrast to mice with a deletion of the entire GR gene that die immediately after birth from severe respiratory distress syndrome.[11,12] Thus, protein–protein interaction and subsequent modulation of other signaling pathways may be critical to sustain extremely diverse effects of glucocorticoids on broad arrays of tissues/organs via the single receptor molecule, GR. Furthermore, this activity may be particularly important in suppressing the immune function and inflammation by glucocorticoids.[11,13] A substantial part of the effects of glucocorticoids on the immune system may be explained by the interaction between GR and nuclear factor-κB (NF-κB), activator protein-1 (AP-1), and probably signal transducers and activators of transcription (STATs).[8,13–15] GR is also known to influence the transcriptional activity of other transcription factors, such as CREB, CAAT/enhancer-binding protein (C/EBP), Nur77, p53, hepatocyte nuclear factor (HNF)-6, GATA-1, Oct-1 and-2, and nuclear factor (NF)-1.[16–23]

We have recently found that GR interacts with one of the orphan nuclear receptors, the chicken ovalbumin upstream promoter–transcription factor II, and mutually affects each other's transcriptional activity.[24] This orphan nuclear receptor plays an important role in glucose, fatty acid, cholesterol, and xenobiotic metabolism, as well as embryonic development.[25] In this review, we describe details of our recent results of physical/functional interaction between GR and COUP-TFII. We speculate that their interaction may play a critical role in the glucocorticoid action on the intermediary metabolism.

STRUCTURE AND ACTION OF GR

The human GR (hGR), a single polypeptide chain of 777 amino acid residues, is a member of the steroid/sterol/thyroid/retinoid/orphan receptor superfamily of nuclear transactivating factors, with over 150 members currently cloned and characterized across species.[26] Together with the mineralocorticoid, progesterone, estrogen, and androgen receptors, GR forms the steroid receptor subfamily. Steroid receptors display a modular structure comprising five to six regions (A–F), with the N-terminal A/B region, also called the immunogenic domain, and the C and E regions corresponding to the DNA-binding (DBD) and ligand-binding (LBD) domains, respectively (FIG. 1).[27] The hGR cDNA was isolated by expression cloning in 1985.[28] The gene of the hGR consists of 9 exons and is located on chromosome 5.[27] It encodes two 3′ splicing variants, GRα and β, from alternative use of a different

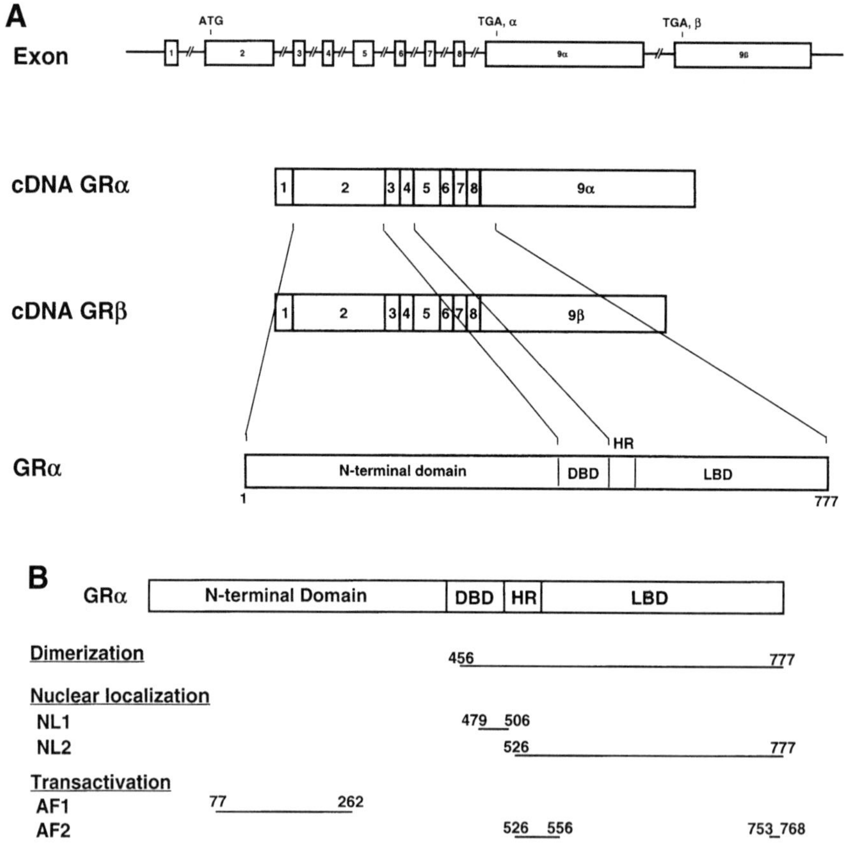

FIGURE 1. (**A** and **B**) Genomic structure and domains of linearized GRα and GRβ molecules. GR, glucocorticoid receptor; HR, hinge region; DBD, DNA-binding domain; LBD, ligand-binding domain; NL, nuclear localization signal; AF-1 and -2, activation function 1 and 2. Unique portions of GRα and GRβ are indicated in grey color.

terminal exon 9α and β, respectively. The hGRα encodes a 777 amino acid protein, while the hGRβ contains 742 amino acids. The first 727 amino acids from the N-terminus are identical in both isoforms. hGRα possesses an additional 50 amino acids, while hGRβ encodes an additional 15 nonhomologous amino acids in its C-terminus. The molecular weights of each receptor isoforms are 97 and 94 kDa, respectively. hGRα is the classic GR that binds to glucocorticoids and transactivates or transrepresses glucocorticoid-responsive promoters. On the other hand, hGRβ does not bind to glucocorticoids and functions as a weak transdominant inhibitor of GRα transactivation; however, its physiologic and pathologic roles are not well known.[29]

The N-terminal domain of GR contains one of the transactivation domains of the GR, activation function (AF) 1, located at amino acids 77–262.[30,31] The activity of this transactivation domain is ligand-independent. Its core activation domain is localized at amino acids 187–244.[31] The DBD of the hGR corresponds to amino acids 420–480, and contains two zinc finger motifs through which it binds to specific sequences of DNA, the glucocorticoid response elements (GREs).[32] The LBD of the GR is comprises 12 helical structures and changes its tertiary configuration in following LBD, thus creating the AF2 surface, which acts as a lignad-induced transactivation domain of the GR.[33,34]

The GR is located primarily in the cytoplasm of cells, as part of a hetero-oligomeric complex with heat-shock proteins (hsp) 90, 70, and 50 and, possibly, other proteins through interaction via the LBD.[35–38] Upon ligand binding, the GR dissociates from the hsp, homodimerizes, and translocates into the nucleus, where it binds to the hormone response elements in the promoter regions of target genes and/or to other transcription factors.[7] The GR contains two nuclear translocation signals (NL), NL1 and NL2; NL1 contains a classic basic-type nuclear localization signal (NLS) structure that overlaps with and extends along the C-terminal from the DNA binding domain of the GR.[39] The function of NL1 is dependent on the importin α, a component of a nuclear translocation system that is energy-dependent protein import machinery through the nuclear pore. NL2, with an uncharacterized motif, overlaps with almost the entire LBD.

GR exerts its transcriptional activity on its responsive promoters via binding to its recognition sequence, GREs.[6] The GRE-bound GR stimulates the transcription rates of responsive genes by facilitating the formation of the transcription initiation complex, including the RNA polymerase II and its ancillary factors.[6] In addition to these molecules, GR, via its AF 1 and 2 domains, first attracts several proteins and protein complexes, which may bridge the DNA-bound GR and the transcription initiation complex as well as modulate the tightly assembled chromatin structure surrounding the promoter region.[40]

INTERACTION OF GR WITH COUP-TFII

COUP-TFII

The chicken ovalbumin upstream promoter–transcription factor II (COUP-TFII) is a protein of 414 amino acids, and its gene is located on chromosome 15.[41] It is an "orphan receptor" (i.e., its native ligand is unknown) and a member of the nuclear steroid/thyroid hormone receptor superfamily.[25] COUP-TFII, along with its closely related protein COUP-TFI in humans, has been characterized as a family of negative coregulators of gene transcription.[25] COUP-TFII binds to the responsive elements with several different sequences, including AGGTCA direct repeats with different spacings. COUP-TFII most strongly interacts with a motif with one base-pair spacing between AGGTCA sequences (called DR1).[25,42,43] Since several other receptors, such as the vitamin D receptor (VDR), thyroid hormone receptor (TR), retinoic acid receptor (RAR), the retinoid X receptor (RXR), the peroxisome proliferators–activated receptors (PPARs), and the orphan nuclear receptor, hepatocyte nuclear factor-4 (HNF-4), bind direct repeats of AGGTCA and use these sequences as their responsive elements, COUP-TFs exert their negative regulatory function by competing for the common response element.[25] COUP-TFs can also actively silence the basal transcription machinery of their target promoters by attracting corepressor molecules, such as the silencing mediator for retinoid and thyroid hormone receptors (SMRT) and the nuclear receptor corepressors (NCoRs), and histone diacetylases through direct interaction with its C-terminal 35 amino acids.[25,44] In addition, COUP-TFII, through the same responsive sequences, can stimulate the transcriptional activity of several target genes that catalyze the glucose and cholesterol metabolism.[25,45–47]

Physical Interaction between GR and COUP-TFII

To identify molecules that interact with GR, we performed a LexA-based yeast two-hybrid screening assay using GRα and β LBDs as baits in the human Jarkat cDNA library.[24] We found that both GRα and β LBDs interact with two independent clones that contain the human COUP-TFII coding sequence. This interaction was confirmed by yeast mating and GST pull-down assays. In the latter, we used bacterially produced and purified GST-fused COUP-TFII, together with *in vitro* translated and [^{35}S]-labeled GRα and β. Both GRα and β bound to GST-COUP-TFII in a ligand-independent fashion. Using a set of GST-COUP-TFII fusion fragments, we found that GRα and β bound to COUP-TFII in a region enclosed between amino acids 75 and 163, a portion corresponding to the DBD of this transcription factor[24] (FIG. 2). Similarly, using bacterially produced and purified GRα and β with [^{35}S]-labeled COUP-TFII, we found that COUP-TFII bound to both GRα and β in a region enclosed by amino acids 490 to 502 that is located in the "hinge region" of the receptors[24] (FIG. 2). To further characterize this interaction, we used several GRα mutants, with point mutations in the region 490–502, in the same GST pull-down assay. GRαQ501A and Q502A, in which aspartic acid was replaced with alanine at positions 501 or 502, respectively, were unable to bind COUP-TFII. GRαA458T, which has a threonine instead of an alanine at position 458 and, therefore, is unable to dimerize, was also able to interact with COUP-TFII, indicating that GRα interacts with COUP-TFII as a monomer.[24]

We next performed chromatin immunoprecipitation assays to study the *in vivo* association of GR and COUP-TFII in the context of a natural chromatin-bound promoter. The rat cholesterol 7α-hydroxylase (CYP7A) promoter, which is known to respond to COUP-TFII through two sets of the COUP-TFII–responsive elements, was used in this experiment.[45,46] In H4IIE rat hepatoma cells, GRα was successfully coprecipitated with this promoter in a dexamethasone (DEX)-dependent fashion.[24] In contrast, GRβ was not coprecipitated with this promoter, despite the *in vitro* bind-

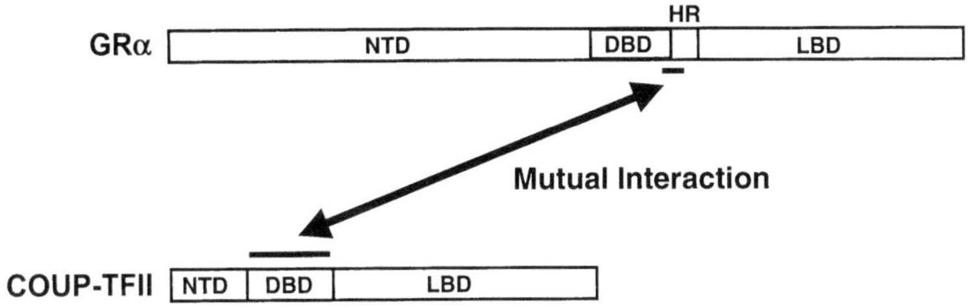

FIGURE 2. Linearized GRα and COUP-TFII and their mutual interaction domains (indicated with *bold line*). NTD, N-terminal domain; DBD, DNA-binding domain; HR, hinge region; LBD, ligand-binding domain.

ing to COUP-TFII. GR and COUP-TFII also formed complexes on the endogenous GREs of the tyrosine aminotransferase promoter. The results were also confirmed using regular immunoprecipitation assays. These results clearly indicate that GR and COUP-TFII form complexes on their responsive promoters *in vivo*.

Reciprocal Transcriptional Modulation of GR and COUP-TFII

The functional aspects of the physical interaction between GR and COUP-TFII were studied in a transient transfection assay using their respective responsive promoters, the MMTV promoter, containing four GREs, and a short fragment of the rat CYP7A promoter that contains two COUP-TFII responsive elements.[45,46] The CYP7A is known as the cholesterol 7α-hydroxylase gene and its product catalyzes the first and rate-limiting step of cholesterol to bile acid in the liver, thus playing a pivotal role in the elimination of cholesterol from the body and maintenance of the cholesterol homeostasis.[48] COUP-TFII suppressed the DEX-stimulated, GR-induced transactivation of the MMTV promoter in a dose-dependent fashion.[24] In contrast, overexpression of GRα enhanced COUP-TFII–stimulated CYP7A promoter activity in a dose-dependent manner. GRβ did not influence COUP-TFII–induced transactivation on the same promoter.

We next examined the mechanisms of the functional interaction between GRα and COUP-TFII. GRα contains two transactivation domains, AF-1 and AF-2, whose activities are supported by a specific interaction with several coactivators and chromatin modulators.[40] We examined the contribution of each of these domains on GR-induced enhancement of COUP-TFII transactivation. GRα(Δ77–261), which is devoid of the AF-1 transactivation domain, completely lost its enhancing effect on the COUP-TFII–induced transcriptional activity of the CYP7A promoter in a glucocorticoid-dependent fashion, while GRα(1–550), which has the AF-1 but not the AF-2 domain, still enhanced COUP-TFII–induced transactivation in a glucocorticoid-independent fashion.[24] These results suggest that the AF-1 domain of GR plays an important role in GR-induced enhancement of COUP-TFII transactivation (FIG. 3).

COUP-TFII is known to actively suppress the transcriptional activity of several promoters by attracting the corepressor SMRT through direct binding via its last 35 amino acids.[44] We thus examined SMRT on COUP-TFII–induced suppression of GR-induced transactivation. Coexpression of SMRT synergistically enhanced the

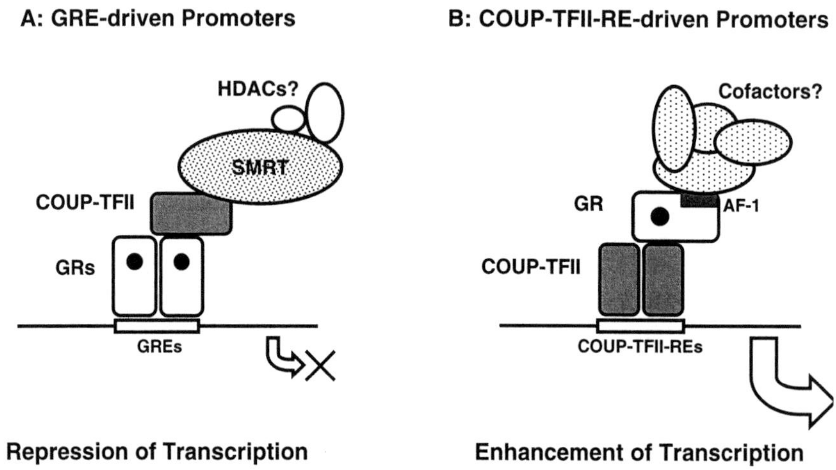

FIGURE 3. Functional interaction of GR and COUP-TFII on their responsive promoters. GR enhances COUP-TFII–induced transactivation through its AF-1 domain probably by attracting AF-1–associating cofactors, while COUP-TFII suppresses GR-induced transactivation in part by attracting SMRT and other corepressors. GRs, glucocorticoid receptors; COUP-TFII, chicken ovalbumin upstream promoter–transcription factor II; GRE, glucocorticoid-responsive element; COUP-TFII-RE, COUP-TFII–responsive element; AF-1, activation function-1; SMRT, silencing mediator for retinoid and thyroid hormone receptors; HDACs, histone deacetylases.

suppressive effect of COUP-TFII on GR transactivation. COUP-TFII(1–380), which is devoid of the SMRT-binding domain, significantly lost the suppressive effect and SMRT did not effectively cooperate with this mutant COUP-TFII.[24] Therefore, COUP-TFII, at least in part, suppressed the GR-induced transcriptional activity by attracting SMRT to the glucocorticoid-responsive promoters. Simplified models of functional interaction of GR and COUP-TFII are shown in FIGURE 3.

Glucose Metabolism: Modulation of Phosphoenolpyruvate Carboxykinase Expression

Glucocorticoids modulate diverse metabolic activities in the liver, adipose tissue, and connective tissues, such as those of carbohydrate, lipid, and amino acid/protein.[1] Changes induced by pharmachologic or stressed-induced doses of glucocorticoids increase circulating concentrations of glucose, free fatty acids, amino acids, and glycerin by facilitating catabolism of fat, protein, and glycogen in peripheral tissues.[1] The liver takes up these compounds and uses them as substrates for gluconeogenesis and subsequent glycogen synthesis. Glucocorticoids influence glucose metabolism by acting at many steps of its metabolic pathways.[1,49] Glucocorticoids increase glycogen synthesis by activating glucokinase, glycogen synthase, and inactivation of glycogen phosphorylase, which helps catalyze glycogen.[49,50] Glucocorticoids also decrease the insulin-stimulated glucose uptake and utilization by peripheral tissues.[51–53] Furthermore, glucocorticoids directly act on the β cells to reduce their insulin secretion.[54–56]

In addition to the above-described effects on glucose metabolism, glucocorticoids also increase hepatic gluconeogenesis, which converts amino acids, fatty acid, and glycerin supplied from peripheral tissues to glucose.[49] Glucocorticoids stimulate this metabolic activity by increasing the production of its rate-limiting enzyme, phosphoenol-pyruvate carboxykinase (PEPCK), by direct stimulation of its promoter activity via GR.[49,57,58]

GR is reported to be associated with the PEPCK promoter through its two GRE-like motifs, termed GR unit 1 and 2 (GRU1 and 2).[47,59–61] In addition, at least three accessory factor-binding elements, called AF-1, 2, and 3, are required for the activation of the PEPCK promoter by GR.[59,60] The factors that bind to these accessory elements include HNF-4/COUP-TFII, HNF-3, and COUP-TFII, respectively.[59–61] The GRU1 and GRU2 correspond to the classical GRE at only 7 and 6 of 12 nucleotides, respectively, and bind GR only with a very low affinity.[59,61] In addition, they are not able to confer glucocorticoid responsiveness if they are connected to the heterologous promoter-driven reported construct.[60] In contrast, mutations of any of the accessory elements significantly reduce the activation of the PEPCK promoter by glucocorticoids.[47,60] Thus, the accessory factors bound on the PEPCK promoter are essential for the GR-induced stimulation of this promoter activity.

Since we found that GR physically interacts with COUP-TFII, we examined the importance of their physical interaction on the PEPCK promoter activity.[24] We found that elimination of COUP-TFII with an antisense to this orphan receptor abolished the glucocorticoid-induced expression of PEPCK mRNA in HepG2 cells.[24] The same treatment also abolished glucocorticoid-induced enhancement of the PEPCK promoter activity. A COUP-TFII–binding defective GR mutant failed to enhance the activity of this promoter. These results indicate that COUP-TFII and its interaction with GR are necessary for the stimulation of the PEPCK promoter and subsequent PEPCK mRNA expression. Since COUP-TFII is one of the cofactors that bind the accessory elements of the PEPCK promoter, GR may be attracted to the PEPCK promoter through COUP-TFII via its protein–protein interaction. Thus, GRU1 and 2 might play a supportive role in attracting GR to the PEPCK promoter.

It has been recently shown that the peroxisome proliferator–activated receptor-γ coactivator-1 (PGC-1), which has a major role in the cellular respiration and adaptive thermogenesis in muscle and brown fat, stimulates the PEPCK expression in hepatocytes, through cooperation with HNF-4.[62,63] PGC-1 is also known to synergistically act with GR on this promoter.[62] Since HNF-4 is one of the molecules that occupy the first accessory elements (AF-1) of the PEPCK promoter and interact with COUP-TFII,[45,62] COUP-TFII and GR might also have functional/physical interaction with this coactivator to stimulate the PEPCK expression.

Cholesterol Metabolism and Apolipoprotein Synthesis

Glucocorticoids increase serum cholesterol levels and modulate serum concentration of the low-density (LDL), very-low-density (VLDL), and high-density (HDL) lipoproteins by affecting various steps of their synthesis/clearance/degradation pathways.[49,64] For example, glucocorticoids stimulate cholesterol synthesis by inducing the expression of its rate-limiting enzyme, HMG-CoA reductase, *in vitro*.[65]

Glucocorticoids also suppress the expression of the LDL receptor on the hepatocyte that may help increase serum LDL levels.[66] In addition, glucocorticoids affect the production of apolipoproteins, components of serum lipoproteins that regulate the metabolic process of lipoproteins in the liver and peripheral tissues. Administration of glucocorticoids increases the serum concentration of apolipoprotein A1, a major protein component of HDL,[67] both in normal subjects and cultured cells.[64,68] Glucocorticoids also increase serum concentrations of apolipoprotein E.[64]

COUP-TFII was first identified as a protein that binds to regulatory elements of the apolipoprotein A1 gene, and for this reason it was originally called ARP-1 (apolipoprotein A1 regulatory protein).[69] Through binding to its promoter and change in its transcriptional activity, COUP-TFII reduces the transcription of the apolipoprotein A1 gene.[69] Since both glucocorticoids and COUP-TFII regulate the expression of apolipoprotein A1, it is likely that glucocorticoids increase the expression of this protein through COUP-TFII via protein–protein interactions between GR and COUP-TFII. Since COUP-TFII also binds to the promoter region of the very-low-density apolipoprotein II, apolipoprotein CII, B, and AII genes, it regulates their transcriptional activity and modulates the lipoprotein lipase expression;[69–71] glucocorticoids may indirectly affect their expressions through COUP-TFII, and hence further regulate the metabolism of other lipoproteins.

In addition to the above-described apolipoprotein metabolism, glucocorticoids also stimulate the bile acid synthesis[72,73] by stimulating the expression of its rate-limiting enzyme cholesterol 7α-hydroxylase, a gene product of the CYP7A.[72] The COUP-TFII binds to the promoter region of the CYP7A and stimulates its transcriptional activity.[46] Since we showed that glucocorticoids enhanced this effect of COUP-TFII on the CYP7A promoter,[24] it is possible that glucocorticoids regulate the expression of this enzyme through protein–protein interaction between the GR and promoter-bound COUP-TFII.

Xenobiotic Metabolism: Regulation of the CYP3A Expression

Glucocorticoids have important effects on the xenobiotic metabolism in the liver, which catalyzes oxidative reactions of clinically used compounds and environmental chemicals.[74] The CYP3A, a microsomal cytochrome P450 monooxygenase enzyme, plays a major role in this reaction.[74] This molecule is the most abundant form in the enzymes that catalyze xenobiotic metabolism.[74] It is also known that several clinically important chemicals, such as pregnane compounds, amilorides, imigazole, PPARγ agonists, and phenobarbital, induce CYP3A expression.[74] Glucocorticoids also stimulate the transcription rate of this enzyme.[75]

Among the CYP3A family enzymes, the rat CYP3A23 gene has been most extensively examined in the regulation of its gene expression by glucocorticoids.[76–79] The promoter region of this enzyme has three regions—DEX-RE1, 2, and Site A—which support inducibility of CYP3A23 by glucocorticoids.[77] COUP-TFII and pregnane X receptor (PXR) bind the former two sites, while HNF-4 associates with the last site.[74,79–82] The mechanism of glucocorticoid-induced regulation of this promoter is complicated—at low concentrations, glucocorticoids stimulate the expression of the PXR protein and indirectly stimulate the activity of the CYP3A23 promoter, while at higher concentrations, glucocorticoids directly stimulate the promoter by acting as ligands for PXR.[83] Interestingly, COUP-TFII regulates this promoter activity by binding to the same region of this promoter that PXR targets.[74,83] Thus, it is highly likely that glucocorticoids additionally regulate this promoter activity via associating with promoter-bound COUP-TFII, and their interaction might, in some part, explain the regulatory mechanisms of the CYP3A expression by glucocorticoids.

SUMMARY

We showed that GR and COUP-TFII physically interact and mutually influence each other's transcriptional activity. This interaction may be important for the expression of PEPCK, which is a rate-limiting enzyme of hepatic gluconeogenesis. GR and COUP-TFII may also act cooperatively on the induction of apolipoproteins and CYP7A and CYP3A, which are important enzymes that catalyze bile acid synthesis and xenobiotic metabolism, respectively. These observations indicate that COUP-TFII may play an important role in the glucocorticoid effects on the intermediary metabolism, such as glucose, cholesterol, and catabolism of chemical compounds. Since knockout animals of GR or COUP-TFII have become available, the biologic importance of their interaction should be tested *in vivo* in these animals. Modulation of their interaction may be also important for the pathologic states that impair glucose and cholesterol metabolism, such as diabetes mellitus and dyslipidemia. Further studies are required to determine the implications of COUP-TFII to these pathologic states.

ACKNOWLEDGMENT

We appreciate Dr. Evangelia Charmandari for critically reading the manuscript.

REFERENCES

1. ORTH, D.N. & W.J. KOVACS. 1998. The adrenal cortex. *In* Williams Textbook of Endocrinology. J.D. Wilson, D.W. Foster, H.M. Kronenberg & P.R. Larsen, Eds.: 517–664. W.B. Saunders Co. Philadelphia.
2. MUNCK, A., P.M. GUYRE & N.J. HOLBROOK. 1984. Physiological functions of glucocorticoids in stress and their relation to pharmacological actions. Endocr. Rev. **5:** 25–44.
3. CLARK, J.K., W.T. SCHRADER & B.W. O'MALLEY. 1992. Mechanism of steroid hormones. *In* Williams Textbook of Endocrinology. J.D. Wilson & D.W. Foster, Eds.: 35–90. W.B. Saunders Co. Philadelphia.
4. BOUMPAS, D.T., G.P. CHROUSOS, R.L. WILDER, *et al.* 1993. Glucocorticoid therapy for immune-mediated diseases: basic and clinical correlates. Ann. Intern. Med. **119:** 1198–1208.
5. KINO, T. & G.P. CHROUSOS. 2001. Glucocorticoid and mineralocorticoid resistance/hypersensitivity syndromes. J. Endocrinol. **169:** 437–445.
6. BEATO, M. & A. SANCHEZ-PACHECO. 1996. Interaction of steroid hormone receptors with the transcription initiation complex. Endocr. Rev. **17:** 587–609.
7. BAMBERGER, C.M., H.M. SCHULTE & G.P. CHROUSOS. 1996. Molecular determinants of glucocorticoid receptor function and tissue sensitivity to glucocorticoids. Endocr. Rev. **17:** 245–261.
8. CALDENHOVEN, E., J. LIDEN, S. WISSINK, *et al.* 1995. Negative cross-talk between RelA and the glucocorticoid receptor: a possible mechanism for the antiinflammatory action of glucocorticoids. Mol. Endocrinol. **9:** 401–412.

9. McKay, L.I. & J.A. Cidlowski. 2000. CBP (CREB binding protein) integrates NF-kappaB (nuclear factor-kappaB) and glucocorticoid receptor physical interactions and antagonism. Mol. Endocrinol. **14:** 1222–1234.
10. Nissen, R.M. & K.R. Yamamoto. 2000. The glucocorticoid receptor inhibits NFkappaB by interfering with serine-2 phosphorylation of the RNA polymerase II carboxy-terminal domain. Genes Dev. **14:** 2314–2329.
11. Reichardt, H.M., K.H. Kaestner, J. Tuckermann, *et al.* 1998. DNA binding of the glucocorticoid receptor is not essential for survival. Cell **93:** 531–541.
12. Cole, T.J., J.A. Blendy, A.P. Monaghan, *et al.* 1995. Targeted disruption of the glucocorticoid receptor gene blocks adrenergic chromaffin cell development and severely retards lung maturation. Genes Dev. **9:** 1608–1621.
13. Reichardt, H.M., J.P. Tuckermann, M. Gottlicher, *et al.* 2001. Repression of inflammatory responses in the absence of DNA binding by the glucocorticoid receptor. EMBO J. **20:** 7168–7173.
14. Stocklin, E., M. Wissler, F. Gouilleux & B. Groner. 1996. Functional interactions between Stat5 and the glucocorticoid receptor. Nature **383:** 726–728.
15. Schule, R., P. Rangarajan, S. Kliewer, *et al.* 1990. Functional antagonism between oncoprotein c-Jun and the glucocorticoid receptor. Cell **62:** 1217–1226.
16. Imai, E., J.N. Miner, J.A. Mitchell, *et al.* 1993. Glucocorticoid receptor-cAMP response element-binding protein interaction and the response of the phosphoenolpyruvate carboxykinase gene to glucocorticoids. J. Biol. Chem. **268:** 5353–5356.
17. Chang, T.J., B.M. Scher, S. Waxman & W. Scher. 1993. Inhibition of mouse GATA-1 function by the glucocorticoid receptor: possible mechanism of steroid inhibition of erythroleukemia cell differentiation. Mol. Endocrinol. **7:** 528–542.
18. Pierreux, C.E., J. Stafford, D. Demonte, *et al.* 1999. Antiglucocorticoid activity of hepatocyte nuclear factor-6. Proc. Natl. Acad. Sci. USA **96:** 8961–8966.
19. Prefontaine, G.G., M.E. Lemieux, W. Giffin, *et al.* 1998. Recruitment of octamer transcription factors to DNA by glucocorticoid receptor. Mol. Cell. Biol. **18:** 3416–3430.
20. Sengupta, S., J.L. Vonesch, C. Waltzinger, *et al.* 2000. Negative cross-talk between p53 and the glucocorticoid receptor and its role in neuroblastoma cells. EMBO J. **19:** 6051–6064.
21. Kusk, P., S. John, G. Fragoso, *et al.* 1996. Characterization of an NF-1/CTF family member as a functional activator of the mouse mammary tumor virus long terminal repeat 5′ enhancer. J. Biol. Chem. **271:** 31269–31276.
22. Philips, A., M. Maira, A. Mullick, *et al.* 1997. Antagonism between Nur77 and glucocorticoid receptor for control of transcription. Mol. Cell. Biol. **17:** 5952–5959.
23. Boruk, M., J.G. Savory & R.J. Hache. 1998. AF-2-dependent potentiation of CCAAT enhancer binding protein beta-mediated transcriptional activation by glucocorticoid receptor. Mol. Endocrinol. **12:** 1749–1763.
24. De Martino, M.U., N. Bhattachryya, S. Alesci, *et al.* 2004. The glucocorticoid receptor (GR) and the orphan nuclear receptor chicken ovalbumin upstream promoter-transcription factor II (COUP-TFII) interact with and mutually affect each other's transcriptional activities: implications for intermediary metabolism. Mol. Endocrinol. In press.
25. Qiu, Y., V. Krishnan, F.A. Pereira, *et al.* 1996. Chicken ovalbumin upstream promoter-transcription factors and their regulation. J. Steroid Biochem. Mol. Biol. **56:** 81–85.
26. Mangelsdorf, D.J., C. Thummel, M. Beato, *et al.* 1995. The nuclear receptor superfamily: the second decade. Cell **83:** 835–839.
27. Kino, T., A. Vottero, E. Charmandari & G.P. Chrousos. 2002. Familial/sporadic glucocorticoid resistance syndrome and hypertension. Ann. N.Y. Acad. Sci. **970:** 101–111.
28. Hollenberg, S.M., C. Weinberger, E.S. Ong, *et al.* 1985. Primary structure and expression of a functional human glucocorticoid receptor cDNA. Nature **318:** 635–641.
29. Vottero, A. & G.P. Chrousos. 1999. Glucocorticoid receptor beta: view I. Trends Endocrinol. Metab. **10:** 333–338.

30. ALMLOF, T., A.E. WALLBERG, J.A. GUSTAFSSON & A.P. WRIGHT. 1998. Role of important hydrophobic amino acids in the interaction between the glucocorticoid receptor tau 1-core activation domain and target factors. Biochemistry **37:** 9586–9594.
31. WARNMARK, A., J.A. GUSTAFSSON & A.P. WRIGHT. 2000. Architectural principles for the structure and function of the glucocorticoid receptor tau 1 core activation domain. J. Biol. Chem. **275:** 15014–15018.
32. LUISI, B.F., W.X. XU, Z. OTWINOWSKI, *et al.* 1991. Crystallographic analysis of the interaction of the glucocorticoid receptor with DNA.1H NMR studies of the glucocorticoid receptor DNA-binding domain: sequential assignments and identification of secondary structure elements. Nature **352:** 497–505.
33. BOURGUET, W., P. GERMAIN & H. GRONEMEYER. 2000. Nuclear receptor ligand-binding domains: three-dimensional structures, molecular interactions and pharmacological implications. Trends Pharmacol. Sci. **21:** 381–388.
34. BLEDSOE, R.K., V.G. MONTANA, T.B. STANLEY, *et al.* 2002. Crystal structure of the glucocorticoid receptor ligand binding domain reveals a novel mode of receptor dimerization and coactivator recognition. Cell **110:** 93–105.
35. OWENS-GRILLO, J.K., K. HOFFMANN, K.A. HUTCHISON, *et al.* 1995. The cyclosporin A-binding immunophilin CyP-40 and the FK506-binding immunophilin hsp56 bind to a common site on hsp90 and exist in independent cytosolic heterocomplexes with the untransformed glucocorticoid receptor. J. Biol. Chem. **270:** 20479–20484.
36. HUTCHISON, K.A., L.F. STANCATO, J.K. OWENS-GRILLO, *et al.* 1995. The 23-kDa acidic protein in reticulocyte lysate is the weakly bound component of the hsp foldosome that is required for assembly of the glucocorticoid receptor into a functional heterocomplex with hsp90. J. Biol. Chem. **270:** 18841–18847.
37. DENIS, M., J.A. GUSTAFSSON & A.C. WIKSTROM. 1988. Interaction of the Mr = 90,000 heat shock protein with the steroid-binding domain of the glucocorticoid receptor. J. Biol. Chem. **263:** 18520–18523.
38. CZAR, M.J., R.H. LYONS, M.J. WELSH, *et al.* 1995. Evidence that the FK506-binding immunophilin heat shock protein 56 is required for trafficking of the glucocorticoid receptor from the cytoplasm to the nucleus. Mol. Endocrinol. **9:** 1549–1560.
39. SAVORY, J.G., B. HSU, I.R. LAQUIAN, *et al.* 1999. Discrimination between NL1- and NL2-mediated nuclear localization of the glucocorticoid receptor. Mol. Cell. Biol. **19:** 1025–1037.
40. MCKENNA, N.J., R.B. LANZ & B.W. O'MALLEY. 1999. Nuclear receptor coregulators: cellular and molecular biology. Endocr. Rev. **20:** 321–344.
41. WANG, L.H., S.Y. TSAI, R.G. COOK, *et al.* 1989. COUP transcription factor is a member of the steroid receptor superfamily. Nature **340:** 163–166.
42. MALIK, S. & S. KARATHANASIS. 1995. Transcriptional activation by the orphan nuclear receptor ARP-1. Nucleic Acids Res. **23:** 1536–1543.
43. BUTLER, A.J. & M.G. PARKER. 1995. COUP-TF II homodimers are formed in preference to heterodimers with RXR alpha or TR beta in intact cells. Nucleic Acids Res. **23:** 4143–4150.
44. SHIBATA, H., Z. NAWAZ, S.Y. TSAI, *et al.* 1997. Gene silencing by chicken ovalbumin upstream promoter-transcription factor I (COUP-TFI) is mediated by transcriptional corepressors, nuclear receptor-corepressor (N-CoR) and silencing mediator for retinoic acid receptor and thyroid hormone receptor (SMRT). Mol. Endocrinol. **11:** 714–724.
45. STROUP, D. & J.Y. CHIANG. 2000. HNF4 and COUP-TFII interact to modulate transcription of the cholesterol 7alpha-hydroxylase gene (CYP7A1). J. Lipid Res. **41:** 1–11.
46. STROUP, D., M. CRESTANI & J.Y. CHIANG. 1997. Orphan receptors chicken ovalbumin upstream promoter transcription factor II (COUP-TFII) and retinoid X receptor (RXR) activate and bind the rat cholesterol 7alpha-hydroxylase gene (CYP7A). J. Biol. Chem. **272:** 9833–9839.
47. STAFFORD, J.M., M. WALTNER-LAW & D.K. GRANNER. 2001. Role of accessory factors and steroid receptor coactivator 1 in the regulation of phosphoenolpyruvate carboxykinase gene transcription by glucocorticoids. J. Biol. Chem. **276:** 3811–3819.
48. DAVIS, R.A., J.H. MIYAKE, T.Y. HUI & N.J. SPANN. 2002. Regulation of cholesterol-7alpha-hydroxylase: BAREly missing a SHP. J. Lipid Res. **43:** 533–543.

49. MILLER, W.L. & G.P. CHROUSOS. 2001. The adrenal cortex. *In* Endocrinology & Metabolism. P. Felig & L.A. Frohman, Eds.: 387–524. McGraw-Hill. New York.
50. STALMANS, W. & M. LALOUX. 1979. Glucocorticoids and hepatic glycogen metabolism. Monogr. Endocrinol. **12:** 517–533.
51. OLEFSKY, J.M. 1975. Effect of dexamethasone on insulin binding, glucose transport, and glucose oxidation of isolated rat adipocytes. J. Clin. Invest. **56:** 1499–1508.
52. SAKODA, H., T. OGIHARA, M. ANAI, *et al.* 2000. Dexamethasone-induced insulin resistance in 3T3-L1 adipocytes is due to inhibition of glucose transport rather than insulin signal transduction. Diabetes **49:** 1700–1708.
53. WEINSTEIN, S.P., T. PAQUIN, A. PRITSKER & R.S. HABER. 1995. Glucocorticoid-induced insulin resistance: dexamethasone inhibits the activation of glucose transport in rat skeletal muscle by both insulin- and non-insulin-related stimuli. Diabetes **44:** 441–445.
54. OGAWA, A., J.H. JOHNSON, M. OHNEDA, *et al.* 1992. Roles of insulin resistance and beta-cell dysfunction in dexamethasone-induced diabetes. J. Clin. Invest. **90:** 497–504.
55. PHILIPPE, J. & M. MISSOTTEN. 1990. Dexamethasone inhibits insulin biosynthesis by destabilizing insulin messenger ribonucleic acid in hamster insulinoma cells. Endocrinology **127:** 1640–1645.
56. GRILL, V., M. ALVARSSON & S. EFENDIC. 1992. Dexamethasone treatment fails to increase arginine-induced insulin release in healthy subjects with low insulin response. Diabetologia **35:** 367–371.
57. COUFALIK, A.H. & C. MONDER. 1981. Stimulation of gluconeogenesis by cortisol in fetal rat liver in organ culture. Endocrinology **108:** 1132–1137.
58. VAN DE WERVE, G., A. LANGE, C. NEWGARD, *et al.* 2000. New lessons in the regulation of glucose metabolism taught by the glucose 6-phosphatase system. Eur. J. Biochem. **267:** 1533–1549.
59. WANG, J.C., P.E. STROMSTEDT, T. SUGIYAMA & D.K. GRANNER. 1999. The phosphoenolpyruvate carboxykinase gene glucocorticoid response unit: identification of the functional domains of accessory factors HNF3 beta (hepatic nuclear factor-3 beta) and HNF4 and the necessity of proper alignment of their cognate binding sites. Mol. Endocrinol. **13:** 604–618.
60. STAFFORD, J.M., J.C. WILKINSON, J.M. BEECHEM & D.K. GRANNER. 2001. Accessory factors facilitate the binding of glucocorticoid receptor to the phosphoenolpyruvate carboxykinase gene promoter. J. Biol. Chem. **276:** 39885–39891.
61. HALL, R.K., F.M. SLADEK & D.K. GRANNER. 1995. The orphan receptors COUP-TF and HNF-4 serve as accessory factors required for induction of phosphoenolpyruvate carboxykinase gene transcription by glucocorticoids. Proc. Natl. Acad. Sci. USA **92:** 412–416.
62. YOON, J.C., P. PUIGSERVER, G. CHEN, *et al.* 2001. Control of hepatic gluconeogenesis through the transcriptional coactivator PGC-1. Nature **413:** 131–138.
63. RHEE, J., Y. INOUE, J.C. YOON, *et al.* 2003. Regulation of hepatic fasting response by PPARgamma coactivator-1alpha (PGC-1): requirement for hepatocyte nuclear factor 4alpha in gluconeogenesis. Proc. Natl. Acad. Sci. USA **100:** 4012–4017.
64. ETTINGER, W.H., JR. & W.R. HAZZARD. 1988. Prednisone increases very low density lipoprotein and high density lipoprotein in healthy men. Metabolism **37:** 1055–1058.
65. AVIGAN, J. 1977. Studies on the effects of hormones on cholesterol synthesis in mammalian cells in culture. Expo. Annu. Biochim. Med. **33:** 1–11
66. AL RAYYES, O., A. WALLMARK & C.H. FLOREN. 1997. Additive inhibitory effect of hydrocortisone and cyclosporine on low-density lipoprotein receptor activity in cultured HepG2 cells. Hepatology **26:** 967–971.
67. VARMA, V.K., T.K. SMITH, M. SORCI-THOMAS & W.H. ETTINGER, JR. 1992. Dexamethasone increases apolipoprotein A-I concentrations in medium and apolipoprotein A-I mRNA abundance from Hep G2 cells. Metabolism **41:** 1075–1080.
68. PARKER, C.R., JR., P.C. MACDONALD, B.R. CARR & J.C. MORRISON. 1987. The effects of dexamethasone and anencephaly on newborn serum levels of apolipoprotein A-1. J. Clin. Endocrinol. Metab. **65:** 1098–1101.
69. LADIAS, J.A. & S.K. KARATHANASIS. 1991. Regulation of the apolipoprotein AI gene by ARP-1, a novel member of the steroid receptor superfamily. Science **251:** 561–565.

70. BEEKMAN, J.M., J. WIJNHOLDS, I.J. SCHIPPERS, *et al.* 1991. Regulatory elements and DNA-binding proteins mediating transcription from the chicken very-low-density apolipoprotein II gene. Nucleic Acids Res. **19:** 5371–5377.
71. LADIAS, J.A., M. HADZOPOULOU-CLADARAS, D. KARDASSIS, *et al.* 1992. Transcriptional regulation of human apolipoprotein genes ApoB, ApoCIII, and ApoAII by members of the steroid hormone receptor superfamily HNF-4, ARP-1, EAR-2, and EAR-3. J. Biol. Chem. **267:** 15849–15860.
72. PRINCEN, H.M., P. MEIJER & B. HOFSTEE. 1989. Dexamethasone regulates bile acid synthesis in monolayer cultures of rat hepatocytes by induction of cholesterol 7 alpha-hydroxylase. Biochem. J. **262:** 341–348.
73. ELLIS, E., B. GOODWIN, A. ABRAHAMSSON, *et al.* 1998. Bile acid synthesis in primary cultures of rat and human hepatocytes. Hepatology **27:** 615–620.
74. QUATTROCHI, L.C. & P.S. GUZELIAN. 2001. Cyp3A regulation: from pharmacology to nuclear receptors. Drug Metab. Dispos. **29:** 615–622.
75. HUNT, C.M., P.B. WATKINS, P. SAENGER, *et al.* 1992. Heterogeneity of CYP3A isoforms metabolizing erythromycin and cortisol. Clin. Pharmacol. Ther. **51:** 18–23
76. HUSS, J.M. & C.B. KASPER. 1998. Nuclear receptor involvement in the regulation of rat cytochrome P450 3A23 expression. J. Biol. Chem. **273:** 16155–16162.
77. HUSS, J.M., S.I. WANG, A. ASTROM , *et al.* 1996. Dexamethasone responsiveness of a major glucocorticoid-inducible CYP3A gene is mediated by elements unrelated to a glucocorticoid receptor binding motif. Proc. Natl. Acad. Sci. USA **93:** 4666–4670.
78. QUATTROCHI, L.C., C.B. YOCKEY, J.L. BARWICK & P.S. GUZELIAN. 1998. Characterization of DNA-binding proteins required for glucocorticoid induction of CYP3A23. Arch. Biochem. Biophys. **349:** 251–260.
79. KLIEWER, S.A., J.T. MOORE, L.WADE, *et al.* 1998. An orphan nuclear receptor activated by pregnanes defines a novel steroid signaling pathway. Cell **92:** 73–82.
80. LECLUYSE, E.L. 2001. Pregnane X receptor: molecular basis for species differences in CYP3A induction by xenobiotics. Chem. Biol. Interact. **134:** 283–289.
81. OGINO, M., K. NAGATA, M. MIYATA & Y. YAMAZOE. 1999. Hepatocyte nuclear factor 4-mediated activation of rat CYP3A1 gene and its modes of modulation by apolipoprotein AI regulatory protein I and v-ErbA-related protein 3. Arch. Biochem. Biophys. **362:** 32–37.
82. HUSS, J.M., S.I. WANG & C.B. KASPER. 1999. Differential glucocorticoid responses of CYP3A23 and CYP3A2 are mediated by selective binding of orphan nuclear receptors. Arch. Biochem. Biophys. **372:** 321–332.
83. HUSS, J.M. & C.B. KASPER. 2000. Two-stage glucocorticoid induction of CYP3A23 through both the glucocorticoid and pregnane X receptors. Mol. Pharmacol. **58:** 48–57.

Modulation of Glucocorticoid Receptor Function via Phosphorylation

NAIMA ISMAILI[a] AND MICHAEL J. GARABEDIAN[a,b]

Departments of Microbiology[a] and Urology,[b] New York University School of Medicine, 550 First Avenue, New York, New York 10016, USA

ABSTRACT: The glucocorticoid receptor (GR) is phosphorylated at multiple serine residues in a hormone-dependent manner. It has been suggested that GR phosphorylation affects turnover, subcellular trafficking, or the transcriptional regulatory functions of the receptor, yet the contribution of individual GR phosphorylation sites to the modulation of GR activity remains enigmatic. This review critically evaluates the literature on GR phosphorylation and presents more recent work on the mechanism of GR phosphorylation from studies using antibodies that recognize GR only when it is phosphorylated. In addition, we present support for the notion that GR phosphorylation modifies protein–protein interactions, which can stabilize the hypophosphorylated form of the receptor in the absence of ligand, as well as facilitate transcriptional activation by the hyperphosphorylation of GR via cofactor recruitment upon ligand binding. Finally, we propose that GR phosphorylation also participates in the non-genomic activation of cytoplasmic signaling pathways evoked by GR. Thus, GR phosphorylation is a versatile mechanism for modulating and integrating multiple receptor functions.

KEYWORDS: GR; phosphorylation; transcription; kinases

INTRODUCTION

Glucocorticoids are produced by the adrenal gland and have a wide range of functions including the regulation of glucose, fat and protein metabolism, cell growth and differentiation; they have effects on mood and cognitive function as well as anti-inflammatory and immunosuppressive actions.[1] Glucocorticoids exert their action through binding to the glucocorticoid receptor (GR), a member of the nuclear receptor (NR) family. The ligand-free form of GR is inactive and held in the cytoplasm through interactions with the heat shock protein 90 (hsp90)-chaperone complex. GR conformation is modified upon hormone binding, allowing its release from the hsp90-complex and its translocation to the nucleus. In this compartment, ligand-activated GR binds as a homodimer to specific DNA motifs termed glucocorticoid response elements (GREs) and activates the transcription of GRE-containing genes. GR can also repress gene transcription via protein–protein interaction independent

Address for correspondence: Michael J. Garabedian, Depts. of Microbiology and Urology, NYU School of Medicine, 550 First Avenue, New York, NY 10016. Voice: 212-263-7662; fax: 212-263-8276.
garabm01@med.nyu.edu

Ann. N.Y. Acad. Sci. 1024: 86–101 (2004).
doi: 10.1196/annals.1321.007

of DNA recognition.[2] In addition, GR activates cell signaling pathways, such as the phosphatidylinositol 3-kinase (PI3K) pathway and its downstream effector kinase Akt, an important regulator of cell cycle progression and cell survival, in a rapid, nontranscriptional manner.[3]

Similar to other steroid receptors, GR is a transcriptional regulatory protein characterized by three distinct functional domains.[4] The N-terminal region harbors a hormone-independent transcriptional activation function (AF-1). The central part of the protein contains a highly conserved zinc finger DNA binding domain (DBD), a nuclear localization signal, and a dimerization motif. The C-terminus includes a ligand-binding domain (LBD) with an additional transcriptional activation function (AF-2). Of these three functional domains, the N-terminal AF-1 region is the most variable among NRs in terms of length and sequence similarities. This variability might be responsible for gene regulation specificity of individual NRs.[5] In addition, this domain is the major target for ligand-dependent phosphorylation at multiple serine residues.[6] There is mounting evidence that posttranslational modification of GR by phosphorylation modulates receptor function.

The literature abounds with examples of protein phosphorylation leading to regulation and/or alteration of protein function, such as enzymatic activity (choline acetyl transferase[7]), subcellular localization (cdc25C;[8] CDH1;[9] SWI5[10]), DNA binding (Ets1;[11] SRF[12]), protein–protein interactions (p53;[13] CDK inhibitor[14]), or ubiquitination (IκBα[15,16]). Furthermore, in the particular case of the multifunctional protein p53, phosphorylation at distinct residues results in differential regulatory functions.[17] For example, the phosphorylation status of residue serine 20 of p53 enhances its interaction with MDM2, while that of residues serines 116 and 127 impairs transcription and MDM2-mediated protein degradation.[13]

GR as well as other NRs are phosphorylated upon hormone treatment. In recent years, progress has been made in understanding how GR phosphorylation affects its activity, subcellular localization, and turnover.

CHARACTERIZATION OF GR PHOSPHORYLATION

Interest in phosphorylation stemmed originally from studies aimed at understanding the mechanisms underlying the binding affinity of GR to its cognate ligand. Based essentially on phosphatase treatment studies of GR from tissue lysates, a phosphorylation–dephosphorylation mechanism was first proposed to be responsible for determining the level of active receptor in the cell based on the observation that GR can be inactivated by dephosphorylation and that only the phosphorylated form of the molecule is capable of binding to the steroid ligand.[18,19] In 1984, Kurl and Jacob provided direct evidence for GR phosphorylation.[20] They showed that incubation of a purified preparation of GR from rat liver cytosol with γ ^{32}P-ATP and Mg^{2+} resulted in transfer of the radiolabeled phosphate to the receptor protein. Additional biochemical studies showed agonist-dependent GR phosphorylation *in vivo*.[21–24] Identification and cloning of mouse and rat GR cDNAs,[25,26] in combination with phosphopeptide mapping of the receptor protein, allowed for a more detailed analysis of GR phosphorylation. Thus, in 1991, Bodwell *et al.* showed that mouse GR is phosphorylated on seven sites whether overexpressed in Chinese hamster ovary cells (Wcl2) or endogenously expressed in WEHI-7 mouse thymoma

Rat	Human	Mouse	Consensus	Kinases
S134	S113	S122	S/T(P)-X-X-E/D	CKII
S162	S141	S150	none	?
T171	*A150*	T159	non polar-X-S/T(P)-P	MAPK/GSK3
S224	S203	S212	S/T(P)-P-X-R/K	CDK
S232	S211	S220	S/T(P)-P-X-R/K	CDK
S246	S226	S234	non polar-X-S/T(P)-P	MAPK
S329	S308	S315	none	?

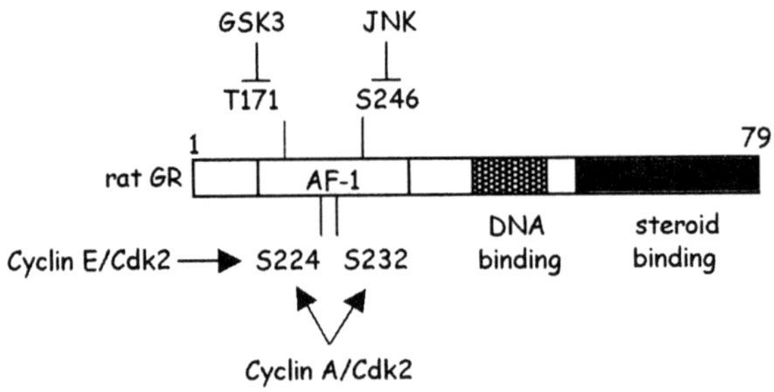

FIGURE 1. Residues phosphorylated on the glucocorticoid receptors. Shown are the sites of phosphorylation identified on the mouse GR and their equivalent positions in the rat and human receptor. The consensus phosphorylation site and the putative protein kinase expected to target the motif are shown. Rat GR T171 (mouse GR T159) is not conserved in human GR. A schematic representation of the rat GR sites phosphorylated and the kinases that modify them *in vitro*. Threonine 171 is phosphorylated by GSK3, and S246 is phosphorylated by JNK. S224 is phosphorylated by cyclin E/CDK2 and cyclinA/CDK2, and S232 is phosphorylated by cyclinA/CDK2.

cells.[27] Solid-phase sequencing of mouse GR revealed that most of the phosphorylated residues identified reside in the N-terminal AF-1 domain and are mainly serines at positions 122, 150, 212, 220, 234, and 315, except for a threonine at position 159. Sequence analysis of the rat and human GR relative to mouse GR showed that the phosphorylated serine residues and their surrounding sequences are conserved in all three species, except for threonine 159 and serine 315 which have no corresponding site in human GR (FIG. 1). Peptide mapping and mutagenesis studies on rat GR led to identification of phosphorylation sites (T171, S224, S232, and S246) corresponding to a subset of the ones described in mouse GR (T159, S212, S220, and S234, respectively).[28] As described for mouse GR, the serines in the amino-terminal region of human GR are the predominant sites of phosphorylation when the receptor is expressed ectopically in yeast. The mutant human GR lacking all five sites (S131, S141, S203, S211, S226) shows dramatic reduction in phosphorylation levels.[29]

It has been shown that rat GR T171 and S246 are constitutively phosphorylated, while phosphorylation at S224 and S232 (S203 and S211 in human GR) is dependent on hormone treatment.[28,30] Although specific residues become phosphorylated in the presence of agonists, the overall charge of phosphorylated GR is not altered as shown by nonequilibrium isoelectric focusing analysis, indicating that heterogeneous phosphorylated forms of GR coexist in the cell.[31] Precise studies of these subpopulations using mutagenesis analysis or biochemical methods are, however, limiting because individual proteins carrying specific phosphorylation sites are indistinguishable. In order to understand the contribution of different phosphorylation subpopulations to GR activities *in vivo*, we have recently developed antibodies that recognize specifically human GR phosphorylated at S203, S211, or S226[32] (Dang and Garabedian, in preparation). These antibodies react with human, mouse, and rat phospho-isoforms, reflecting the conservation of these receptor phosphorylation sites among species. In addition, the antibodies are extremely versatile in that they react with specific GR phospho-isoforms by immunoblotting, immunohistochemistry, immuno-fluorescence, and immunoprecipitation. Using a battery of agonists and antagonists, in combination with kinetic studies of phosphorylation at the different sites, we have been able to demonstrate the presence of different subpopulations of human GR, and we conclude that phosphorylation of human GR is a dynamic process—phosphorylation at S203 is present at a basal level in the absence of ligand and becomes stronger in the presence of hormone, whereas phopho-S211 is strictly agonist dependent. However, upon hormone treatment, S211 phosphorylation increases at a more substantial and sustained, albeit slower, rate than phosphorylation at S203.

PROTEIN KINASES IMPLICATED IN GR N-TERMINAL PHOSPHORYLATION

Early studies on the phosphorylation of GR from rat liver cytosol suggested the involvement of endogenous protein kinases.[33] Sequence analysis of mouse, rat, and human GR indicates conservation of both the phosphorylation sites and the adjacent sequences. Rat GR S224 (human GR S203 and mouse GR S212), S232 (human GR S211, mouse GR S220), and S246 (human GR S226 and mouse GR S234) are followed by a proline, thus matching the consensus motif recognized by either MAPKs or CDKs.

Indeed, the two major kinase families that seem to be responsible for a subset of GR N-terminal phosphorylation are (1) the mitogen-activated protein kinases (MAPK) and (2) the cyclin-dependent kinases (CDK). The superfamily of MAPKs consists of three main members—the extracellular signal-regulated protein kinases (ERKs), the c-Jun N-terminal kinases (JNKs), and the p38 family of kinases. MAPKs respond to a variety of cellular stimuli such as growth factors, stress events, cytokines, and mitogens. When activated, MAPKs mediate a wide range of cellular processes including gene transcription, chromatin remodeling, apoptosis, and inflammation, to mention a few. The MAPKs mediate phosphorylation of serine or threonine residues lying within the consensus sequence: nonpolar-X-S/T(P)-P.

As with the MAPKs, the CDK family is large and it phosphorylates proteins on serine or threonine residues within the consensus sequence: S/T(P)-P-X-R/K. These kinases are activated (1) by associating with cyclin subunits, (2) through binding of inhibitory polypeptides (e.g., p21 and p27), and (3) by activating or inhibiting phosphorylation reactions.[34]

To determine whether GR is, as predicted, a genuine substrate for CDKs and MAPKs, GR phosphorylation was analyzed using a simple *in vitro* kinase assay.[28] This showed that rat GR S246 (human GR S226) is phosphorylated by MAPK while S224 (human GR S203) and S232 (human GR S211) are modified by the cyclinA/CDK2 complex. S224, but not S232, can also be phosphorylated by cyclinE/CDK2 using this *in vitro* system. When S224 is mutated to alanine, phosphorylation of S232 by cyclinA/CDK2 complex is reduced, suggesting interdependency between these sites. In addition, it appears that the actions of individual family members are varied. For example, cyclin B/CDC2 induces mitosis; cyclin E/CDK2 controls DNA replication; cyclin D is synthesized in response to growth factors; CDK5 is expressed in postmitotic neurons; and cyclinH/CDK7 is part of the TFIIH complex. Conceivably, each cyclin/CDK complex phosphorylates a distinct set of specific target proteins or target residues. Evidence from our laboratory suggests that cyclinA/CDK2 efficiently phosphorylates the receptor at S203 and perhaps modifies S211 more efficiently when S203 is phosphorylated. Our genetic analysis suggests that both G_1 and G_2 cyclin/CDK complexes affect receptor transcriptional enhancement to different extents *in vivo*. Thus, phosphorylation of the receptor appears to be influenced by multiple cyclin/CDK complexes. These findings suggest that individual GR phosphorylation sites could potentially be modified by multiple kinases, depending on distinct cell signaling *in vivo*. The fact that MAPK phosphorylates the estrogen receptor on S118 in a ligand-independent manner, whereas cyclinH/CDK7 mediates hormone-induced phosphorylation of S118, supports this possibility.[35]

In another study, we narrowed down the MAPKs responsible for rat GR S246 phosphorylation to JNK and ERK, and we reported that inhibition of GR-mediated transcriptional activation *in vivo* depends on receptor phosphorylation at rat GR S246 (human GR S226) by JNK, but not ERK.[36] In agreement with our findings, Itoh *et al.* showed that activated JNK phosphorylates human GR at S226 and enhances its nuclear export after withdrawal of hormone, leading most likely to downregulation of its activity.[37]

In vivo analysis of individual phosphorylation sites of human GR using phospho-specific antibodies shows a more intricate interplay between the different phosphorylation sites. First, mutation of human GR S203 to alanine only slightly impaired phosphorylation of S211 in U2OS-hGR cells, further strengthening our proposal that multiple kinases might be recruited at any given site in response to specific cellular events.[32] Second, a serine-to-alanine substitution at S203 increased S226 phosphorylation and *vice versa* (Wang, Dang, and Garabedian, in preparation), indicating that one site has a greater tendency to be phosphorylated when the other is dephosphorylated. One possibility is that the phosphorylation at one site creates "steric hindrance" and thus impedes the access of the kinase to the other site. Alternatively, phosphorylation at one site could result in the recruitment of a phosphatase that prevents hyperphosphorylation at the other site. Third, there may be a temporal order of phosphorylation whereby S203 is phosphorylated prior to S211 and S226 phosphorylation. This is consistent with phosphopeptide mapping of the rat GR that identified phospho-S232 (human GR S211) only in association with phospho-S224 (human GR S203), but never by itself. Interestingly, the order of phosphorylation at these sites parallels the temporal order of expression of their cognate kinases—the cyclin E/CDK2, which phosphorylates S224 but not S232, appears in late G_1, whereas cyclinA/CDK2, which phosphorylates bothS224 and S232, is active during G_1/S phase of the cell cycle.

Cell cycle–dependent induction by GR of endogenous genes, such as tyrosine aminotransferase (TAT),[38] alkaline phosphatase,[39] EGF receptor,[40] and metallothionein-I (MT-I), or stably transfected MMTV promoter, was reported.[41] However, while cells were glucocorticoid responsive during the G_1 and S phases, cells in G_2 as well as in M phase were reported to be completely resistant to glucocorticoids despite the fact that GR phosphorylation was more pronounced at this phase in the absence of ligand.[41,42] In addition, inhibition of GR transcriptional regulation seemed to be confined to activation, because repression by GR was not affected in G_2 cells.[43] These observations led to the speculation that GR hyperphosphorylation might be responsible for resistance to glucocorticoids during the G_2/M phase, possibly by hindering association of GR with its DNA targets. However, a recent report by Abel *et al.* indicated that mitotic repression of GR-induced transcription on either endogenous genes or stably integrated MMTV-GFP gene was due to general chromatin condensation, and not to modification of GR.[44] The discrepancy of their findings with previous reports was attributed to the use in the latter of Hoechst 33342 (HOE) to synchronize cells in G_2, a compound they found to interfere with GR-dependent transcription independent of the cell cycle.

PROTEIN PHOSPHATASES MODULATING GR PHOSPHORYLATION AND ACTIVITY

Little is known about the phosphatases that might be regulating GR phosphorylation. It has been reported that after exiting the nucleus upon hormone withdrawal, GR is unable to reenter the nucleus when the cells are treated with okadaic acid, a cell-permeable inhibitor of serine/threonine protein phosphatases.[45] Under these conditions, GR becomes associated with hsp90 but its ability to bind the hormone is not accompanied by its subsequent release from the hsp90 complex.[46] It was suggested that dephosphorylation events are required for GR dissociation from hsp90 and its subsequent translocation from the cytoplasm to the nucleus. The effects of okadaic acid were attributed initially to inhibition of phosphatases PP1 and/or PP2.[45] It is now known that additional phosphatases are sensitive to this compound including PP4 and PP5.[47] The latter, a serine-threonine protein phosphatase, associates through its tetratricopeptide repeats with hsp90 and is therefore indirectly associated with the unliganded GR via hsp90.[48,49] Suppression of PP5 expression increases the association of GR with its cognate DNA-binding sequence. In addition, PP5 induces GR transcriptional activity in the absence of hormone and increases hormone-dependent GR activity by 10-fold.[50] These findings are consistent with the effect of okadaic acid treatment on GR transcriptional activity in a study by Somers and DeFranco.[51] Moreover, GR nuclear accumulation in the absence of glucocorticoids was observed when the activity of PP5 was suppressed,[52] further suggesting that PP5 might be involved in dephosphorylating GR at some sites. Based on the properties of PP5, it is tempting to suggest that it might play a role in regulating GR phosphorylation and modulating its association with hsp90, its nucleocytoplasmic shuttling, its DNA binding, and transcriptional regulatory activities. Detailed analysis of GR phosphorylation in the absence of PP5 and the effect on GR localization and activity, using RNA interference against PP5 and GR phospho-specific antibodies, would shed more light on the involvement of PP5 in these processes.

GR PHOSPHORYLATION AND RECEPTOR TRANSCRIPTIONAL ACTIVITY

To examine whether the phosphorylation status of GR contributes to its transcriptional activity, Mason and Housley analyzed the effects of serine-to-alanine substitutions on the activation of GR transcription from a reporter gene (MMTV-LTR-CAT).[53] Unexpectedly, mutations at all seven phosphorylation sites exhibited only a modest decrease in activity compared to the wild type. Consistent with this finding, a later study by Webster *et al.* showed that a combination of single or multiple mutations had little effect on GR transcriptional activity from a transfected MMTV reporter gene.[54] In contrast, all but one of these phosphorylation mutants had a significant effect on GR transcriptional activity from a minimal promoter containing simple glucocorticoid response elements (GRE_2-E1b-CAT), indicating that the effect of phosphorylation on GR activity is promoter-specific. Notably, mutations at S203 and/or S211 (in human numbering), exhibited more than a 50% decrease in transcriptional activity, while S226A had little effect. Similar results were described for human GR mutated at these sites by Almlof *et al.* S203A and S211A mutants showed a decrease in transcriptional activity in a reconstituted GR signaling system in yeast while S226A displayed a twofold increase in activity, indicating that phosphorylation of this site decreases GR activity.[29] The effects of these particular mutants on transcription are in agreement with the effects on GR activity observed in either MAPK- or CDK-deficient yeast mutants. GR activity increased in yeast strains altered in the human MAPK homologues FUS3 and KSS1, indicating that phosphorylation of S226 inhibits transcription.[28] Conversely, GR activity decreased in yeast strains altered in the human CDK homologue CDC28, indicating that phosphorylation of S203 and S211 is important for activity.[28]

Different approaches for studying phosphorylation of human GR with phospho-specific antibodies against phospho-GR-S203 or phospho-GR-S211 have demonstrated that differentially phosphorylated GR isoforms localize to unique subcellular compartments *in vivo*. In the absence of hormone, some receptors are phosphorylated at S203 at basal levels while others remain unphosphorylated. In the presence of hormone, a subpopulation of GR that is phosphorylated at both sites is detected in the cytoplasm whereas a second population that is phosphorylated at S211, but not S203, was localized to the nucleus and strongly correlates with GR transcriptional activation. The S203 phosphorylated isoforms of GR seem to be confined to the cytoplasm and to the perinuclear region. Chromatin immunoprecipitation assays (ChIP) to study recruitment of different phospho-isoforms of GR at the endogenous tyrosine aminotransferase (TAT) promoter demonstrate the presence of the phospho-S211, but not the phospho-S203, GR isoforms at this promoter (Blind and Garabedian, in preparation). This supports the observation that the GR phospho-S203 form does not seem to accumulate in the nucleus and is not recruited to DNA. However, the S203A GR mutant shows a 50% decrease in transcriptional activity compared to the wild-type GR in transfection assays. Considering that the S203A mutation does not substantially impede phosphorylation of S211. This same mutation, however, is accompanied by an increase in phosphorylation at S226, which in turn may affect GR transcriptional enhancement (Wang and Garabedian, in preparation). Whether these effects are inherent to the function of phospho-GR S203 at the cytoplasmic-nuclear junction or are due to the interdependency of the phosphorylation sites, the regulation of GR function(s) through specific phosphorylation events appears to be quite complex.

ROLE OF PHOSPHORYLATION IN PROTEIN–PROTEIN INTERACTIONS—IMPLICATIONS FOR GR ACTIVITY

The transcriptional activity of steroid receptors is determined, in part, by interactions with proteins termed coactivators and corepressors.[55,56] Different factors have been reported to interact with GR through the N-terminal or the C-terminal activation domains AF-1 and AF-2.[57]

We have identified two proteins, DRIP150 and TSG101, that interact with the GR N-terminal transcriptional activation domain. These proteins appear to function as a GR coactivator and a corepressor, respectively.[58] DRIP150 is a subunit of the mediator complex and will be discussed below.[58–61] TSG101 was originally found in a screen designed to identify tumor-suppressor protein and was discovered by its ability to neoplastically transform mouse 3T3 fibroblasts when either deficient or overexpressed.[62] TSG101 transcripts with deletions in coding regions, attributed to alternative splicing, were found in different carcinomas,[63,64] although truncated forms of TSG101 were also identified in nonneoplastic cells.[65] Thus, the role of TSG101 as a tumor suppressor remains controversial.

TSG101 has been implicated in a variety of cellular functions, including transcriptional regulation,[58,66,67] cell growth and cycling,[68] modulation of the MDM2/p53 feedback control loop,[69] and the release of HIV-1 from cells;[70] it localizes to the perinuclear region of the cell.[71] TSG101 is a homologue of ubiquitin-conjugating (E2) enzymes, but, because its N-terminal ubiquitin-conjugating E2 variant (UEV) domain lacks a cysteine residue required for thioester bond formation with ubiquitin, it is inactive as a ubiquitin conjugase.[72,73] Nevertheless, the TSG101 protein can bind to ubiquitin and affect ubiquitin-dependent proteolysis.[74] Importantly, the interaction with GR and TSG101 appears to be modulated by phosphorylation. GR phosphorylation mutants with serine-to-alanine (A) and serine-to-glutamic acid (E) substitutions, which mimic "nonphosphorylated" and "phosphorylated" states of GR, were tested for their ability to interact with TSG101 using the yeast two-hybrid interaction assay. Interestingly, the nonphosphorylated form of GR (GR S203A/S211A) showed enhanced interaction with TSG101, as compared with the wild-type GR or the GR S203E/S211E phospho-mimetic derivative, suggesting that TSG101 interacts *more* favorably with GR when it is not phosphorylated (Blind, Hilltelman and Garabedian, unpublished observation). Given that TSG101 is involved in protein stability and has a preferential association with the hypophosphorylated form of GR, its effect on GR stability in mammalian cells was analyzed. A significant accumulation of GR protein is detected in the GR S203A/S211A double mutant in the absence of ligand when TSG101 is overexpressed, whereas no increase in the wild-type GR or GR S203E/

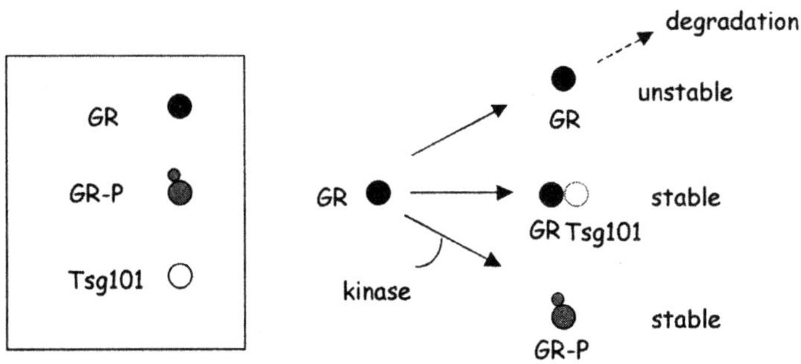

FIGURE 2. TSG101 stabilizes the ligand-free and hypophosphorylated form of GR. A model whereby TSG101 preferentially associates with the nonphosphorylated form of the ligand-free GR and protects it from degradation, while the nonphosphorylated, non-TSG101 bound form of the receptor is degraded. We posit that GR phosphorylation can also stabilize the receptor independent of TSG101 binding. Our findings suggest that GR phosphorylation and TSG101 binding to the nonphosphorylated GR are redundant mechanisms that operate to prevent the ligand-free form of the receptor from being degraded. This may represent a "fail safe" mechanism to maintain GR levels *in vivo.*

S211E stability was observed under the same conditions. The opposite effect is observed when TSG101 expression is silenced using specific siRNAs, rendering the hypophosphorylated form of GR unstable (Ismaili and Garabedian, in preparation). Thus, in the absence of ligand, TSG101 is recruited preferentially to the nonphosphorylated pool of GR and protects it from degradation (FIG. 2). The significance of this finding to the repressive effect of TSG101 on GR transcriptional activity remains unclear. Nevertheless, the effect of TSG101 on the hypophosphorylated GR may represent a novel method to establish steady-state receptor levels in various cell types by modulating GR phosphorylation and/or TSG101 levels.

DRIP150 activates GR transcriptional activity both in yeast and in mammalian cells through the AF-1 region that harbors most of the phosphorylation sites. Many examples of transcription factors that recruit coactivators upon phosphorylation have been described.[75–77] It is therefore conceivable that GR phosphorylation could similarly enhance protein–protein interactions. Consequently, GR phosphorylation– site mutants were tested for their abilities to interact with DRIP150 using the yeast two-hybrid interaction assay. In contrast to TSG101, GR phosphorylation enhances DRIP150-GR association, indicating that this protein–protein interaction is also dependent on phosphorylation status of GR (Blind, Hilltelman, and Garabedian, unpublished observation). Thus, when GR is hyperphosphorylated, this facilitates recruitment of DRIP150, which results in enhancement of GR transcriptional activity.

EFFECT OF PHOSPHORYLATION ON GR STABILITY

The ubiquitin–proteasome pathway is responsible for degradation of a plethora of short-lived proteins.[78] It is known that GR is downregulated upon hormone binding.[79,80] Inhibition of the proteasome by MG-132 or β-lactone eliminates GR downregulation and leads to enhanced GR transcriptional activation. In addition, ubiquitination of GR was detected,[81] suggesting that the ubiquitin–proteasome pathway is also involved in GR degradation. GR has been shown to associate with UBC9, an E2 ubiquitin ligase,[33,82] and with MDM2, an E3 ubiquitin ligase, although the significance of these interactions is not well understood. Recent reports described the presence of a negative crosstalk between GR and p53 and a ligand-dependent interaction between p53 and GR that results in their degradation by MDM2.[83,84] MDM2 is a ubiquitin ligase for p53 and for itself.[85,86] It forms an autoregulatory loop with p53 by binding to its N-terminal region, thus inhibiting its transcriptional activity and increasing its degradation by the proteosome. It was also shown that GR downregulation by estradiol is due to an estrogen receptor–specific upregulation of MDM2, further implicating MDM2 in GR degradation.[87] Recently, Lin *et al.* demonstrated that AR phosphorylation by Akt triggers AR ubiquitination by MDM2 and subsequent degradation by the proteosome.[88] This illustrates the importance of phosphorylation on steroid hormone receptor turnover. It is not clear however, in the case of GR, whether phosphorylation is also a signal for its downregulation by MDM2 or other factors. Earlier studies reported that the GR antagonist RU486, like the receptor agonist dexamethasone, promotes downregulation of GR. However, because RU486 does not apparently change the phosphorylation state of the receptor yet induces downregulation, phosphorylation alone may not be the signal for GR protein degradation.[89] Nevertheless, Webster *et al.* showed that mutations of the phosphorylation sites of GR extend its half-life and abolish its downregulation following treatment with ligand, suggesting that GR phosphorylation status determines its stability.[54]

EFFECT OF p27^{Kip1} ON GR PHOSPHORYLATION AND TRANSCRIPTIONAL ACTIVATION

We have also recently examined GR transcriptional regulation, receptor phosphorylation, and glucocorticoid-dependent growth inhibition using primary mouse embryonic fibroblasts (MEFs) derived from p27-deficient mice.[90] p27 is an important negative regulator of CDK activity, which affects cell cycle entry and exit, as well as the phosphorylation of many substrates, including GR. Many anti-mitogenic signals lead to the accumulation of p27, which in turn inhibits CDK activity and leads to cell cycle arrest. Since total CDK2 activity is elevated in p27-deficient cells and GR is a substrate for CDK2 phosphorylation *in vitro*, we have found that GR phosphorylation at two putative CDK sites, as well as receptor transcriptional enhancement, are elevated in p27$^{-/-}$ MEFs relative to control cells. This increased GR transcriptional activation appears to be mediated through the GR N-terminus. Coexpression of the GR N-terminal coactivator DRIP150 further enhanced GR-dependent transcriptional activation in the p27$^{-/-}$ MEFs as compared to wild-type MEFs. Thus, DRIP150-induction of GR transcriptional activation appears enhanced by the loss of p27 expression, suggesting this change facilitates the formation of a functional GR-DRIP150 complex through alterations in GR phosphorylation.

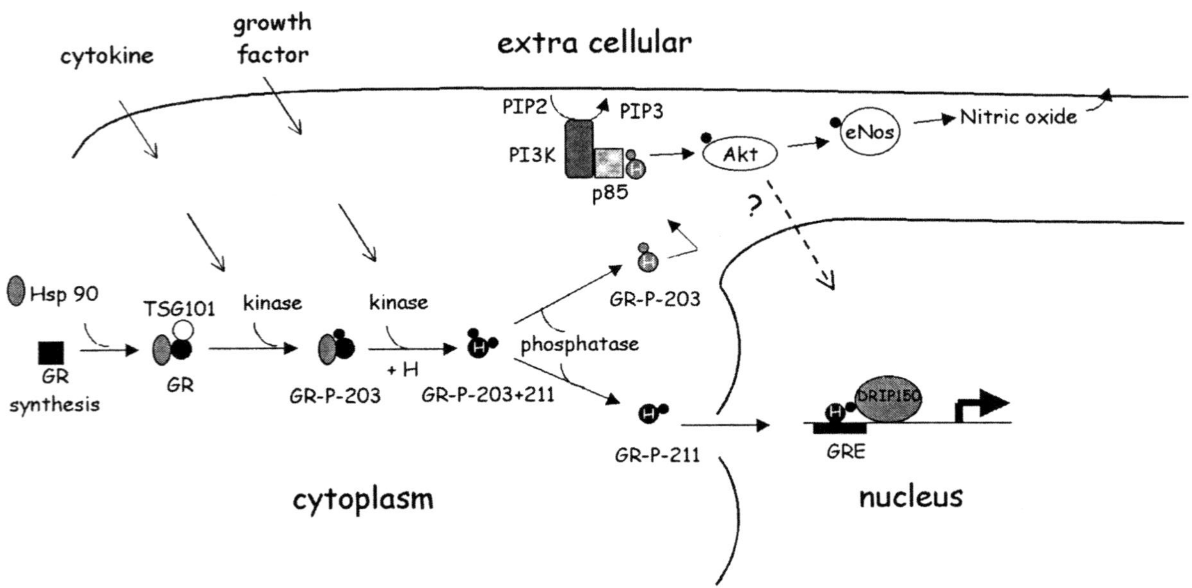

FIGURE 3. A model for the modulation of nuclear and nonnuclear actions of GR by differential phosphorylation. GR is synthesized and becomes associated with hsp90. The nonphosphorylated form of GR is also associated with TSG101, which prevents it from being degraded. Phosphorylation of GR would reduce GR:TSG101 association. Upon agonist binding, GR is phosphorylated at multiple sites by distinct protein kinase that can be regulated via extracellular signals, such as growth factors (GF). Phosphatases then selectively dephosphorylate GR. Particular GR phospho-isoform (e.g., S203-P) would associate, for example, with the regulatory p85 subunit of PI3K and stimulate the formation of PIP3, leading to the phosphorylation and activation of Akt, which enhances nitric oxide release through phosphorylation of eNOS. In the nucleus, a different GR phospho-isoform (e.g., S211-P) binds DNA and regulates gene expression through interaction with coactivator proteins, such as DRIP150. Potential crosstalk between the GR nonnuclear activation of Akt and nuclear GR function is also possible and is represented by a *dashed line* with open arrowhead and question mark to the left.

EFFECT OF GR PHOSPHORYLATION ON NONTRANSCRIPTIONAL ACTIVITY OF GR

In addition to its customary role in transcription, GR has recently been reported to activate the PI3K-Akt pathway in a rapid, nontranscriptional manner. The PI3K-Akt pathway is an important regulator of cell cycle progression and cell survival.[91] Dexamethasone-bound GR activates PI3K and Akt. The Akt activation is blocked by cotreatment withPI3K inhibitors, but not by inhibitors of transcription. This effect is mediated by GR, as cells that lack endogenous GR, such as COS7, do not activate PI3K or Akt upon dexamethasone stimulation. However, transient expression of GR or the dimerization defective GR (A548T), which is unable to activate transcription, reconstituted Akt activation in response to dexamethasone in COS7 cells.[3] The mechanism of PI3K activation by GR is unknown but may involve interaction of GR with the p85 regulatory subunit of PI3K. Interestingly, the mineralocorticoid receptor (MR) is not capable of activating the PI3K pathway. Because the GR and MR ligand-binding and DNA binding domains are highly conserved, while their N-termini are divergent, it is likely that the GR N-terminus is responsible for the rapid nontranscriptional induction of PI3K activity via interaction with p85. We therefore speculate that the association of a particular GR phospho-isoform with the regulatory p85 subunit of PI3K stimulates the formation of PIP3, leading to the phosphorylation and activation of Akt, which enhances nitric oxide release through phosphorylation of eNOS. In the nucleus, a different GR phospho-isoform binds DNA and regulates gene expression through interaction with coactivator proteins. Potential crosstalk between the GR nontranscriptional activation of Akt and nuclear GR function is also possible and may be an important mechanism of signal integration by GR via phosphorylation (FIG. 3).

SUMMARY

Phosphorylation of GR appears to facilitate a host of receptor functions including transcriptional activation, stability, and subcellular localization and recycling. Receptor phosphorylation may also play a role in the nongenomic effects of GR. Thus, it appears that GR phosphorylation functions to direct and refine receptor activity in response to particular physiological needs. The ability of the receptor to respond to extracellular signals via phosphorylation permits plasticity in receptor action that, in conjunction with the steroid ligands, may be crucial in coordinating the various growth and metabolic processes managed by glucocorticoids.

ACKNOWLEDGMENTS

We thank Raymond Blind, Zhen Wang, and Thoa Dang for communicating unpublished findings and Zhen Wang, Men-Jean Lee, and Thoa Dang for critically reading the manuscript. This work was supported by grants from the National Institutes of Health and the American Cancer Society.

REFERENCES

1. BLUM, A. & E. MASER. 2003. Enzymology and molecular biology of glucocorticoid metabolism in humans. Prog. Nucleic Acids Res. Mol. Biol. **75:** 173–216.
2. YAMAMOTO, K.R. 1995. Multilayered control of intracellular receptor function. Harvey Lecture, 1995. **91:** 1–19.
3. LIMBOURG, F.P. & J.K. LIAO. 2003. Nontranscriptional actions of the glucocorticoid receptor. J. Mol. Med. **81:** 168-174.
4. BEATO, M. & J. KLUG. 2000. Steroid hormone receptors: an update. Hum. Reprod. Update **6**: 225–236.
5. LE RICOUSSE, S. *et al.* 1996. Glucocorticoid and progestin receptors are differently involved in the cooperation with a structural element of the mouse mammary tumor virus promoter. Proc. Natl. Acad. Sci. USA **93**: 5072–5077.
6. BODWELL, J.E. *et al.* 1998. Glucocorticoid receptor phosphorylation: overview, function and cell cycle-dependence. J. Steroid Biochem. Mol. Biol. **65**: 91–99.
7. DOBRANSKY, T. & R.J. RYLETT. 2003. Functional regulation of choline acetyltransferase by phosphorylation. Neurochem. Res. **28**: 537–542.
8. TAKIZAWA, C.G. & D.O. MORGAN. 2000. Control of mitosis by changes in the subcellular location of cyclin-B1-Cdk1 and Cdc25C. Curr. Opin. Cell Biol. **12**: 658–665.
9. ZHOU, Y. *et al.* 2003. Nuclear localization of the cell cycle regulator CDH1 and its regulation by phosphorylation. J. Biol. Chem. **278**: 12530–12536.
10. JANS, D.A. & S. HUBNER. 1996. Regulation of protein transport to the nucleus: central role of phosphorylation. Physiol. Rev. **76**: 651–685.
11. DITTMER, J. 2003. The biology of the Ets1 proto-oncogene. Mol. Cancer **2**(1): p. 29.
12. GILLE, H. *et al.* 1996. Phosphorylation-dependent formation of a quaternary complex at the c-fos SRE. Mol. Cell. Biol. **16:** 1094–1102.
13. WEI, G., G. LIU & X. LI. 2003. Identification of two serine residues important for p53 DNA binding and protein stability. FEBS Lett. **543:** 16–20.
14. NASH, P. *et al.* 2001. Multisite phosphorylation of a CDK inhibitor sets a threshold for the onset of DNA replication. Nature **414:** 514–521.
15. MANTOVANI, F. & L. BANKS. 2003. Regulation of the discs large tumor suppressor by a phosphorylation-dependent interaction with the beta-TrCP ubiquitin ligase receptor. J. Biol. Chem. **278:** 42477–42486.
16. CRAIG, K.L. & M. TYERS. 1999. The F-box: a new motif for ubiquitin dependent proteolysis in cell cycle regulation and signal transduction. Prog. Biophys. Mol. Biol. **72:** 299–328.
17. APPELLA, E. & C.W. ANDERSON. 2000. Signaling to p53: breaking the posttranslational modification code. Pathol. Biol. (Paris) **48:** 227–245.
18. WESTLY, H.J. & K.W. KELLEY. 1987. Down-regulation of glucocorticoid and beta-adrenergic receptors on lectin-stimulated splenocytes. Proc. Soc. Exp. Biol. Med. **185:** 211–218.
19. NIELSEN, C.J., J.J. SANDO & W.B. PRATT. 1977. Evidence that dephosphorylation inactivates glucocorticoid receptors. Proc. Natl. Acad. Sci. USA **74:** 1398–1402.
20. KURL, R.N. & S.T. JACOB. 1984. Phosphorylation of purified glucocorticoid receptor from rat liver by an endogenous protein kinase. Biochem. Biophys. Res. Commun. **119:** 700–705.
21. SANCHEZ, E.R. *et al.* 1987. Glucocorticoid receptor phosphorylation in mouse L-cells. J. Steroid Biochem. **27:** 215–225.
22. TIENRUNGROJ, W. *et al.* 1987. Glucocorticoid receptor phosphorylation, transformation, and DNA binding. J. Biol. Chem. **262:** 17342–17349.
23. ORTI, E. *et al.* 1989. A dynamic model of glucocorticoid receptor phosphorylation and cycling in intact cells. J. Steroid Biochem. **34:** 85–96.
24. DALMAN, F.C. *et al.* 1988. Localization of phosphorylation sites with respect to the functional domains of the mouse L cell glucocorticoid receptor. J. Biol. Chem. **263:** 12259–12267.
25. DANIELSEN, M., J.P. NORTHROP & G.M. RINGOLD. 1986. The mouse glucocorticoid receptor: mapping of functional domains by cloning, sequencing and expression of wild-type and mutant receptor proteins. EMBO J. **5:** 2513–2522.

26. MIESFELD, R. *et al.* 1984. Characterization of a steroid hormone receptor gene and mRNA in wild-type and mutant cells. Nature **312:** 779–781.
27. BODWELL, J.E. *et al.* 1991. Identification of phosphorylated sites in the mouse glucocorticoid receptor. J. Biol. Chem. **266:** 7549–7555.
28. KRSTIC, M.D. *et al.* 1997. Mitogen-activated and cyclin-dependent protein kinases selectively and differentially modulate transcriptional enhancement by the glucocorticoid receptor. Mol. Cell. Biol. **17:** 3947–3954.
29. ALMLOF, T., A.P. WRIGHT & J.A. GUSTAFSSON. 1995. Role of acidic and phosphorylated residues in gene activation by the glucocorticoid receptor. J. Biol. Chem. **270:** 17535–17540.
30. POCUCA, N. *et al.* 1998. Using yeast to study glucocorticoid receptor phosphorylation. J. Steroid Biochem. Mol. Biol. **66:** 303–318.
31. BODWELL, J.E. *et al.* 1996. Glucocorticoid receptors: ATP and cell cycle dependence, phosphorylation, and hormone resistance. Am. J. Respir. Crit. Care Med. **154:** S2–6.
32. WANG, Z., J. FREDERICK & M.J. GARABEDIAN. 2002. Deciphering the phosphorylation "code" of the glucocorticoid receptor in vivo. J. Biol. Chem. **277:** 26573–26580.
33. KAUL, S. *et al.* 2002. Ubc9 is a novel modulator of the induction properties of glucocorticoid receptors. J. Biol. Chem. **277:** 12541–12549.
34. NIGG, E.A. 1995. Cyclin-dependent protein kinases: key regulators of the eukaryotic cell cycle. Bioessays **17:** 471–480.
35. CHEN, D. *et al.* 2000. Activation of estrogen receptor alpha by S118 phosphorylation involves a ligand-dependent interaction with TFIIH and participation of CDK7. Mol. Cell **6:** 127–137.
36. ROGATSKY, I., S.K. LOGAN & M.J. GARABEDIAN. 1998. Antagonism of glucocorticoid receptor transcriptional activation by the c-Jun N-terminal kinase. Proc. Natl. Acad. Sci. USA **95:** 2050–2055.
37. ITOH, M. *et al.* 2002. Nuclear export of glucocorticoid receptor is enhanced by c-Jun N-terminal kinase-mediated phosphorylation. Mol. Endocrinol. **16:** 2382–2392.
38. MARTIN, D.W., JR., G.M. TOMKINS & M.A. BRESLER. 1969. Control of specific gene expression examined in synchronized mammalian cells. Proc. Natl. Acad. Sci. USA **63:** 842–849.
39. GRIFFIN, M.J. & R.H. BOTTOMLEY. 1969. Regulation of alkaline phosphatase in HeLa clones of differing modal chromosome number. Ann. N.Y. Acad. Sci. **166:** 417–432.
40. FANGER, B.O., R.A. CURRIE & J.A. CIDLOWSKI. 1986. Regulation of epidermal growth factor receptors by glucocorticoids during the cell cycle in HeLa S3 cells. Arch. Biochem. Biophys. **249:** 116–125.
41. HSU, S.C., M. QI & D.B. DEFRANCO. 1992. Cell cycle regulation of glucocorticoid receptor function. EMBO J. **11:** 3457–3468.
42. HU, J.M., J.E. BODWELL & A. MUNCK. 1994. Cell cycle-dependent glucocorticoid receptor phosphorylation and activity. Mol. Endocrinol. **8:** 1709–1713.
43. HSU, S.C. & D.B. DEFRANCO. 1995. Selectivity of cell cycle regulation of glucocorticoid receptor function. J. Biol. Chem. **270:** 3359–3364.
44. ABEL, G.A. *et al.* 2002. Activity of the GR in G2 and mitosis. Mol. Endocrinol. **16:** 1352–1366.
45. DEFRANCO, D.B. *et al.* 1991. Protein phosphatase types 1 and/or 2A regulate nucleocytoplasmic shuttling of glucocorticoid receptors. Mol. Endocrinol. **5:** 1215–1228.
46. GALIGNIANA, M.D. *et al.* 1999. Inhibition of glucocorticoid receptor nucleocytoplasmic shuttling by okadaic acid requires intact cytoskeleton. J. Biol. Chem. **274:** 16222–16227.
47. KLOEKER, S. *et al.* 2003. Parallel purification of three catalytic subunits of the protein serine/threonine phosphatase 2A family (PP2A(C), PP4(C), and PP6(C)) and analysis of the interaction of PP2A(C) with alpha4 protein. Protein Expr. Purif. **31:** 19–33.
48. CHEN, M.S. *et al.* 1996. The tetratricopeptide repeat domain of protein phosphatase 5 mediates binding to glucocorticoid receptor heterocomplexes and acts as a dominant negative mutant. J. Biol. Chem. **271:** 32315–32320.
49. RUSSELL, L.C. *et al.* 1999. Identification of conserved residues required for the binding of a tetratricopeptide repeat domain to heat shock protein 90. J. Biol. Chem. **274:** . 20060–20063.

50. ZUO, Z. *et al.* 1999. Ser/Thr protein phosphatase type 5 (PP5) is a negative regulator of glucocorticoid receptor-mediated growth arrest. Biochemistry **38:** 8849–8857.
51. SOMERS, J.P. & D.B. DEFRANCO. 1992. Effects of okadaic acid, a protein phosphatase inhibitor, on glucocorticoid receptor-mediated enhancement. Mol. Endocrinol. **6**: 26–34.
52. DEAN, D.A. *et al.* 2001. Serine/threonine protein phosphatase 5 (PP5) participates in the regulation of glucocorticoid receptor nucleocytoplasmic shuttling. BMC Cell Biol. **2:** 6.
53. MASON, S.A. & P.R. HOUSLEY. 1993. Site-directed mutagenesis of the phosphorylation sites in the mouse glucocorticoid receptor. J. Biol. Chem. **268:** 21501–21504.
54. WEBSTER, J.C. *et al.* 1997. Mouse glucocorticoid receptor phosphorylation status influences multiple functions of the receptor protein. J. Biol. Chem. **272:** 9287–9293.
55. ROSENFELD, M.G. & C.K. GLASS. 2001. Coregulator codes of transcriptional regulation by nuclear receptors. J. Biol. Chem. **276:** 36865–36868.
56. HERMANSON, O., C.K. GLASS & M.G. ROSENFELD. 2002. Nuclear receptor coregulators: multiple modes of modification. Trends Endocrinol. Metab. **13:** 55–60.
57. JENKINS, B.D., C.B. PULLEN & B.D. DARIMONT. 2001. Novel glucocorticoid receptor coactivator effector mechanisms. Trends Endocrinol. Metab. **12:** 122–126.
58. HITTELMAN, A.B. *et al.* 1999. Differential regulation of glucocorticoid receptor transcriptional activation via AF-1-associated proteins. EMBO J. **18:** 5380–5388.
59. BOUBE, M. *et al.* 2002. Evidence for a mediator of RNA polymerase II transcriptional regulation conserved from yeast to man. Cell **110:** 143–151.
60. ITO, M. & R.G. ROEDER. 2001. The TRAP/SMCC/Mediator complex and thyroid hormone receptor function. Trends Endocrinol. Metab. **12:** 127–134.
61. RACHEZ, C. & L.P. FREEDMAN. 2001. Mediator complexes and transcription. Curr. Opin. Cell Biol. **13:** 274–280.
62. LI, L. & S.N. COHEN. 1996. Tsg101: a novel tumor susceptibility gene isolated by controlled homozygous functional knockout of allelic loci in mammalian cells. Cell **85:** 319–329.
63. LEE, M.P. & A.P. FEINBERG. 1997. Aberrant splicing but not mutations of TSG101 in human breast cancer. Cancer Res. **57:** 3131–3134.
64. SUN, Z. *et al.* 1997. Frequent abnormalities of TSG101 transcripts in human prostate cancer. Oncogene **15:** 3121–3125.
65. CHANG, J.G. *et al.* 1999. Analysis of TSG101 tumour susceptibility gene transcripts in cervical and endometrial cancers. Br. J. Cancer **79:** 445–450.
66. WATANABE, M. *et al.* 1998. A putative tumor suppressor, TSG101, acts as a transcriptional suppressor through its coiled-coil domain. Biochem. Biophys. Res. Commun. **245:** 900–905.
67. SUN, Z. *et al.* 1999. Tumor susceptibility gene 101 protein represses androgen receptor transactivation and interacts with p300. Cancer **86:** 689–696.
68. OH, H. *et al.* 2002. Negative regulation of cell growth and differentiation by TSG101 through association with p21(Cip1/WAF1). Proc. Natl. Acad. Sci. USA **99:** 5430–5435.
69. LI, L. *et al.* 2001. A TSG101/MDM2 regulatory loop modulates MDM2 degradation and MDM2/p53 feedback control. Proc. Natl. Acad. Sci. USA **98:** 1619–1624.
70. GARRUS, J.E. *et al.* 2001. Tsg101 and the vacuolar protein sorting pathway are essential for HIV-1 budding. Cell **107:** 55–65.
71. XIE, W., L. LI & S.N. COHEN. 1998. Cell cycle-dependent subcellular localization of the TSG101 protein and mitotic and nuclear abnormalities associated with TSG101 deficiency. Proc. Natl. Acad. Sci. USA **95:** 1595–1600.
72. KOONIN, E.V. & R.A. ABAGYAN. 1997. TSG101 may be the prototype of a class of dominant negative ubiquitin regulators. Nat. Genet. **16:** 330–331.
73. PONTING, C.P., Y.D. CAI & P. BORK. 1997. The breast cancer gene product TSG101: a regulator of ubiquitination? J. Mol. Med. **75:** 467–469.
74. DUPRE, S., C. VOLLAND & R. HAGUENAUER-TSAPIS. 2001. Membrane transport: ubiquitylation in endosomal sorting. Curr. Biol. **11:** R932–934.
75. LI, Q.J. *et al.* 2003. MAP kinase phosphorylation-dependent activation of Elk-1 leads to activation of the co-activator p300. EMJO J. **22:** 281–291.

76. JANKNECHT, R., N.J. WELLS & T. HUNTER. 1998. TGF-beta-stimulated cooperation of smad proteins with the coactivators CBP/p300. Genes Dev. **12:** 2114–2119.
77. HAMMER, G.D. *et al.* 1999. Phosphorylation of the nuclear receptor SF-1 modulates cofactor recruitment: integration of hormone signaling in reproduction and stress. Mol. Cell **3:** 521–526.
78. PICKART, C.M. 2000. Ubiquitin biology: an old dog learns an old trick. Nat. Cell Biol. **2:** E139–141.
79. CIDLOWSKI, J.A. & N.B. CIDLOWSKI. 1982. Glucocorticoid receptors and the cell cycle: evidence that the accumulation of glucocorticoid receptors during the S phase of the cell cycle is dependent on ribonucleic acid and protein synthesis. Endocrinology **110:** 1653–1662.
80. HOECK, W., S. RUSCONI & B. GRONER. 1989. Down-regulation and phosphorylation of glucocorticoid receptors in cultured cells. Investigations with a monospecific antiserum against a bacterially expressed receptor fragment. J. Biol. Chem. **264:** 14396–14402.
81. WALLACE, A.D. & J.A. CIDLOWSKI. 2001. Proteasome-mediated glucocorticoid receptor degradation restricts transcriptional signaling by glucocorticoids. J. Biol. Chem. **276:** 42714–42721.
82. GOTTLICHER, M. *et al.* 1996. Interaction of the Ubc9 human homologue with c-Jun and with the glucocorticoid receptor. Steroids **61:** 257–262.
83. SENGUPTA, S. *et al.* 2000. Negative cross-talk between p53 and the glucocorticoid receptor and its role in neuroblastoma cells. EMBO J. **19:** 6051–6064.
84. SENGUPTA, S. & B. WASYLYK. 2001. Ligand-dependent interaction of the glucocorticoid receptor with p53 enhances their degradation by Hdm2. Genes Dev. **15:** 2367–2380.
85. ARGENTINI, M., N. BARBOULE & B. WASYLYK. 2001. The contribution of the acidic domain of MDM2 to p53 and MDM2 stability. Oncogene **20:** 1267–1275.
86. HONDA, R. & H. YASUDA. 2000. Activity of MDM2, a ubiquitin ligase, toward p53 or itself is dependent on the RING finger domain of the ligase. Oncogene **19:** 1473–1476.
87. KINYAMU, H.K. & T.K. ARCHER. 2003. Estrogen receptor-dependent proteasomal degradation of the glucocorticoid receptor is coupled to an increase in mdm2 protein expression. Mol. Cell. Biol. **23:** 5867–5881.
88. LIN, H.K. *et al.* 2002. Phosphorylation-dependent ubiquitylation and degradation of androgen receptor by Akt require Mdm2 E3 ligase. EMBO J. **21:** 4037–4048.
89. HOECK, W. & B. GRONER. 1990. Hormone-dependent phosphorylation of the glucocorticoid receptor occurs mainly in the amino-terminal transactivation domain. J. Biol. Chem. **265:** 5403–5408.
90. WANG, Z. & M.J. GARABEDIAN. 2003. Modulation of glucocorticoid receptor transcriptional activation, phosphorylation and growth inhibition by p27KIP1. J. Biol. Chem. **278:** 50897–50901.
91. HAFEZI-MOGHADAM, A. *et al.* 2002. Acute cardiovascular protective effects of corticosteroids are mediated by non-transcriptional activation of endothelial nitric oxide synthase. Nat. Med. **8:** 473–479.

The Origin and Functions of Multiple Human Glucocorticoid Receptor Isoforms

NICK Z. LU AND JOHN A. CIDLOWSKI

The Laboratory of Signal Transduction, Molecular Endocrinology Group, National Institute of Environmental Health Sciences, National Institutes of Health, Department of Health and Human Services, Research Triangle Park, North Carolina 27709, USA

ABSTRACT: Glucocorticoid hormones are necessary for life and are essential in all aspects of human health and disease. The actions of glucocorticoids are mediated by the glucocorticoid receptor (GR), which binds glucocorticoid hormones and regulates gene expression, cell signaling, and homeostasis. Decades of research have focused on the mechanisms of action of one isoform of GR, GRα. However, in recent years, increasing numbers of human GR (hGR) isoforms have been reported. Evidence obtained from this and other laboratories indicates that multiple hGR isoforms are generated from one single hGR gene via mutations and/or polymorphisms, transcript alternative splicing, and alternative translation initiation. Each hGR protein, in turn, is subject to a variety of posttranslational modifications, and the nature and degree of posttranslational modification affect receptor function. We summarize here the processes that generate and modify various hGR isoforms with a focus on those that impact the ability of hGR to regulate target genes. We speculate that unique receptor compositions and relative receptor proportions within a cell determine the specific response to glucocorticoids. Unchecked expression of some isoforms, for example hGRβ, has been implicated in various diseases.

KEYWORDS: glucocorticoid receptor isoforms; alternative splicing; phosphorylation; ubiquitination; receptor mobility

INTRODUCTION

Glucocorticoids are essential for proper embryogenesis, development, growth, and survival.[1,2] In addition, glucocorticoids are broadly used as therapeutics in acute and chronic treatment of asthma,[3,4] rheumatoid arthritis,[5] degenerative osteoarthritis,[6] ulcerative colitis,[7] eosinophilic gastritis,[8] transplant rejection,[9] complications from acquired immunodeficiency syndromes,[10] as well as many other inflammatory and immune diseases. Furthermore, glucocorticoids have also been applied effectively as chemotherapeutic agents in the treatment of cancers, especially cancers of

Address for correspondence: John A. Cidlowski, The Laboratory of Signal Transduction, Molecular Endocrinology Group, National Institute of Environmental Health Sciences, NIH, Department of Health and Human Services, 111 Alexander Drive, Research Triangle Park, NC 27709. Voice: 919-541-1564; fax: 919-541-1367.
cidlowski@niehs.nih.gov

**Ann. N.Y. Acad. Sci. 1024: 102–123 (2004). © 2004 New York Academy of Sciences.
doi: 10.1196/annals.1321.008**

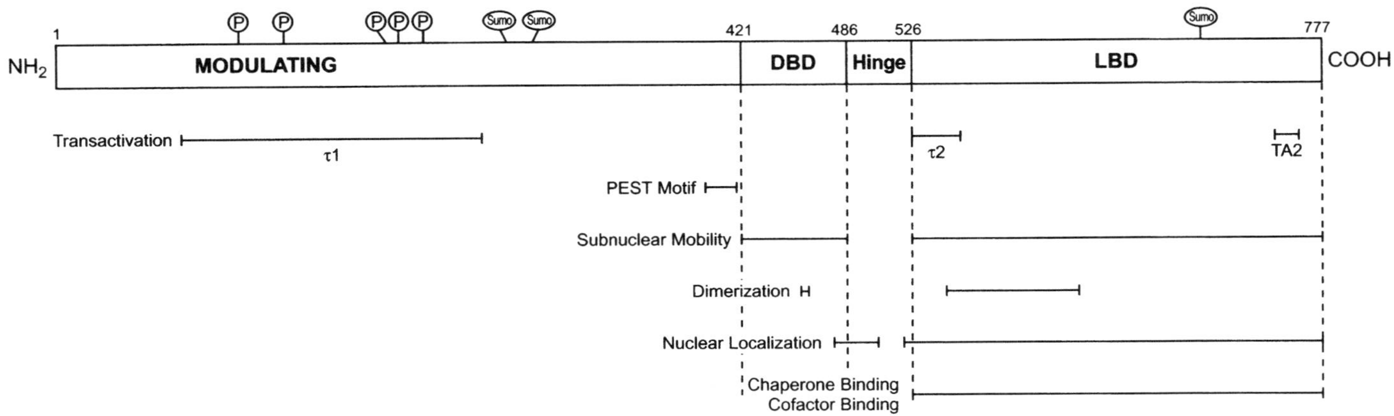

FIGURE 1. Domains and modified residues of the human glucocorticoid receptor (hGR). The N-terminal modulating domain contains the sequence for the main transactivation activity, τ1, while τ2 and an additional transactivation activity (TA2) reside in the ligand binding domain (LBD). Several functional domains of hGR overlap with each other. For example, portions of the DNA binding domain (DBD) are critical for ligand-dependent nuclear translocation of the receptor and receptor dimerization. P, phosphorylation sites; Sumo, sumoylation sites.

hematological origins,[1] including Hodgkin's lymphoma, acute lymphoblastic leukemia, and multiple myeloma. Despite the extensive clinical usage of glucocorticoids in the clinic, the mechanisms underlying the remarkable diversity of the glucocorticoid receptor (GR) function are poorly understood.

GR, along with related steroid receptors such as estrogen receptor (ER), progestin receptor (PR), androgen receptor (AR), and mineralocorticoid receptor (MR), likely emerged through a series of gene duplication events from a common ancestral receptor some 400 million years ago.[11] Similar to other steroid receptors, GR protein has a modular structure.[12–16] From amino terminus to carboxyl terminus are the amino acid sequences for the transactivation domain 1 (τ1 or TA1), DNA binding domain (DBD), hinge region, and ligand binding domain (LBD, FIG. 1). Additional transactivation domains embedded in the LBD, the τ2 and TA2, are less potent in autonomous transactivation activity than the τ1 domain. Correctly folded GR presents "pockets" for cognate hormone recognition[17] and motifs for recognizing specific DNA sequences termed glucocorticoid response element (GREs) on target genes.[18] Somewhat overlapping with the aforementioned major domains are additional regions that may allow interdomain interactions, e.g., between TA1 and DBD.[19] These regions also facilitate interactions between GR and other proteins, including chaperones that are involved in the compartmentalization and trafficking of the receptor[20–22] and coregulators that control the efficacy of the receptor function.[23–29] In addition, these regions may also facilitate the heterodimerization of GR isoforms[17,30] and direct interactions between GR and other transcription factors that may expand the potential gene targets of GR.[31–41]

In this article, we describe the processes that generate multiple human GR (hGR) isoforms from a single gene, including alternative RNA splicing, alternative translation initiation, and gene mutations. Also summarized is recent evidence for post-translational modifications of GR proteins, with the emphasis on phosphorylation and ubiquitination as well as the consequences of these modification processes on receptor function. Finally, we present novel observations on the intranuclear movement of hGR as a result of selective ligand binding.

COMPLEXITIES WITHIN THE hGR GENE

Only one GR gene has been identified in every species examined to date. The hGR gene is located on chromosome 5q31-32 and comprises over 140 kb of nucleotides, less than 2% (~2.5 kb) of which are exons.[42–47] There are 9 exons in the hGR gene (FIG. 2): exon 1 (~116–981 bp) is a leader sequence; exon 2 (1,197 bp) contains the coding sequence for τ1 at the amino terminal; exons 3 (167 bp) and 4 (117 bp) code for the first and second zinc-finger motif in the DBD, respectively; exons 5 (280 bp), 6 (145 bp), 7 (131 bp), and 8 (158 bp) code for τ2 and a large portion of the LBD; and exon 9 (4,108 bp) contains coding sequences for the two alternative carboxyl termini of the LBD, α and β, and their respective 3′ untranslated regions. Remarkable homology has been found within the splice junctions of exons for the DBD and LBD among GR and related steroid hormone receptors, such as PR, AR, and ER.[45] Divergence, however, exists between GR and less-related nuclear receptors, such as thyroid hormone receptors (TR) and vitamin D receptors. These findings suggest that, evolutionarily, three parallel branches of receptors for steroids,

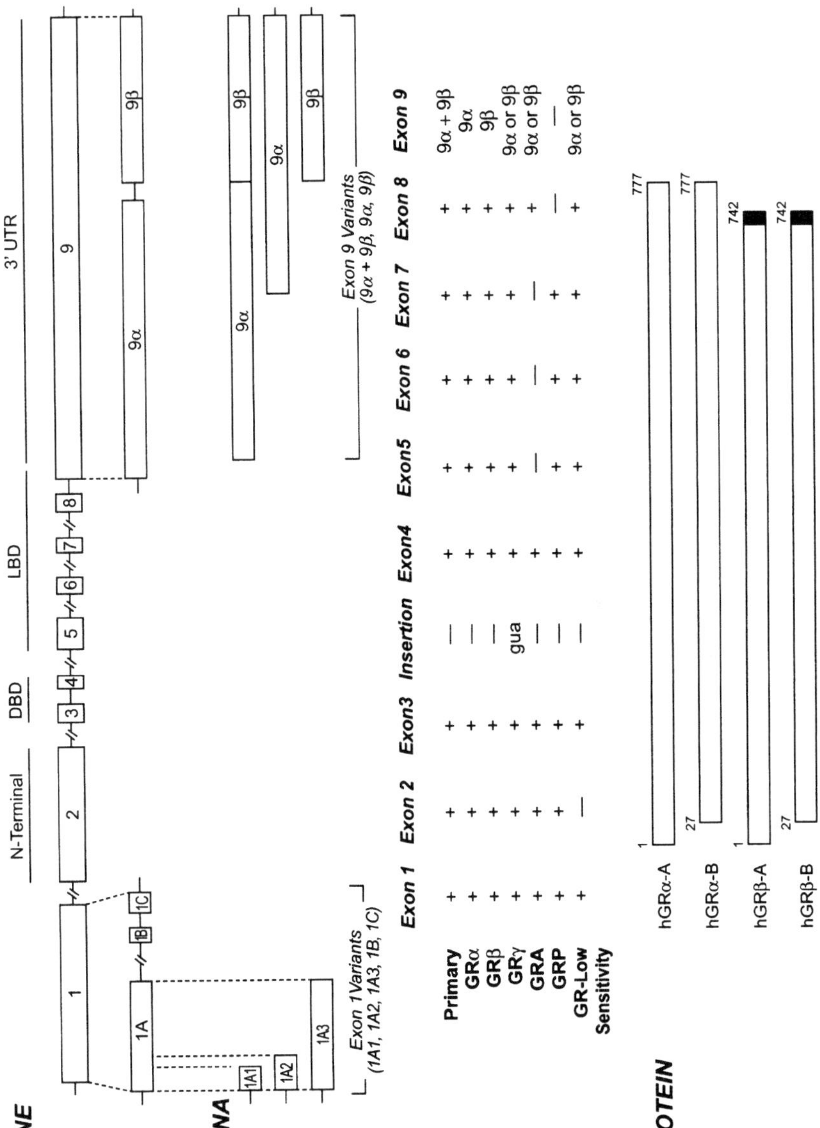

FIGURE 2. *See following page for legend.*

vitamin D, and thyroid hormones split early from a common ancestral receptor. The least homologous sequences and intron-exon organization among steroid receptors are found in exons 1 and 2.

Exon 1 of hGR exists in three forms, 1A, 1B, and 1C, each of which is driven by a distinct proximal promoter.[47–49] Thus, promoter 1A, approximately 31 kb upstream of the first start codon, drives the expression of exon 1A (~212–981 bp); promoter 1B drives the expression of exon 1B (~116 bp) approximately 5 kb upstream of the first start codon; whereas promoter 1C drives the expression of exon 1C (184 bp) approximately 4 kb upstream of the first start codon. None of the promoter regions of the hGR gene contain a consensus TATA or CAAT box, but all contain multiple GC islands, reflecting the necessity for constitutive expression of hGR. Multiple transcription factors have been reported to bind to various sites on the hGR promoters, including four Sp1 and three YY1 binding sites in promoter 1B[50,51] and six Sp1, one AP-2, one NF-κB and one YY1 sites in promoter 1C.[47–52] This vast array of transcription factors seems to ensure constitutive expression of hGR under a variety of physiological conditions. However, the expression level of hGR transcripts containing each species of exon 1 may also be regulated in a cell type–specific as well as developmental state–dependent manner. For example, exons 1B and 1C are ubiquitous although expression levels differ in various cells and tissues.[47] The hGR transcripts containing exon 1A3, one of the three 1A exons (see below), are more abundant in cancer cells of hematopoietic lineage than in cancer cells from the bone, liver, lung, or breast. In addition, hormonal factors also may regulate hGR promoter usage. For example, promoter 1A, but not 1B or 1C, contains an element identified as a noncanonical GRE, through which the expression of the exon 1A3 is upregulated in CEM-C7 T cells but, interestingly, downregulated in IM-9 B-lymphoma cells by dexamethasone.[47]

Although the expression level of exon 1 is highly regulated, the first exon of the hGR transcript is not a component of the coding region due to an in-frame stop codon at the very beginning of exon 2, only 9 bp upstream of the first start codon. However, the terminology of “5′ untranslated region” may be a misnomer since, in the mouse GR exon 1, at least one potential open reading frame exists in the leader sequence. This 5′ open reading frame may be translated into a small peptide of

---

FIGURE 2. Organization of the human glucocorticoid receptor (hGR) gene and diversification of hGR transcripts and proteins. The hGR gene contains nine exons (numbered in boxes), coding various regions of the receptor, such as the DNA binding domain (DBD), ligand binding domain (LBD), and untranslated regions (UTR). Alternative promoter usage and mRNA alternative splicing generate multiple hGR transcripts. For example, at least five exon 1 variants, 1A1, 1A2, 1A3, 1B, and 1C, can join exon 2, yielding transcripts containing various 5′ leader sequences. In addition, alternative splicing of exon 9 generates mRNAs coding for hGRα or hGRβ. Furthermore, alternative splicing can also result in the insertion of an additional codon (GRγ), exon skipping (GRA), or exon deletion (GRP), generating receptor isoforms with blunted activity. All mRNA variants, except GR–low sensitivity (mouse lymphoma cells), have been detected in human cells and tissues. Compositions of exon 9 have been confirmed experimentally for primary transcript of hGR, hGRα, and hGRβ, and predicted for the other mRNA variants. During translation of transcripts containing both AUG^1 and AUG^{27}, the number of GR proteins is doubled by alternative translation initiation. The labels on hGR proteins refer to amino acids in the full-length hGRα receptor and are from references listed in the text.

8.5 kDa,[53] which is thought to regulate the translation efficiency of the full-length mouse GR protein.[54] Interestingly, transcripts of many potent regulatory proteins, such as cytokines, growth factors, kinases, and transcription factors, similar to GR, often contain a 5′ leader sequence.[55] The length of the leader sequence has been correlated with the translational efficiency for some transcripts; however, such information about hGR is lacking.

The mouse GR gene also has three promoter regions homologous to hGR,[56] and in cancer cells, such as S-49 lymphoma cells, four or five promoters maybe active.[57] In humans, the versatility of the GR gene in directing the expression of hGR receptors can be demonstrated not only by alternative promoter usage, but also by numerous polymorphisms within the gene. Over a hundred natural single nucleotide polymorphisms have been documented in the hGR database (http://www.ncbi.nlm.nih.gov:80/SNP/snp_ref.cgi?locusId=2908). Although few of these polymorphic loci are correlated with human diseases, individual differences in glucocorticoid responses may very well be attributed to sequence substitution within the hGR gene.[58] Furthermore, scores of additional natural mutations in the hGR gene have been recorded in instances of glucocorticoid resistance, as defined by the decreased therapeutic effects of glucocorticoid drugs in patients after prolonged usage or the absence of ACTH suppression by dexamethasone challenge. The former type of glucocorticoid resistance frequently occurs selectively in tissues exposed to glucocorticoids (upper airways, for example, when inhaling agents are used for asthma) while the latter usually indicates a generalized dysfunction. In either scenario, mutations within the hGR gene are often the cause. Key amino acid changes in the τ1 region, LBD, DBD, or mutations leading to truncated proteins have all been identified to cause glucocorticoid resistance.[58]

ALTERNATIVE SPLICING OF hGR TRANSCRIPTS

As described above, three promoters drive the expression of at least three exons 1 (1A, 1B, or 1C). Exon 1A, through alternative splicing, produces three additional isoforms,[47] 1A1 (~212 bp), 1A2 (~308 bp), and 1A3 (981 bp, FIG. 2). Sequences of exons 1A1, 1A2, and 1A3 are identical towards the 5′ end whereas three distinct splicing donor sites at the 3′ end join with the common acceptor sites on exon 2, increasing the potential number of hGR transcripts to five, i.e., 1A1, 1A2, 1A3, 1B, and 1C. Additional alternative splicing events exist and affect the coding region of hGR as well. For example, at the carboxyl terminus of the hGR primary transcript, exon 9 comprises the originally defined exon 9α (2,475 bp), an intron of 155 bp, and exon 9β (1,478 bp).[45,46] This large exon can be alternatively spliced to join exon 8, generating hGRα and hGRβ (FIG. 2).

hGRβ

Amino acid sequence analysis revealed that hGRα and hGRβ isoforms are identical from the amino terminus through amino acid 727 but diverge beyond this position, with hGRα having an additional 50 amino acids and hGRβ having an additional, non homologous, 15 amino acids. The existence of the hGRβ isoform was predicted ever since the cloning of the hGR cDNA;[42] however, only hGRα appeared to bind hormone and induce the expression of a glucocorticoid-responsive reporter gene in a hormone-dependent manner.[42,59] In contrast, the hGRβ isoform has been largely ignored because early studies reported that the recombinant hGRβ does not bind hormones and does not activate glucocorticoid-responsive promoters.[42,59] Recent years have seen a spur of interest in hGRβ since this isoform was found to have widespread tissue expression.[46,60,61] hGRβ acts as a dominant negative inhibitor for hGRα transcriptional regulation and, importantly,[46,60] increased hGRβ expression has been correlated with several diseases related to glucocorticoid resistance.[41,62–67]

Using Northern blot or reverse transcription PCR analyses, both hGRβ and hGRα mRNAs have been detected in multiple adult and fetal tissues, including the heart, brain, placenta, lung, liver, skeletal muscle, and pancreas.[46,60] To investigate the expression of hGRβ protein, we have produced an antipeptide, hGRβ-specific antibody termed BShGR.[61] This antibody has been made against the unique 15-amino acid peptide at the carboxyl terminus of hGRβ and recognizes both the native and denatured conformations of hGRβ, but it does not crossreact with hGRα. Using BShGR on Western blots and in immunoprecipitation experiments, we have also detected hGRβ protein in a variety of human cell lines and tissues. A second hGRβ antibody has been recently produced in a separate laboratory and has also been used to confirm the wide distribution pattern of hGRβ.[62]

In the absence of ligand, hGRα resides in the cytoplasmic compartment, forming a complex with molecular chaperones like hsp90. When treated with cognate hormones, it is released from the molecular complex in the cytoplasm and translocates to the nucleus. In support of this two-step translocation model, we previously have shown that, in HeLa-S3 cells, hGRα translocates from the cytoplasm to the nucleus in a hormone-dependent manner.[46,68] However, in marked contrast to hGRα, hGRβ has been found largely in the nucleus, independent of glucocorticoid treatment.[46] Further detailed analysis has demonstrated that within tissues, hGRβ is expressed at high levels in a cell type–specific manner.[61] For example, hGRβ protein is expressed abundantly in the epithelial cells lining the terminal bronchiole of the lung, forming the outer layer of Hassall's corpuscle in the thymus, and lining the bile duct in the liver. In contrast, thymic lymphocytes and other epithelial cells in these tissues show very little immunoreactivity. Moderate immunoreactivity has also been observed in hepatocytes. These studies indicate that relative levels of hGRα and hGRβ may vary considerably among different cells. Thus, ratios of hGRα and hGRβ proteins determined from whole tissues or organs do not necessarily reflect the ratio of hGRα and hGRβ within an individual cell.

The molecular difference between hGRα and hGRβ lies within the 3′ end of the LBD. The 50 amino acids at the carboxyl terminus of hGRα are replaced with 15 unique amino acids in hGRβ. With few exceptions,[69,70] amino acid changes in the hGRα LBD often result in a reduction or complete loss of hormone binding.[59,71–73] In agreement with previous reports,[42,59,60] we observed that this natural carboxyl terminus modification prevents agonist binding to hGRβ protein.[46] Similar observations have been reported for other steroid receptors. For example, the truncated version of the human PR-B, missing the carboxyl terminal 42 amino acids, does not bind progesterone or the synthetic agonist R5020 but does bind the antiprogestin RU486.[74] This finding suggests that amino acids at the extreme carboxyl terminus of the human PR are critical for agonist but not antagonist binding. To date, we have not found ligands that bind to hGRβ.

Consequently, independent of steroid treatment, hGRβ seems to be transcriptionally inactive on reporter genes studied thus far. However, hGRβ can bind GREs in the promoter regions of target genes.[60] In addition, hGRβ binds GRE-containing DNA with a greater capacity than hGRα in the absence of glucocorticoids.[75] Glucocorticoid treatment enhances hGRα binding, but not hGRβ, to DNA. hGRβ interacts with hsp90, which, in the nucleus, may facilitate the release of chromatin-bound hGR.[76] Remarkably, hGRβ inhibits the hGRα-mediated activation of several reporter genes in a dose-dependent fashion.[46,60] Furthermore, hGRβ represses the activity of endogenous hGRα.[75] In contrast, the ability of the PR or AR to activate reporter genes is only weakly affected by hGRβ, demonstrating that the dominant negative activity of hGRβ is specific for hGRα. In addition, hGRβ also inhibits hGRα-mediated repression of NF-κB– and AP-1–responsive promoters.[46,77]

The molecular basis for the dominant negative activity of hGRβ has been recently elucidated.[78] Molecular modeling of the wild type and mutant hGRα and hGRβ has delineated a possible structural basis for the lack of hormone binding and the dominant negative actions of hGRβ. The absence of helix 12 in the LBD is neither necessary nor sufficient for the dominant negative phenotype of hGRβ. Using a series of truncated hGRα mutants and sequential mutagenesis, our laboratory has generated a series of hGRα/β hybrids. We have demonstrated that two residues within the unique 15 amino acids of hGRβ are responsible for the dominant negative activity. In addition, hGRα and hGRβ have been found to physically associate with each other as heterodimers,[75] which may hinder the formation of the transcriptionally active hGRα homodimers. Thus, the physiological significance of hGRβ may reside in its ability to antagonize the function of hGRα. High levels of hGRβ would confer glucocorticoid resistance, and low levels of hGRβ would confer hypersensitivity to glucocorticoids.

Thus, it is of great interest to determine the factors that regulate the relative expression levels of hGRα and hGRβ: the identification of such factors would elucidate potential treatment targets for hGR-related diseases. Webster and colleagues have recently reported that in HeLa-S3 cells TNFα treatment selectively enhances the steady-state levels of the hGRβ protein isoform, making hGRβ the predominant endogenous receptor isoform over hGRα.[52] Similar results have also been observed following treatment of human CEM-C7 lymphoid cells with TNFα or IL-1. TNFα and IL-1 are both pro-inflammatory agents whose actions can be counteracted by glucocorticoids. The increase in hGRβ protein expression correlates with the development of glucocorticoid resistance. For example, increases of hGRβ levels have been reported in T cells in the airway, peripheral blood mononuclear cells, and in tuberculin-induced inflammatory lesions in glucocorticoid-insensitive asthmatics.[41,62–65] Elevated expression of hGRβ in peripheral blood mononuclear cells from patients with rheumatoid arthritis has also been correlated with glucocorticoid resistance.[66] In another report, high levels of hGRβ are found in 10 of 12 patients with glucocorticoid-resistant colitis.[67] Additionally, Hauk and colleagues have demonstrated that isolated peripheral blood mononuclear cells, when stimulated with various superantigens, become insensitive to glucocorticoids: this insensitivity is believed to be the result of an increased expression of hGRβ.[63] In a separate study, incubation of myoblasts with 50–1,000 nM of cortisol resulted in a dose-dependent decline in hGRα expression and a dose-dependent increase in hGRβ expression.[79] These studies underscore the importance of hGRβ in causing diseases and suggest that a strong correlation exists between the expression level of hGRβ, relative to hGRα, and resistance to glucocorticoids.

Glucocorticoid insensitivity is observed not only in disease states but also during normal physiological processes. For example, we and others have shown that genes that are positively regulated by hGRα are unresponsive to glucocorticoids during the G2 phase of the cell cycle.[80,81] During development, the chicken retina is resistant to circulating glucocorticoids before embryonic day 6, but thereafter becomes progressively more sensitive even though the level of hGRα does not change significantly over this time period.[82] In each case, cell cycle or developmentally regulated induction of hGRβ might account for the temporary resistance. Indeed, alternative splicing is often regulated in a cell type– and developmental state–specific fashion, or in response to specific cellular signals. Information on the regulation of hGR alternative splicing events is scarce, although a recent report suggests that serine arginine-rich protein p30 is involved in directing alternative splicing of hGR pre-mRNA to hGRβ in neutrophils.[83]

Other members of the nuclear receptor superfamily, TRα for example, bear resemblance to the alternative splicing pattern of hGR. Through alternative splicing of the last exon, TRα generates two receptor isoforms, TRα1 and TRα2, that differ at the carboxyl terminus.[84,85] The TRα2 isoform does not bind thyroid hormones and represses the transcriptional activity of TRα1 by competing with TRα1 for binding to the thyroid hormone responsive elements.[85] These data imply that a wide range of hormone responses can be achieved by varying the ratio of receptor isoforms within a cell. The carboxyl terminal sequence of rat GR is homologous to hGR with both α and β isoforms being produced.[86] In contrast, mouse GR has exon 9α but no 9β.[87]

Other hGR Splice Variants

Several other hGR splice variants have been detected in tissues and in certain cancer cell lines. For example, hGRγ transcripts (FIG. 2) have been detected where a 3-bp sequence from the intron separating exons 3 and 4 is retained, yielding an in-frame single amino acid insertion between the two zinc-fingers in DBD.[88] This isoform of hGR is also widely expressed and represents 4–8% of total hGR message in various tissues. Interestingly, however, hGRγ exhibits only about half of the transcriptional activity of hGRα. Recently, the level of hGRγ has been correlated with glucocorticoid resistance in childhood acute lymphoblastic leukemia.[89] A similar insertion at this splice junction has also been detected in the mouse GR,[90] rainbow trout GR,[91,92] as well as human MR.[93]

An additional splice variant of GR has been reported in glucocorticoid-resistant mouse lymphoma cells. This isoform lacks the entire exon 2 that encodes the amino terminal τ1 region, labeled as GR–low sensitivity in FIGURE 2.[94–96] One other splice variant of the GR, GRP, has retained the intron between exons 7 and 8, thus missing the appropriate exons 8 and 9.[97] The GRP variant is expressed at a high level in glucocorticoid-resistant myeloma patients.[98] In the same patient group, another splice variant missing the entire sequences of exons 5, 6, and 7, and thus a significant portion of the LBD, has also been identified and termed GRA.[97] Both GRP and GRA have been determined to contribute to glucocorticoid resistance.

ALTERNATIVE INITIATION OF hGR TRANSLATION

During translation of hGR transcripts, ribosome entry occurs at the 5′ end of the hGR message. Sequential addition of amino acids occurs after the recognition of the first start codon and concludes when the ribosomes encounter the stop codon at the 3′ end of hGR transcripts, yielding the full-length 777 amino acid peptide. However, translation reinitiation occurs at codon AUG^{27} in hGR transcripts, generating a receptor peptide of 751 amino acids that lacks the first 26 amino acids from the full-length hGR.[99] In the original paper that describes these two isoforms (FIG. 2), the 94 kDa full-length receptor translated from hGRα transcript is named hGRα-A and the 91 kDa protein is named hGRα-B.[99]

A survey of eukaryotic mRNAs has revealed that alternative start codon usage, also termed ribosomal leaky scanning, may occur in as many as 5% of the transcribed messages.[100] During translation, suboptimal nucleotide context in the proximity of the first start codon promotes weak ribosomal binding. Additional ribosomes are therefore permitted to scan for binding sites downstream of the first start codon. When the weak context at the AUG^{1} of hGRα was replaced with a consensus sequence that facilitates optimal interaction between ribosomes and mRNA, the production of hGRα-B can be diminished.[99]

hGRα-A and hGRα-B exhibit similar ligand-dependent translocation from the cytoplasm to the nucleus.[99] Interestingly, hGRα-B, in transient transfection experiments, activates reporter genes to a greater extent than hGRα-A. This is in agreement with the notion that ribosomal leaky scanning, instead of reflecting "sloppiness" of the translation machinery, deliberately produces potent regulators of cell function.[55] Whether hGRα-A and hGRα-B are differentially expressed in a tissue-specific manner and how their expressions are regulated are topics under investigation. Potentially, A and B isoforms derived from various hGR transcripts through alternative translation initiation may diversify the hGR receptor family exponentially. For example, the hGRβ transcript, which contains both AUG^{1} and AUG^{27}, produces both hGRβ-A and hGRβ-B isoforms (unpublished results).

POSTTRANSLATIONAL MODIFICATIONS OF hGR

Mature hGR proteins are covalently modified by various processes, which further modulate the transcription regulation activity of the receptors. For example, three consensus sumoylation sites (FIG. 1) have been identified within the hGR peptide sequence, and this modification process seems to affect receptor activity.[101–103] In addition, nitrosylation at cysteine residues on GR likely decreases ligand binding and may disable glucocorticoids from exerting anti-inflammatory effects during fatal septic shock.[104] In this section, we discuss in detail two other posttranslational modification processes that are directly linked with GRα-A function: phosphorylation and ubiquitination. The information on posttranslational modification of other GR isoforms is scarce.

Phosphorylation

Like most other nuclear receptors, mature GR proteins are phosphorylated. When activated by agonists, GR becomes hyperphosphorylated on several of the eight residues at the amino terminus of the receptor. Identification of the phosphorylated residues on GR required the heroic efforts of several laboratories.[105] Receptor proteins were radiolabeled with [^{32}P] *in vivo*, purified, and subjected to digestion by trypsin at optimal conditions. Incomplete hydrolysis may introduce overlapping, thus confounding signals, whereas overdigestion may cut the peptide into fragments too small to be sufficiently purified, thereby increasing the number of misses. Tryptic peptides were then separated by HPLC, the phosphate content of each fraction measured, and the amino acids sequenced. Eight phosphorylated residues directly identified in the mouse GR are serines 122, 150, 212, 220, 234, 315, and 412, and threonine 159. Five corresponding amino acids in hGR are serines 113, 141, 203, 211, and 226 (FIG. 1). There are no counterparts in hGR for the other phosphorylated residues identified in the mouse GR, and this difference may underlie species-specific receptor functions.[106]

Since the phosphorylated residues of GR are concentrated in the τ1 region of the receptor, significant changes of receptor transactivation activity were anticipated when receptor phosphorylation was disrupted in receptors containing serine/threonine to alanine substitutions. However, it was not until after a series of target promoters were surveyed that the profile of transcription activity regulation by phosphorylation was revealed. For example, in COS-1 cells, replacing all eight phosphorylated residues in mouse GR does not alter the receptor's ability to induce a reporter gene driven by the mouse mammary tumor virus promoter.[107] In contrast, the non-phosphorylated receptor exhibits only 25–50% of the transactivation activity of the phosphorylated receptor on a simple GRE2-driven reporter. Thus, phosphorylation enhances the transactivation activity in a gene-specific manner. Different degrees of receptor phosphorylation, therefore, may extend the range of the gene regulatory capability of GR. Wang and colleagues, using antibodies that recognize phosphorylated Ser211 on hGR, have demonstrated a positive correlation between the amount of Ser211 phosphorylation and transactivation activity of hGR.[108] It is not known whether GR phosphorylation status affects GRE-independent regulation of gene expression.

Factors that facilitate GR phosphorylation include agonists such as dexamethasone and triamcinolone, but not antagonists, such as RU486.[109] In addition, low amount of basal phosphorylation on GR, observed during the DNA synthesis phase of the cell cycle but not during the mitotic phase, assists GR hyperphosphorylation.[109] Therefore, agonist treatment stimulates the degree of GR phosphorylation during S phase but not G2/M phase. Correspondingly, cells synchronized at S phase are glucocorticoid sensitive but cells synchronized at G2/M phase are glucocorticoid resistant.[110] This insight may assist with the designing of efficient chemotherapy regimens that could potentially overcome glucocorticoid resistance in some patients.

In the presence of agonists, phosphorylated GR has a half-life of 8–9 h whereas the half-life of non-phosphorylated GR is about 32 h.[107] This observation is in agreement with findings that agonist-activated GR has a shorter half-life than un-liganded receptors.[111–119] Thus, transcriptionally active GR exhibits a fast turnover rate. However, a slow turnover rate does not necessarily correlate with low receptor activity.

Ubiquitination

We recently reported that the mouse GR is degraded through the ubiquitin-proteasome pathway.[120] Protein phosphorylation facilitates E2 ubiquitin-conjugating enzymes and/or E3 ubiquitin-ligase to recognize target proteins and covalently link the 76 amino acid ubiquitin to lysine residue(s).[121–123] Proteins tagged with poly-ubiquitin are trafficked primarily to the multiprotein complex known as the proteasome for degradation.[124–126] GR has been shown to interact with an E2-conjugating protein[127] and two E3-ligase proteins.[128]

A number of proteins rapidly degraded through the proteasome pathway contain PEST regions, which contain the amino acids Pro (P), Glu (E), Ser (S), and Thr (T). Hallmarks of PEST regions include phosphorylation sites, stretches of hydrophilic amino acids, and Lys, Arg, and His residues.[129,130] Analysis of mouse GR using a PEST-FIND program revealed a PEST motif from amino acids 407–426 (FIG. 1). This region has a PEST-FIND score of +18.3; on a scale from −50 to +50 a value above +5 is indicative of a possible functional PEST motif.[129] For example, two proteins known to be degraded by the proteasome, IκBα and FOS, have PEST scores of 5.9 and 10.1, respectively.[131–133] In addition, PEST-FIND analysis calculated a score of +18.3 for the rat GR and +16.1 for hGR, suggesting that PEST motifs in GR are conserved among mammals. Ser-412 within the mouse PEST region is a site of ligand-dependent phosphorylation.[105,134]

Pretreatment of COS-1 cells expressing mouse GR with proteasome inhibitor, MG-132, effectively blocks GR protein downregulation (degradation) by the glucocorticoid dexamethasone.[120] Furthermore, direct evidence for ubiquitination of the GR has been obtained by immunoprecipitation of cellular extracts from proteasome-impaired cells. MG-132 also blocks agonist-induced degradation of ERα, PR, as well as the aryl hydrocarbon receptor.[135,136] Interestingly, both MG-132 and a second proteasome inhibitor, β-lactone, significantly enhance the transactivation activity of transfected mouse GR as well as endogenous hGR in HeLa cells. Mutation of Lys426 within the PEST element abrogates ligand-dependent downregulation of the mouse GR protein and simultaneously enhances GR-induced transcriptional activation. MG-132 does not affect the receptor level or the transcriptional activity of K426A mutant mouse GR.[120]

Thus, when the turnover rate of GR is decreased by proteasomal inhibition, GR activity is increased. Inhibition of degradation also enhances the transcription regulatory activity of other transcription factors, such as the aryl hydrocarbon receptor, Sp1, and p53.[137–139] In contrast, proteasomal inhibition decreased ligand-induced transcriptional activity of ERα or TRα.[137,140] The causal features common to each category of relationship between turnover rate and activity are not known although it has been suggested that the formation of an ERα coactivator complex may be disrupted by proteasome inhibitors.[137] In addition, consensus PEST motifs are not present in either ERα or TRα and proteasomal activity may be necessary to produce a transcriptionally active form of TRα.[140] As discussed above, phosphorylation shortens the half-life but enhances the transactivation activity of GR. However, the long-lived GR proteins in the presence of MG132 exhibit increased transactivation activity. Together, these data support the notion that the amount of available phosphorylated GR, instead of receptor turnover rate, determines receptor activity. Recently, the phosphorylation status of Ser211 has been suggested as a biomarker for hGR activity *in vivo*.[108]

GR phosphorylation occurs within 5–10 min of hormone addition and the half-maximal rate ($t_{1/2}$) for GR dephosphorylation is 90–120 min.[109] Phosphorylated GR, in the presence of agonist, has a half-life of 9 h, before being trafficked to the proteasome for degradation. Intriguing questions about the activation and degradation of GR remain to be answered. How does an activated GR molecule navigate through the cell nucleus where the genome resides? Furthermore, what signals terminate the usage of an individual receptor?

LIGAND-DEPENDENT GR TRAFFICKING IN THE NUCLEUS

In the absence of hormone, GRα resides predominantly in the cytoplasm of cells, forming a multiprotein complex with two molecules of hsp90 and several additional proteins.[141] Ligand binding induces conformational changes that are followed by the release of GRα from the chaperones and translocation of the receptor into the nucleus.[142] Most current experiments using transiently expressed fluorescent proteins (GFP or YFP) tagged receptors suggest that complete nuclear translocation of GFP-hGRα occurs within 30 min after ligand addition.[143] Similar data on nuclear translocation have been obtained for additional GFP-tagged receptors in the steroid receptor family as well.[144–148]

In the nucleus, the agonist-bound GFP-hGRα has been reported to be organized into discrete foci,[143] which is consistent with earlier results from immunocytochemical studies on endogenous GR.[149] A similar punctate distribution in the nucleus has also been found for agonist-bound GFP-tagged ERα,[150,151] AR,[144–146] MR,[147] vitamin D receptor,[148] and TRβ.[152] Treatment with an antagonist does not result in foci formation of GR,[143] AR,[144–146] or MR,[147] although additional studies are needed to determine whether this characteristic is strictly limited to certain antagonists. ERα antagonists induce a less pronounced punctate receptor distribution than agonists.[150,151] Nuclear GR foci take shape within 15 min of agonist treatment, but it is not completely understood whether all of these foci colocalize with transcription initiation sites.[153] Active transcription complexes are assembled in an orderly fashion where activated GR initiates the recruitment of RNA polymerase II, the cofactors GRIP-1,[154] SRC1, and CBP (which contains inherent acetyltransferase activity), BRG1 (a chromatin remodeler), and other transcription factors such as NFI and AP-2.[155]

Additional nuclear GR have been reported to colocalize with the nuclear matrix. The nuclear matrix is the non-chromatin elements of the nuclear structure readily observed under an electron microscope.[156,157] A main constituent protein, hnRNP U,[158,159] interacts with GR.[160] The rat GR τ2 region contains a nuclear matrix–targeting signal that facilitates the interaction between GR and hnRNP U.[161] Furthermore, overexpression of hnRNP U inhibits GR-induced transactivation.[161,162] This observation is consistent with the previous finding that ligand-bound GR is more resistant to high salt extraction from the nucleus than non-liganded GR.[163]

The relationship between chromatin-bound GR and nuclear matrix–bound GR was then examined in a series of elegant experiments, GR release from and redocking onto chromatin was visualized.[153] The recycling process is rather rapid with a half maximal rate ($t_{1/2}$) of 5 s. Our recent photobleaching experiments indicate that the mobility of nuclear hGR is highly dependent on the ligand that occupies the receptor.[164] For example, YFP-hGRα-A in the nucleus is less mobile when activated by triamcinolone acetonide ($t_{1/2} = 2.38$ s) than by cortisone ($t_{1/2} = 0.97$ s). The affinity of hGRα-A for the former ligand is more than 10-fold higher than that for the latter. The positive correlation between ligand affinity and the ligand's ability to decelerate nuclear hGRα-A seems to be true when a panel of GR ligands was tested.[164] The structural determinants of hGRα-A mobility have been mapped as well.[164] Both the DBD and LBD of the receptor are required for the ligand-induced decrease in receptor mobility. Interestingly, the proteasome inhibitor MG132 immobilizes a subpopulation of non-liganded receptors.[165] This immobilization can be blocked by high- affinity dexamethasone but not by low-affinity cortisone. Thus, the range of GR mobility and function is extended further by a vast array of natural and synthetic ligands.

PERSPECTIVE

Efforts from many laboratories including ours provide convincing evidence that many forms of hGR exist in various physiological and pathological states. Gene regulation activities as well as expression levels of various hGR isoforms differ in various *in vitro* and *in vivo* systems. Different hGR isoforms may contribute to tissue specificity and differences of glucocorticoid responsivity among individuals. Similarly, changes in posttranslational modification status and in the relative proportion of receptor isoforms within a cell or tissue may result in dysfunction of GR-mediated physiology. The majority of our knowledge concerning how GR is modified and how modified GR regulates various gene targets has been generated from studies on the GRα isoform. GR activation requires binding of cognate ligands. About 5 to 10 min after agonist addition, GR phosphorylation is stimulated, followed by the assembly of the receptor-mediated transcription complex within target gene-containing chromatins. Before the receptors are directed to the proteasome for degradation, GR in the nucleus may be shuffled among multiple target sites rapidly (in seconds) to impact the genome. Posttranslational modification status of an individual receptor could very likely be a determinant of receptor usage. The latest development in methodology will greatly improve our understanding of the modification status and function of GR. For example, the number of phosphorylation residues on PR has been recently updated from seven to fourteen.[166] Although the discovery of multiple GR isoforms represents a step closer to understanding the pivotal roles of GR in health and disease, continuous endeavors from investigators in different laboratories will be needed to unveil how the diversity of GR relates to its function and how these mechanisms can be utilized in developing effective treatment regimens for GR-related diseases.

REFERENCES

1. Barnes, P.J. 1998. Anti-inflammatory actions of glucocorticoids: molecular mechanisms. Clin. Sci. (Lond.) **94:** 557–572.
2. Sapolsky, R.M., L. Romero & A.U. Munck. 2000. How do glucocorticoids influence stress responses? Integrating permissive, suppressive, stimulatory, and preparative actions. Endocr. Rev. **21:** 55–89.
3. Corrigan, C.J., P.H. Brown, N.C. Barnes, *et al.* 1991. Glucocorticoid resistance in chronic asthma. Glucocorticoid pharmacokinetics, glucocorticoid receptor characteristics, and inhibition of peripheral blood T cell proliferation by glucocorticoids *in vitro*. Am. Rev. Respir. Dis. **144:** 1016–1025.

4. BARNES, P.J., A.P. GREENING & G.K. CROMPTON. 1995. Glucocorticoid resistance in asthma. Am. J. Respir. Crit. Care Med. **152:** S125–140.
5. KIRKHAM, B.W., M.M. CORKILL, S.C. DAVISON & G.S. PANAYI. 1991. Response to glucocorticoid treatment in rheumatoid arthritis: *in vitro* cell mediated immune assay predicts *in vivo* responses. J. Rheumatol. **18:** 821–825.
6. DI BATTISTA, J.A., M. ZHANG, J. MARTEL-PELLETIER, *et al.* 1999. Enhancement of phosphorylation and transcriptional activity of the glucocorticoid receptor in human synovial fibroblasts by nimesulide, a preferential cyclooxygenase 2 inhibitor. Arthritis Rheum. **42:** 157–166.
7. LICHTIGER, S., D.H. PRESENT, A. KORNBLUTH, *et al.* 1994. Cyclosporine in severe ulcerative colitis refractory to steroid therapy. N. Engl. J. Med. **330:** 1841–1845.
8. QUAN, S.F., J.B. SEDGWICK, M.V. NELSON & W.W. BUSSE. 1993. Corticosteroid resistance in eosinophilic gastritis—relation to *in vitro* eosinophil survival and interleukin 5. Ann. Allergy **70:** 256–260.
9. LANGHOFF, E., J. LADEFOGED, B.K. JAKOBSEN, *et al.* 1986. Recipient lymphocyte sensitivity to methylprednisolone affects cadaver kidney graft survival. Lancet **1:** 1296–1297.
10. NORBIATO, G., M. BEVILACQUA, T. VAGO, *et al.* 1992. Cortisol resistance in acquired immunodeficiency syndrome. J. Clin. Endocrinol. Metab. **74:** 608–613.
11. THORNTON, J.W. 2001. Evolution of vertebrate steroid receptors from an ancestral estrogen receptor by ligand exploitation and serial genome expansions. Proc. Natl. Acad. Sci. USA **98:** 5671–5676.
12. EVANS, R.M. 1988. The steroid and thyroid hormone receptor superfamily. Science **240:** 889–895.
13. HOLLENBERG, S.M. & R.M. EVANS. 1988. Multiple and cooperative trans-activation domains of the human glucocorticoid receptor. Cell **55:** 899–906.
14. CARSON-JURICA, M.A., W.T. SCHRADER & B.W. O'MALLEY. 1990. Steroid receptor family: structure and functions. Endocr. Rev. **11:** 201–220.
15. KUMAR, R. & E.B. THOMPSON. 1999. The structure of the nuclear hormone receptors. Steroids **64:** 310–319.
16. YAMAMOTO, K.R. 1985. Steroid receptor regulated transcription of specific genes and gene networks. Annu. Rev. Genet. **19:** 209–252.
17. BLEDSOE, R.K., V.G. MONTANA, T.B. STANLEY, *et al.* 2002. Crystal structure of the glucocorticoid receptor ligand binding domain reveals a novel mode of receptor dimerization and coactivator recognition. Cell **110:** 93–105.
18. LUISI, B.F., W.X. XU, Z. OTWINOWSKI, *et al.* 1991. Crystallographic analysis of the interaction of the glucocorticoid receptor with DNA. Nature **352:** 497–505.
19. KUMAR, R., I.V. BASKAKOV, G. SRINIVASAN, *et al.* 1999. Interdomain signaling in a two-domain fragment of the human glucocorticoid receptor. J. Biol. Chem. **274:** 24737–24741.
20. PICARD, D. & K.R. YAMAMOTO. 1987. Two signals mediate hormone-dependent nuclear localization of the glucocorticoid receptor. EMBO J. **6:** 3333–3340.
21. DALMAN, F.C., L.C. SCHERRER, L.P. TAYLOR, *et al.* 1991. Localization of the 90-kDa heat shock protein-binding site within the hormone-binding domain of the glucocorticoid receptor by peptide competition. J. Biol. Chem. **266:** 3482–3490.
22. WIKSTROM, A.C., C. WIDEN, A. ERLANDSSON, *et al.* 2002. Cytosolic glucocorticoid receptor-interacting proteins. Ernst Schering Res. Found. Workshop: 177–196.
23. BEATO, M. & A. SANCHEZ-PACHECO. 1996. Interaction of steroid hormone receptors with the transcription initiation complex. Endocr. Rev. **17:** 587–609.
24. COLLINGWOOD, T.N., F.D. URNOV & A.P. WOLFFE. 1999. Nuclear receptors: coactivators, corepressors and chromatin remodeling in the control of transcription. J. Mol. Endocrinol. **23:** 255–275.
25. MCKENNA, N.J., R.B. LANZ & B.W. O'MALLEY. 1999. Nuclear receptor coregulators: cellular and molecular biology. Endocr. Rev. **20:** 321–344.
26. GLASS, C. K. & M. G. ROSENFELD. 2000. The coregulator exchange in transcriptional functions of nuclear receptors. Genes Dev. **14:** 121–141.
27. WALLBERG, A.E., A. WRIGHT & J.A. GUSTAFSSON. 2000. Chromatin-remodeling complexes involved in gene activation by the glucocorticoid receptor. Vitam. Horm. **60:** 75–122.

28. JENKINS, B.D., C.B. PULLEN & B.D. DARIMONT. 2001. Novel glucocorticoid receptor coactivator effector mechanisms. Trends Endocrinol. Metab. **12:** 122–126.
29. KUMAR, R., J.C. LEE, D.W. BOLEN & E.B. THOMPSON. 2001. The conformation of the glucocorticoid receptor af1/tau1 domain induced by osmolyte binds co-regulatory proteins. J. Biol. Chem. **276:** 18146–18152.
30. DAHLMAN-WRIGHT, K., A. WRIGHT, J.A. GUSTAFSSON & J. CARLSTEDT-DUKE. 1991. Interaction of the glucocorticoid receptor DNA-binding domain with DNA as a dimer is mediated by a short segment of five amino acids. J. Biol. Chem. **266:** 3107–3112.
31. MACDONALD, R.G. & J.A. CIDLOWSKI. 1981. Glucocorticoid-stimulated protein degradation in lymphocytes: quantitation by sodium dodecyl sulfate-polyacrylamide gel electrophoresis. Arch. Biochem. Biophys. **212:** 399–410.
32. YANG-YEN, H.F., J.C. CHAMBARD, Y.L. SUN, *et al.* 1990. Transcriptional interference between c-Jun and the glucocorticoid receptor: mutual inhibition of DNA binding due to direct protein-protein interaction. Cell **62:** 1205–1215.
33. SCHULE, R., P. RANGARAJAN, S. KLIEWER, *et al.* 1990. Functional antagonism between oncoprotein c-Jun and the glucocorticoid receptor. Cell **62:** 1217–1226.
34. JONAT, C., H.J. RAHMSDORF, K.K. PARK, *et al.* 1990. Antitumor promotion and antiinflammation: down-modulation of AP-1 (Fos/Jun) activity by glucocorticoid hormone. Cell **62:** 1189–1204.
35. CALDENHOVEN, E., J. LIDEN, S. WISSINK, *et al.* 1995. Negative cross-talk between RelA and the glucocorticoid receptor: a possible mechanism for the antiinflammatory action of glucocorticoids. Mol. Endocrinol. **9:** 401–412.
36. MANGELSDORF, D.J., C. THUMMEL, M. BEATO, *et al.* 1995. The nuclear receptor superfamily: the second decade. Cell **83:** 835–839.
37. SCHEINMAN, R.I., A. GUALBERTO, C.M. JEWELL, *et al.* 1995. Characterization of mechanisms involved in transrepression of NF-kappa B by activated glucocorticoid receptors. Mol. Cell Biol. **15:** 943–953.
38. BAMBERGER, C.M., H.M. SCHULTE & G.P. CHROUSOS. 1996. Molecular determinants of glucocorticoid receptor function and tissue sensitivity to glucocorticoids. Endocr. Rev. **17:** 245–261.
39. STOCKLIN, E., M. WISSLER, F. GOUILLEUX & B. GRONER. 1996. Functional interactions between Stat5 and the glucocorticoid receptor. Nature **383:** 726–728.
40. MCKAY, L.I. & J.A. CIDLOWSKI. 1999. Molecular control of immune/inflammatory responses: interactions between nuclear factor-kappa B and steroid receptor-signaling pathways. Endocr. Rev. **20:** 435–459.
41. WEBSTER, J.C. & J.A. CIDLOWSKI. 1999. Mechanisms of glucocorticoid-receptor-mediated repression of gene expression. Trends Endocrinol. Metab. **10:** 396–402.
42. HOLLENBERG, S.M., C. WEINBERGER, E.S. ONG, *et al.* 1985. Primary structure and expression of a functional human glucocorticoid receptor cDNA. Nature **318:** 635–641.
43. FRANCKE, U. & B.E. FOELLMER. 1989. The glucocorticoid receptor gene is in 5q31-q32 [corrected]. Genomics **4:** 610–612.
44. THERIAULT, A., E. BOYD, S.B. HARRAP, *et al.* 1989. Regional chromosomal assignment of the human glucocorticoid receptor gene to 5q31. Hum. Genet. **83:** 289–291.
45. ENCIO, I.J. & S.D. DETERA-WADLEIGH. 1991. The genomic structure of the human glucocorticoid receptor. J. Biol. Chem. **266:** 7182–7188.
46. OAKLEY, R.H., M. SAR & J.A. CIDLOWSKI. 1996. The human glucocorticoid receptor beta isoform. Expression, biochemical properties, and putative function. J. Biol. Chem. **271:** 9550–9559.
47. BRESLIN, M.B., C.D. GENG & W.V. VEDECKIS. 2001. Multiple promoters exist in the human GR gene, one of which is activated by glucocorticoids. Mol. Endocrinol. **15:** 1381–1395.
48. ZONG, J., J. ASHRAF & E.B. THOMPSON. 1990. The promoter and first, untranslated exon of the human glucocorticoid receptor gene are GC rich but lack consensus glucocorticoid receptor element sites. Mol. Cell Biol. **10:** 5580–5585.
49. NOBUKUNI, Y., C.L. SMITH, G.L. HAGER & S.D. DETERA-WADLEIGH. 1995. Characterization of the human glucocorticoid receptor promoter. Biochemistry **34:** 8207–8214.
50. BRESLIN, M.B. & W.V. VEDECKIS. 1998. The human glucocorticoid receptor promoter upstream sequences contain binding sites for the ubiquitous transcription factor, Yin Yang 1. J. Steroid Biochem. Mol. Biol. **67:** 369–381.

51. Nunez, B.S. & W.V. Vedeckis. 2002. Characterization of promoter 1B in the human glucocorticoid receptor gene. Mol. Cell. Endocrinol. **189:** 191–199.
52. Webster, J.C., R.H. Oakley, C.M. Jewell & J.A. Cidlowski. 2001. Proinflammatory cytokines regulate human glucocorticoid receptor gene expression and lead to the accumulation of the dominant negative beta isoform: a mechanism for the generation of glucocorticoid resistance. Proc. Natl. Acad. Sci. USA **98:** 6865–6870.
53. Diba, F., C.S. Watson & B. Gametchu. 2001. 5′UTR sequences of the glucocorticoid receptor 1A transcript encode a peptide associated with translational regulation of the glucocorticoid receptor. J. Cell. Biochem. **81:** 149–161.
54. Chen, F., C.S. Watson & B. Gametchu. 1999. Association of the glucocorticoid receptor alternatively-spliced transcript 1A with the presence of the high molecular weight membrane glucocorticoid receptor in mouse lymphoma cells. J. Cell. Biochem. **74:** 430–446.
55. Kozak, M. 2002. Pushing the limits of the scanning mechanism for initiation of translation. Gene **299:** 1–34.
56. Strahle, U., A. Schmidt, G. Kelsey, *et al.* 1992. At least three promoters direct expression of the mouse glucocorticoid receptor gene. Proc. Natl. Acad. Sci. USA **89:** 6731–6735.
57. Chen, F., C.S. Watson & B. Gametchu. 1999. Multiple glucocorticoid receptor transcripts in membrane glucocorticoid receptor-enriched S-49 mouse lymphoma cells. J. Cell. Biochem. **74:** 418–429.
58. Bray, P.J. & R.G. Cotton. 2003. Variations of the human glucocorticoid receptor gene (NR3C1): pathological and *in vitro* mutations and polymorphisms. Hum. Mutat. **21:** 557–568.
59. Giguere, V., S.M. Hollenberg, M.G. Rosenfeld & R.M. Evans. 1986. Functional domains of the human glucocorticoid receptor. Cell **46:** 645–652.
60. Bamberger, C.M., A.M. Bamberger, M. de Castro & G.P. Chrousos. 1995. Glucocorticoid receptor beta, a potential endogenous inhibitor of glucocorticoid action in humans. J. Clin. Invest. **95:** 2435–2441.
61. Oakley, R.H., J.C. Webster, M. Sar, *et al.* 1997. Expression and subcellular distribution of the beta-isoform of the human glucocorticoid receptor. Endocrinology **138:** 5028–5038.
62. de Castro, M., S. Elliot, T. Kino, *et al.* 1996. The non-ligand binding beta-isoform of the human glucocorticoid receptor (hGR beta): tissue levels, mechanism of action, and potential physiologic role. Mol. Med. **2:** 597–607.
63. Hauk, P.J., Q.A. Hamid, G.P. Chrousos & D.Y. Leung. 2000. Induction of corticosteroid insensitivity in human PBMCs by microbial superantigens. J. Allergy Clin. Immunol. **105:** 782–787.
64. Hamid, Q.A., S.E. Wenzel, P.J. Hauk, *et al.* 1999. Increased glucocorticoid receptor beta in airway cells of glucocorticoid-insensitive asthma. Am. J. Respir. Crit. Care Med. **159:** 1600–1604.
65. Sousa, A.R., S.J. Lane, J.A. Cidlowski, *et al.* 2000. Glucocorticoid resistance in asthma is associated with elevated *in vivo* expression of the glucocorticoid receptor beta-isoform. J. Allergy Clin. Immunol. **105:** 943–950.
66. Chikanza, I.C. 2002. Mechanisms of corticosteroid resistance in rheumatoid arthritis: a putative role for the corticosteroid receptor beta isoform. Ann. N.Y. Acad. Sci. **966:** 39–48.
67. Honda, M., F. Orii, T. Ayabe, *et al.* 2000. Expression of glucocorticoid receptor beta in lymphocytes of patients with glucocorticoid-resistant ulcerative colitis. Gastroenterology **118:** 859–866.
68. Cidlowski, J.A., D.L. Bellingham, F.E. Powell-Oliver, *et al.* 1990. Novel antipeptide antibodies to the human glucocorticoid receptor: recognition of multiple receptor forms *in vitro* and distinct localization of cytoplasmic and nuclear receptors. Mol. Endocrinol. **4:** 1427–1437.
69. Warriar, N., C. Yu & M.V. Govindan. 1994. Hormone binding domain of human glucocorticoid receptor. Enhancement of transactivation function by substitution mutants M565R and A573Q. J. Biol. Chem. **269:** 29010–29015.

70. CHAKRABORTI, P.K., M.J. GARABEDIAN, K.R. YAMAMOTO & S.S. SIMONS, JR. 1991. Creation of "super" glucocorticoid receptors by point mutations in the steroid binding domain. J. Biol. Chem. **266:** 22075–22078.
71. DANIELSEN, M., J.P. NORTHROP & G.M. RINGOLD. 1986. The mouse glucocorticoid receptor: mapping of functional domains by cloning, sequencing and expression of wild-type and mutant receptor proteins. EMBO J. **5:** 2513–2522.
72. HOLLENBERG, S.M., V. GIGUERE & R.M. EVANS. 1989. Identification of two regions of the human glucocorticoid receptor hormone binding domain that block activation. Cancer Res. **49:** 2292s–2294s.
73. HURLEY, D.M., D. ACCILI, C.A. STRATAKIS, *et al.* 1991. Point mutation causing a single amino acid substitution in the hormone binding domain of the glucocorticoid receptor in familial glucocorticoid resistance. J. Clin. Invest. **87:** 680–686.
74. VEGETO, E., G.F. ALLAN, W.T. SCHRADER, *et al.* 1992. The mechanism of RU486 antagonism is dependent on the conformation of the carboxy-terminal tail of the human progesterone receptor. Cell **69:** 703–713.
75. OAKLEY, R.H., C.M. JEWELL, M.R. YUDT, *et al.* 1999. The dominant negative activity of the human glucocorticoid receptor beta isoform. Specificity and mechanisms of action. J. Biol. Chem. **274:** 27857–27866.
76. LIU, J. & D.B. DEFRANCO. 1999. Chromatin recycling of glucocorticoid receptors: implications for multiple roles of heat shock protein 90. Mol. Endocrinol. **13:** 355–365.
77. GOUGAT, C., D. JAFFUEL, R. GAGLIARDO, *et al.* 2002. Overexpression of the human glucocorticoid receptor alpha and beta isoforms inhibits AP-1 and NF-kappaB activities hormone independently. J. Mol. Med. **80:** 309–318.
78. YUDT, M.R., C.M. JEWELL, R.J. BIENSTOCK & J.A. CIDLOWSKI. 2003. Molecular origins for the dominant negative function of human glucocorticoid receptor beta. Mol. Cell Biol. **23:** 4319–4330.
79. WHORWOOD, C.B., S.J. DONOVAN, P.J. WOOD & D.I. PHILLIPS. 2001. Regulation of glucocorticoid receptor alpha and beta isoforms and type I 11beta-hydroxysteroid dehydrogenase expression in human skeletal muscle cells: a key role in the pathogenesis of insulin resistance? J. Clin. Endocrinol. Metab. **86:** 2296–2308.
80. FANGER, B.O., R.A. CURRIE & J.A. CIDLOWSKI. 1986. Regulation of epidermal growth factor receptors by glucocorticoids during the cell cycle in HeLa S3 cells. Arch. Biochem. Biophys. **249:** 116–125.
81. HSU, S.C., M. QI & D.B. DEFRANCO. 1992. Cell cycle regulation of glucocorticoid receptor function. EMBO J. **11:** 3457–3468.
82. GOROVITS, R., I. BEN-DROR, L.E. FOX, *et al.* 1994. Developmental changes in the expression and compartmentalization of the glucocorticoid receptor in embryonic retina. Proc. Natl. Acad. Sci. USA **91:** 4786–4790.
83. XU, Q., D.Y. LEUNG & K.O. KISICH. 2003. Serine-arginine-rich protein p30 directs alternative splicing of glucocorticoid receptor pre-mRNA to glucocorticoid receptor beta in neutrophils. J. Biol. Chem. **278:** 27112–27118.
84. MITSUHASHI, T., G.E. TENNYSON & V.M. NIKODEM. 1988. Alternative splicing generates messages encoding rat c-erbA proteins that do not bind thyroid hormone. Proc. Natl. Acad. Sci. USA **85:** 5804–5808.
85. KOENIG, R.J., M.A. LAZAR, R.A. HODIN, *et al.* 1989. Inhibition of thyroid hormone action by a non-hormone binding c-erbA protein generated by alternative mRNA splicing. Nature **337:** 659–661.
86. KORN, S.H., E. KOERTS-DE LANG, G.E. ENGEL, *et al.* 1998. Alpha and beta glucocorticoid receptor mRNA expression in skeletal muscle. J. Muscle Res. Cell. Motil. **19:** 757–765.
87. OTTO, C., H.M. REICHARDT & G. SCHUTZ. 1997. Absence of glucocorticoid receptor-beta in mice. J. Biol. Chem. **272:** 26665–26668.
88. RIVERS, C., A. LEVY, J. HANCOCK, *et al.* 1999. Insertion of an amino acid in the DNA-binding domain of the glucocorticoid receptor as a result of alternative splicing. J. Clin. Endocrinol. Metab. **84:** 4283–4286.
89. BEGER, C., K. GERDES, M. LAUTEN, *et al.* 2003. Expression and structural analysis of glucocorticoid receptor isoform gamma in human leukaemia cells using an isoform-specific real-time polymerase chain reaction approach. Br. J. Haematol. **122:** 245–252.

90. KASAI, Y. 1990. Two naturally-occurring isoforms and their expression of a glucocorticoid receptor gene from an androgen-dependent mouse tumor. FEBS Lett. **274:** 99–102.
91. DUCOURET, B., M. TUJAGUE, J. ASHRAF, *et al.* 1995. Cloning of a teleost fish glucocorticoid receptor shows that it contains a deoxyribonucleic acid-binding domain different from that of mammals. Endocrinology **136:** 3774–3783.
92. TAKEO, J., J. HATA, C. SEGAWA, *et al.* 1996. Fish glucocorticoid receptor with splicing variants in the DNA binding domain. FEBS Lett. **389:** 244–248.
93. BLOEM, L.J., C. GUO & J.H. PRATT. 1995. Identification of a splice variant of the rat and human mineralocorticoid receptor genes. J. Steroid Biochem. Mol. Biol. **55:** 159–162.
94. OKRET, S., Y.W. STEVENS, J. CARLSTEDT-DUKE, *et al.* 1983. Absence in glucocorticoid-resistant mouse lymphoma P1798 of a glucocorticoid receptor domain responsible for biological effects. Cancer Res. **43:** 3127–3131.
95. DIEKEN, E.S., E.U. MEESE & R.L. MIESFELD. 1990. nti glucocorticoid receptor transcripts lack sequences encoding the amino-terminal transcriptional modulatory domain. Mol. Cell Biol. **10:** 4574–4581.
96. IP, M.M., W.K. SHEA, D. SYKES & D.A. YOUNG. 1991. The truncated glucocorticoid receptor in the P1798 mouse lymphosarcoma is associated with resistance to glucocorticoid lysis but not to other glucocorticoid-induced functions. Cancer Res. **51:** 2786–2796.
97. MOALLI, P.A., S. PILLAY, N.L. KRETT & S.T. ROSEN. 1993. Alternatively spliced glucocorticoid receptor messenger RNAs in glucocorticoid-resistant human multiple myeloma cells. Cancer Res. **53:** 3877–3879.
98. KRETT, N.L., S. PILLAY, P.A. MOALLI, *et al.* 1995. A variant glucocorticoid receptor messenger RNA is expressed in multiple myeloma patients. Cancer Res. **55:** 2727–2729.
99. YUDT, M.R. & J.A. CIDLOWSKI. 2001. Molecular identification and characterization of a and b forms of the glucocorticoid receptor. Mol. Endocrinol. **15:** 1093–1103.
100. KOZAK, M. 1984. Compilation and analysis of sequences upstream from the translational start site in eukaryotic mRNAs. Nucleic Acids Res. **12:** 857–872.
101. TIAN, S., H. POUKKA, J.J. PALVIMO & O. A. JANNE. 2002. Small ubiquitin-related modifier-1 (SUMO-1) modification of the glucocorticoid receptor. Biochem. J. **367:** 907–911.
102. LE DREAN, Y., N. MINCHENEAU, P. LE GOFF & D. MICHEL. 2002. Potentiation of glucocorticoid receptor transcriptional activity by sumoylation. Endocrinology **143:** 3482–3489.
103. HOLMSTROM, S., M.E. VAN ANTWERP & J.A. INIGUEZ-LLUHI. 2003. Direct and distinguishable inhibitory roles for SUMO isoforms in the control of transcriptional synergy. Proc. Natl. Acad. Sci. USA **100:** 15758–15763.
104. GALIGNIANA, M.D., G. PIWIEN-PILIPUK & J. ASSREUY. 1999. Inhibition of glucocorticoid receptor binding by nitric oxide. Mol. Pharmacol. **55:** 317–323.
105. BODWELL, J.E., E. ORTI, J.M. COULL, *et al.* 1991. Identification of phosphorylated sites in the mouse glucocorticoid receptor. J. Biol. Chem. **266:** 7549–7555.
106. ROGATSKY, I., C.L. WAASE & M.J. GARABEDIAN. 1998. Phosphorylation and inhibition of rat glucocorticoid receptor transcriptional activation by glycogen synthase kinase-3 (GSK-3). Species-specific differences between human and rat glucocorticoid receptor signaling as revealed through GSK-3 phosphorylation. J. Biol. Chem. **273:** 14315–14321.
107. WEBSTER, J.C., C.M. JEWELL, J.E. BODWELL, *et al.* 1997. Mouse glucocorticoid receptor phosphorylation status influences multiple functions of the receptor protein. J. Biol. Chem. **272:** 9287–9293.
108. WANG, Z., J. FREDERICK & M.J. GARABEDIAN. 2002. Deciphering the phosphorylation "code" of the glucocorticoid receptor *in vivo*. J. Biol. Chem. **277:** 26573–26580.
109. BODWELL, J.E., J.C. WEBSTER, C.M. JEWELL, *et al.* 1998. Glucocorticoid receptor phosphorylation: overview, function and cell cycle-dependence. J. Steroid Biochem. Mol. Biol. **65:** 91–99.

110. KING, K.L. & J.A. CIDLOWSKI. 1998. Cell cycle regulation and apoptosis. Annu. Rev. Physiol. **60:** 601–617.
111. CIDLOWSKI, J.A. & N.B. CIDLOWSKI. 1981. Regulation of glucocorticoid receptors by glucocorticoids in cultured HeLa S3 cells. Endocrinology **109:** 1975–1982.
112. SVEC, F. & M. RUDIS. 1981. Glucocorticoids regulate the glucocorticoid receptor in the AtT-20 cell. J. Biol. Chem. **256:** 5984–5987.
113. MCINTYRE, W.R. & H.H. SAMUELS. 1985. Triamcinolone acetonide regulates glucocorticoid-receptor levels by decreasing the half-life of the activated nuclear-receptor form. J. Biol. Chem. **260:** 418–427.
114. DONG, Y., L. POELLINGER, J.A. GUSTAFSSON & S. OKRET. 1988. Regulation of glucocorticoid receptor expression: evidence for transcriptional and posttranslational mechanisms. Mol. Endocrinol. **2:** 1256–1264.
115. HOECK, W. & B. GRONER. 1990. Hormone-dependent phosphorylation of the glucocorticoid receptor occurs mainly in the amino-terminal transactivation domain. J. Biol. Chem. **265:** 5403–5408.
116. BURNSTEIN, K.L., C.M. JEWELL & J.A. CIDLOWSKI. 1991. Evaluation of the role of ligand and thermal activation of specific DNA binding by *in vitro* synthesized human glucocorticoid receptor. Mol. Endocrinol. **5:** 1013–1022.
117. BURNSTEIN, K.L., D.L. BELLINGHAM, C.M. JEWELL, *et al.* 1991. Autoregulation of glucocorticoid receptor gene expression. Steroids **56:** 52–58.
118. BELLINGHAM, D.L., M. SAR & J.A. CIDLOWSKI. 1992. Ligand-dependent down-regulation of stably transfected human glucocorticoid receptors is associated with the loss of functional glucocorticoid responsiveness. Mol. Endocrinol. **6:** 2090–2102.
119. SILVA, C.M., F.E. POWELL-OLIVER, C.M. JEWELL, *et al.* 1994. Regulation of the human glucocorticoid receptor by long-term and chronic treatment with glucocorticoid. Steroids **59:** 436–442.
120. WALLACE, A.D. & J.A. CIDLOWSKI. 2001. Proteasome-mediated glucocorticoid receptor degradation restricts transcriptional signaling by glucocorticoids. J. Biol. Chem. **276:** 42714–42721.
121. BALDI, L., K. BROWN, G. FRANZOSO & U. SIEBENLIST. 1996. Critical role for lysines 21 and 22 in signal-induced, ubiquitin-mediated proteolysis of I kappa B-alpha. J. Biol. Chem. **271:** 376–379.
122. CHEN, Z., J. HAGLER, V.J. PALOMBELLA, *et al.* 1995. Signal-induced site-specific phosphorylation targets I kappa B alpha to the ubiquitin-proteasome pathway. Genes Dev. **9:** 1586–1597.
123. SKOWYRA, D., D.M. KOEPP, T. KAMURA, *et al.* 1999. Reconstitution of G1 cyclin ubiquitination with complexes containing SCFGrr1 and Rbx1. Science **284:** 662–665.
124. NIRMALA, P.B. & R.V. THAMPAN. 1995. Ubiquitination of the rat uterine estrogen receptor: dependence on estradiol. Biochem. Biophys. Res. Commun. **213:** 24–31.
125. SYVALA, H., A. VIENONEN, Y.H. ZHUANG, *et al.* 1998. Evidence for enhanced ubiquitin-mediated proteolysis of the chicken progesterone receptor by progesterone. Life Sci. **63:** 1505–1512.
126. WARD, C.L., S. OMURA & R.R. KOPITO. 1995. Degradation of CFTR by the ubiquitin-proteasome pathway. Cell **83:** 121–127.
127. KAUL, S., J.A. BLACKFORD, JR., S. CHO & S.S. SIMONS, JR. 2002. Ubc9 is a novel modulator of the induction properties of glucocorticoid receptors. J. Biol. Chem. **277:** 12541–12549.
128. SENGUPTA, S. & B. WASYLYK. 2001. Ligand-dependent interaction of the glucocorticoid receptor with p53 enhances their degradation by Hdm2. Genes Dev. **15:** 2367–2380.
129. RECHSTEINER, M. & S.W. ROGERS. 1996. PEST sequences and regulation by proteolysis. Trends Biochem. Sci. **21:** 267–271.
130. ROGERS, S., R. WELLS & M. RECHSTEINER. 1986. Amino acid sequences common to rapidly degraded proteins: the PEST hypothesis. Science **234:** 364–368.
131. TSURUMI, C., N. ISHIDA, T. TAMURA, *et al.* 1995. Degradation of c-Fos by the 26S proteasome is accelerated by c-Jun and multiple protein kinases. Mol. Cell Biol. **15:** 5682–5687.

132. LIN, R., P. BEAUPARLANT, C. MAKRIS, *et al.* 1996. Phosphorylation of IkappaBalpha in the C-terminal PEST domain by casein kinase II affects intrinsic protein stability. Mol. Cell Biol. **16:** 1401–1409.
133. ERNST, M.K., L.L. DUNN & N. R. RICE. 1995. The PEST-like sequence of I kappa B alpha is responsible for inhibition of DNA binding but not for cytoplasmic retention of c-Rel or RelA homodimers. Mol. Cell Biol. **15:** 872–882.
134. BODWELL, J.E., J.M. HU, E. ORTI & A. MUNCK. 1995. Hormone-induced hyperphosphorylation of specific phosphorylated sites in the mouse glucocorticoid receptor. J. Steroid Biochem. Mol. Biol. **52:** 135–140.
135. ALARID, E.T., N. BAKOPOULOS & N. SOLODIN. 1999. Proteasome-mediated proteolysis of estrogen receptor: a novel component in autologous down-regulation. Mol. Endocrinol. **13:** 1522–1534.
136. NAWAZ, Z., D.M. LONARD, A.P. DENNIS, *et al.* 1999. Proteasome-dependent degradation of the human estrogen receptor. Proc. Natl. Acad. Sci. USA **96:** 1858–1862.
137. LONARD, D.M., Z. NAWAZ, C.L. SMITH & B.W. O'MALLEY. 2000. The 26S proteasome is required for estrogen receptor-alpha and coactivator turnover and for efficient estrogen receptor-alpha transactivation [In Process Citation]. Mol. Cell. **5:** 939–948.
138. DAVARINOS, N.A. & R.S. POLLENZ. 1999. Aryl hydrocarbon receptor imported into the nucleus following ligand binding is rapidly degraded via the cytosplasmic proteasome following nuclear export. J. Biol. Chem. **274:** 28708–28715.
139. ROBERTS, B.J. & M.L. WHITELAW. 1999. Degradation of the basic helix-loop-helix/Per-ARNT-Sim homology domain dioxin receptor via the ubiquitin/proteasome pathway. J. Biol. Chem. **274:** 36351–36356.
140. DACE, A., L. ZHAO, K.S. PARK, *et al.* 2000. Hormone binding induces rapid proteasome-mediated degradation of thyroid hormone receptors. Proc. Natl. Acad. Sci. USA **97:** 8985–8990.
141. PRATT, W.B. 1993. The role of heat shock proteins in regulating the function, folding, and trafficking of the glucocorticoid receptor. J. Biol. Chem. **268:** 21455–21458.
142. JEWELL, C.M., J.C. WEBSTER, K.L. BURNSTEIN, *et al.* 1995. Immunocytochemical analysis of hormone mediated nuclear translocation of wild type and mutant glucocorticoid receptors. J. Steroid Biochem. Mol. Biol. **55:** 135–146.
143. HTUN, H., J. BARSONY, I. RENYI, *et al.* 1996. Visualization of glucocorticoid receptor translocation and intranuclear organization in living cells with a green fluorescent protein chimera. Proc. Natl. Acad. Sci. USA **93:** 4845–4850.
144. SAITOH, M., R. TAKAYANAGI, K. GOTO, *et al.* 2002. The presence of both the amino- and carboxyl-terminal domains in the AR is essential for the completion of a transcriptionally active form with coactivators and intranuclear compartmentalization common to the steroid hormone receptors: a three-dimensional imaging study. Mol. Endocrinol. **16:** 694–706.
145. TOMURA, A., K. GOTO, H. MORINAGA, *et al.* 2001. The subnuclear three-dimensional image analysis of androgen receptor fused to green fluorescence protein. J. Biol. Chem. **276:** 28395–28401.
146. TYAGI, R.K., Y. LAVROVSKY, S.C. AHN, *et al.* 2000. Dynamics of intracellular movement and nucleocytoplasmic recycling of the ligand-activated androgen receptor in living cells. Mol. Endocrinol. **14:** 1162–1174.
147. FEJES-TOTH, G., D. PEARCE & A. NARAY-FEJES-TOTH. 1998. Subcellular localization of mineralocorticoid receptors in living cells: effects of receptor agonists and antagonists. Proc. Natl. Acad. Sci. USA **95:** 2973–2978.
148. RACZ, A. & J. BARSONY. 1999. Hormone-dependent translocation of vitamin D receptors is linked to transactivation. J. Biol. Chem. **274:** 19352–19360.
149. VAN STEENSEL, B., M. BRINK, K. VAN DER MEULEN, *et al.* 1995. Localization of the glucocorticoid receptor in discrete clusters in the cell nucleus. J. Cell Sci. **108:** 3003–3011.
150. HTUN, H., L.T. HOLTH, D. WALKER, J.R. DAVIE & G.L. HAGER. 1999. Direct visualization of the human estrogen receptor alpha reveals a role for ligand in the nuclear distribution of the receptor. Mol. Biol. Cell. **10:** 471–486.

151. STENOIEN, D.L., M.G. MANCINI, K. PATEL, *et al.* 2000. Subnuclear trafficking of estrogen receptor-alpha and steroid receptor coactivator-1. Mol. Endocrinol. **14:** 518–534.
152. BAUMANN, C.T., P. MARUVADA, G.L. HAGER & P.M. YEN. 2001. Nuclear cytoplasmic shuttling by thyroid hormone receptors. multiple protein interactions are required for nuclear retention. J. Biol. Chem. **276:** 11237–11245.
153. MCNALLY, J.G., W.G. MULLER, D. WALKER, *et al.* 2000. The glucocorticoid receptor: rapid exchange with regulatory sites in living cells. Science **287:** 1262–1265.
154. BECKER, M., C. BAUMANN, S. JOHN, *et al.* 2002. Dynamic behavior of transcription factors on a natural promoter in living cells. EMBO Rep. **3:** 1188–1194.
155. MULLER, W.G., D. WALKER, G.L. HAGER & J.G. MCNALLY. 2001. Large-scale chromatin decondensation and recondensation regulated by transcription from a natural promoter. J. Cell Biol. **154:** 33–48.
156. PEDERSON, T. 2000. Half a century of "the nuclear matrix." Mol. Biol. Cell. **11:** 799–805.
157. NICKERSON, J. 2001. Experimental observations of a nuclear matrix. J. Cell. Sci. **114:** 463–474.
158. FACKELMAYER, F.O. & A. RICHTER. 1994. Purification of two isoforms of hnRNP-U and characterization of their nucleic acid binding activity. Biochemistry **33:** 10416–10422.
159. MATTERN, K.A., B.M. HUMBEL, A.O. MUIJSERS, *et al.* 1996. hnRNP proteins and B23 are the major proteins of the internal nuclear matrix of HeLa S3 cells. J. Cell. Biochem. **62:** 275–289.
160. EGGERT, M., J. MICHEL, S. SCHNEIDER, *et al.* 1997. The glucocorticoid receptor is associated with the RNA-binding nuclear matrix protein hnRNP U. J. Biol. Chem. **272:** 28471–28478.
161. TANG, Y., R.H. GETZENBERG, B.N. VIETMEIER, *et al.* 1998. The DNA-binding and tau2 transactivation domains of the rat glucocorticoid receptor constitute a nuclear matrix-targeting signal. Mol. Endocrinol. **12:** 1420–1431.
162. EGGERT, H., M. SCHULZ, F.O. FACKELMAYER, *et al.* 2001. Effects of the heterogeneous nuclear ribonucleoprotein U (hnRNP U/SAF-A) on glucocorticoid-dependent transcription *in vivo*. J. Steroid Biochem. Mol. Biol. **78:** 59–65.
163. CIDLOWSKI, J.A. & A. MUNCK. 1980. Multiple forms of nuclear binding of glucocorticoid-receptor complexes in rat thymocytes. J. Steroid Biochem. **13:** 105–112.
164. SCHAAF, M.J. & J.A. CIDLOWSKI. 2003. Molecular determinants of glucocorticoid receptor mobility in living cells: the importance of ligand affinity. Mol. Cell Biol. **23:** 1922–1934.
165. DEROO, B.J. & T.K. ARCHER. 2001. Glucocorticoid receptor-mediated chromatin remodeling *in vivo*. Oncogene **20:** 3039–3046.
166. KNOTTS, T.A., R.S. ORKISZEWSKI, R.G. COOK, *et al.* 2001. Identification of a phosphorylation site in the hinge region of the human progesterone receptor and additional amino-terminal phosphorylation sites. J. Biol. Chem. **276:** 8475–8483.

Overview of the Actions of Glucocorticoids on the Immune Response

A Good Model to Characterize New Pathways of Immunosuppression for New Treatment Strategies

D. FRANCHIMONT

Gastroenterology Department, Erasme University Hospital, 1070 Brussels, Belgium

ABSTRACT: Glucocorticoids have been used for over 50 years in the treatment of inflammatory and autoimmune diseases and in preventing graft rejection. Today, knowledge of their molecular, cellular, and pharmacological properties allows a better understanding of glucocorticoid-mediated immunosuppression. Glucocorticoids exert both negative and positive effects with a dynamic and bi-directional spectrum of activities on various limbs and components of the immune response. They modulate genes involved in the priming of the innate immune response, while their actions on the adaptive immune response are to suppress cellular (Th1) immunity and promote humoral (Th2) immunity. Interestingly, glucocorticoids can also induce tolerance to specific antigens by influencing dendritic cell maturation and function and promoting the development of regulatory high IL-10–producing T cells. The *ex vivo* therapeutic use of glucocorticoids could therefore represent an adjuvant treatment to cell therapy in autoimmune diseases, avoiding the long-term deleterious adverse effects of glucocorticoids. Thus, the panoramic view of glucocorticoid actions on the immune system provides an interesting model for characterizing important biological pathways of immunosuppression.

KEYWORDS: glucocorticoids; immune response; innate immunity; adaptive immunity; tolerance

INTRODUCTION

Integrating the study of glucocorticoids in the emerging concepts of immunology appears essential to allow a comprehensive understanding of glucocorticoids' actions on the immune response. Many of the studies carried out during the past 50 years have revealed the broad spectrum of actions of glucocorticoids on the immune response with diametrically opposite effects on several components of the immune system. Gathering molecular, cellular, and pharmacological knowledge may help portray the immunosuppression mediated by glucocorticoids (sometimes with over-simplification) and distinguish their positive from their negative immune actions.

Address for correspondence: Denis P. Franchimont, M.D., Department of Gastroenterology, Erasme University Hospital, Free University of Brussels, 808, Lennik Road, 1070 Brussels, Belgium. Voice: 32-2-555-3712; fax: 32-2-555-4697.

Denis.Franchimont@ulb.ac.be

Ann. N.Y. Acad. Sci. 1024: 124–137 (2004). © 2004 New York Academy of Sciences. doi: 10.1196/annals.1321.009

This overview summarizes the mechanism of actions of glucocorticoids on innate and adaptive immunity.

INNATE IMMUNITY

Cell Trafficking and Adhesion Molecules

Intravenous injection of glucocorticoids or acute cortisol release following stress greatly influences white cell trafficking during local or systemic inflammation. The redistribution of T cells, the increased blood leukocytosis, and the profound depletion of eosinophils and basophils result from the complex regulation of leukocyte and endothelial adhesion molecules and chemokines by glucocorticoids. These factors modulate bone marrow release and peripheral tissue infiltration and clearance of leukocytes. The increased neutrophilia and the decrease of eosinophils after glucocorticoid administration are familiar to most clinicians. In fact, glucocorticoids promote the survival and proliferation of neutrophils while they induce apoptosis of eosinophils and basophils.[1–5] Glucocorticoids enhance granulocyte colony–stimulating factor (GM–CSF), which in turn affects the proliferation and expansion of neutrophils. Leukotriene B4 (LTB4) also seems to be a critical survival factor since inhibitors of 5-lipo-oxygenase (5LO) or LTB4 antagonists prevent glucocorticoid-induced neutrophil survival.[6,7] Very interestingly, glucocorticoids upregulate the high-affinity leukocyte LTB4 receptor (BLT1) expression and enhance LTB4-mediated neutrophil survival and chemotaxis.[7] The glucocorticoid-induced neutrophilia is enhanced by increased release of bone-marrow polymorphonuclear cells and inhibition of neutrophil transmigration to inflammatory sites. Inhibition of neutrophil transmigration by glucocorticoids occurs through inhibition of leukocyte adhesion molecules, such as L-selectin.[8,9] L-selectin seems to be critical to glucocorticoid-mediated actions on neutrophil migration.[10,11] The immediate action of glucocorticoids on neutrophils is, however, not related to a direct regulation of L-selectin on circulating peripheral neutrophils.[9] In fact, glucocorticoids promote shedding of L-selectin from neutrophils, preventing their attachment to the endothelial cells. Glucocorticoids enhance lipocortin 1 (annexin 1) expression, which in turn would activate a metalloprotease "sheddase" that frees L-selectin from neutrophils.[12,13] This mechanism has been recently described and accounts for the early changes of neutrophilia during acute stress or glucocorticoid therapy.[14] Also, glucocorticoids decrease L-selectin expression on bone marrow progenitors and differentiating neutrophils.[9] Chemokines released from the inflamed tissue are critically important to activate neutrophils and initiate a program of cell transmigration. Inhibition of neutrophil transmigration by glucocorticoids also occurs through inhibition of chemokine release of IL-8 and other cysteine-X-cysteine (CXC) chemokines.[15–18]

Eosinophils, basophils, and Th2 cells are recruited by chemokines, released by immune and non-immune cells such as epithelial cells or airway smooth muscle cells. The β chemokines, or cysteine-cysteine (CC)-chemokines, enhance adhesion molecule expression on endothelial cells and bind their cognate receptor on eosinophils and basophils to allow them to roll through the vessel wall. In fact, glucocorticoids prevent tissue invasion of eosinophils by inhibiting the release of CC-chemokines, such as eotaxin, eotaxin 2, MCP-4, and RANTES released by bronchial epithelial cells.[19] Simultaneously, while glucocorticoids decrease the expression of some chemokine receptors, they enhance the expression of others on eosinophils further complicating the overall picture of eosinophil trafficking.[20]

The flow and movement of monocytes seem tightly regulated by glucocorticoids via similar mechanisms, namely through the repression of monocyte and endothelial adhesion molecules and the regulation of chemokines and their receptors.[8]

INNATE IMMUNE RESPONSE

Glucocorticoids act on the immune system by both suppressing and stimulating a large number of pro-inflammatory or anti-inflammatory mediators. In many ways, glucocorticoids lead to termination of inflammation by enhancing the clearance of foreign antigens, toxins, microorganisms, and dead cells. They do so by enhancing opsonization and the activity of scavenger systems, and by stimulating macrophage phagocytotic ability and antigen uptake.[1,21–23] Glucocorticoids stimulate the expression of the mannose receptor (MR) or the scavenger receptor CD163, promoting clearance of microorganisms, dead cell bodies and antigens.[24,25] They potentiate IFNγ-induced FcγRI. It is noteworthy that this effect is observed in monocytes but not in polymorphonuclear neutrophils.[26] At the same time, they prevent inflammation from overshooting by suppressing the synthesis of many inflammatory mediators, such as several cytokines and chemokines, prostaglandins, leukotrienes, proteolytic enzymes, free oxygen radicals, and nitric oxide.

A great number of cytokines (including IL-1β, TNFα, IL-6, IL-8, IL-12 and IL-18, etc.) is broadly downregulated by glucocorticoids. Similarly, secretion of many cysteine-cysteine (CC) and cysteine-X-cysteine (CXC) chemokines is strongly suppressed. Interestingly, anti-inflammatory cytokines such as IL-10 and TGFβ (and to some extent, although controversial, IL-1RA) are upregulated by glucocorticoids.[27–31] Although the negative regulation of pro-inflammatory cytokines by glucocorticoids has been clearly demonstrated both *in vitro* and *in vivo*, glucocorticoids enhance the receptor of some of these pro-inflammatory cytokines and chemokines.[30] It is tempting to speculate that they increase receptor expression to enhance the sensitivity of target immune cells to these pro-inflammatory mediators and help these immune cells to terminate inflammation. Soluble or decoy receptors, inhibiting or further enhancing the inflammatory process, are also regulated by glucocorticoids. For example, the decoy receptor IL-1RII, which binds IL-1 without driving any signaling, is enhanced by glucocorticoids.[32] This represents an anti-inflammatory mechanism of action of glucocorticoids.

An intriguing inflammatory mediator is the macrophage inhibitory factor (MIF) whose secretion is triggered by glucocorticoids. Synthesized by anterior pituitary cells and macrophages in response to endotoxin challenge and pro-inflammatory cytokines, such as TNFα, MIF exerts major pro-inflammatory response on macrophages and T cells and overrides the anti-inflammatory and immunosuppressive actions of glucocorticoids.[33,34] In fact, MIF increases lethality while genetic deletion or therapeutic neutralization confers protection to endotoxemia, ARDS, or septic shock.[35,36] That Toll-like receptor (TLR) 4 expression is increased by MIF underscores the essential role of MIF in the macrophage response to endotoxins and gram-negative bacteria.[37] Thus, glucocorticoids enhancing the secretion of a major pro-inflammatory cytokine such as MIF, which counteracts their effects, represent a yin/yang mechanism of the acute phase response and septic shock.

By inhibiting phospholipase A2, inducible cyclooxygenase (COX) 2, and inducible prostaglandin synthase (PGS) 2, they block the release of arachidonic acid and the synthesis of prostaglandins (PGE2, PGH2, and PGD2) and platelet activating factor (PAF); both constitutive COX 1 and PGS 1 remaining unaffected by glucocorticoids.[6,38,39] This explains why glucocorticoids alter late phase of type 1 hypersensitivity. Lipocortin 1 would mediate glucocorticoid inhibition of phospholipase A2. Several lines of evidence suggest that glucocorticoids prevent COX-2 expression through both posttranscriptional and posttranslational mechanisms.[40] Strikingly, glucocorticoids seem to enhance 5-lipooxygenase and 5-lipooxygenase activating protein (FLAP) expression in monocytes and eosinophils.[6,41–43] The overall reduced secretion of leukotrienes by glucocorticoids would therefore be related more to the inhibition of phospholipase A2 expression and activity. Glucocorticoids may also suppress nicotineamide adenine dinucleotide phosphate, the reduced form (NADPH) oxidase, and superoxide dismutase expression and, hence, the production of free oxygen radicals.[44] Nitric oxide appears as an important intra- and intercellular signaling molecule in shaping the innate and adaptive immune response with both detrimental and protective effects. Glucocorticoids suppress inducible nitric oxide synthetase expression, which results in a decrease of nitric oxide release by endothelial cells.[45] This inhibition seems also mediated by the glucocorticoid secondary messenger lipocortin 1 and would prevent early endothelial cell–mediated inflammatory reaction. Thus, glucocorticoids represent a potent anti-inflammatory agent given their homogeneous inhibition on arachidonic acid metabolites, free oxygen radicals, and nitric oxide.

ADAPTIVE IMMUNITY

Antigen Presentation and Adaptive Immune Response

Dendritic cells (DCs) represent the crucial interplay between innate and adaptive immunity. Upon encounter of microorganisms or antigens, tissue-resident DCs rapidly differentiate and migrate to secondary lymphoid organs.[46] These immature DCs demonstrate increased antigen uptake ability and specialized antigen-processing machinery.[47] Glucocorticoids enhance this "immature" phenotype. They improve opsonization and the activity of scavenger systems and stimulate macrophage phagocytosis, pinocytosis ability and antigen uptake.[25] Toll-like receptors mediate the pattern of microorganism antigen recognition in macrophages and both immature and mature DCs.[48–50] Interestingly, glucocorticoids modulate TLR expression, emphasizing their crucial role early in the innate immune response.[30]

During their migration, DCs mature and express major histocompatibility complex (MHC) class II and costimulatory molecules to efficiently present the antigen, as professional APCs (antigen presenting cells), to naïve or memory T cells.[51] A crosstalk between T and DCs through TCR/MHC II–bound antigen, costimulatory molecules, and cytokines allows the development of a T cell–immune response, generating T cell clonal expansion or deletion.[52] After exposure to glucocorticoids, these DCs have a decreased ability to present antigens and elicit a T cell response,

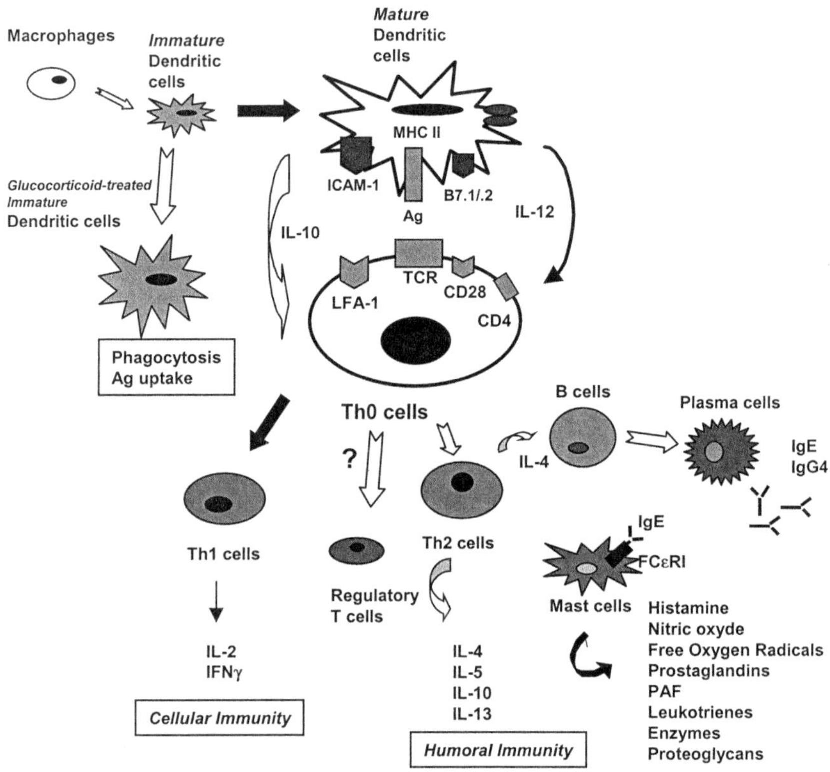

FIGURE 1. Effect of glucocorticoids on dendritic cell maturation and adaptive Th1 and Th2 immune response. *Open arrow*, stimulation by glucocorticoids; *filled arrow*, inhibition by glucocorticoids.

primarily because glucocorticoids prevent the upregulation of MHC class II and co-stimulatory molecules, including B7.2 (CD86), to some extent B7.1 (CD80), CD40, and the ICAM-1/LFA-1 complex[23,25,53] (FIG. 1). However, it is noteworthy that terminally differentiated DCs continue to express these molecules because of their relative resistance to glucocorticoids. The timing of exposure to glucocorticoids appears thus essential during dendritic cell maturation.[54]

Once a T cell immune response has flared up, glucocorticoids may modulate and interfere with its type (FIG. 1). The differentiation of CD4+ T cells into T helper (Th)1 lymphocytes, which drive cellular immunity, or into Th2 lymphocytes, which drive humoral immunity, depends on the type of antigen encountered and the type of cytokines produced during antigen presentation. Indeed, glucocorticoids block IL-12 secretion by monocytes and DCs. Interleukin 12 is critical to the development of Th1-directed cellular immune response and, hence, is the link between innate and cellular immunity.[55,56] On the other hand, glucocorticoids promote Th2 development by enhancing IL-10 secretion by macrophages and, somehow, by dendritic cells. In fact, several studies clearly showed both in mice and humans that the presence of glucocorticoids during the primary immune response enhanced Th2 cytokine and decreased Th1 cytokine secretion by CD4+ lymphocytes upon secondary stimulation.[31,55,57] This effect seemed IL-12–dependent since it was reversed by the addition of exogenous IL-12 to glucocorticoid-treated APCs during the primary stimulation. Moreover, the poor glucocorticoid regulation of costimulatory molecules in mature DCs ruled out any role of costimulation in this process.[22] Recently, terminally differentiated DCs were classified into DC 1 and DC 2, depending on their cytokine secretion profile and their aptitude for forcing the differentiation of naïve T cells into Th1 and Th2 mature lymphocytes (see below). Thus, DC phenotypes may be altered by glucocorticoids into DC 2 program, which will ultimately generate a Th2 immune response and the secretion of Th2 cytokines.

CELLULAR IMMUNE RESPONSE

Interleukin-12 is critical for Th1 lymphocyte differentiation and secretion of Th1 cytokines such as IFNγ and TNFα. Alterations in IL-12 signaling, such as observed in IL-12–/– and IL-12 R–deficient mice, are associated with a defective Th1 immune response. In fact, glucocorticoids not only block IL-12 secretion by monocytes/macrophages and dendritic cells but also suppress IL-12 R β1 and β2 expression on T cells.[58] They also interfere with IL-12 signaling and prevent IL-12–induced signal transducer and activator of transcription (Stat) 4 phosphorylation and Stat4-mediated expression of dependent genes, such as IRF-1.[59] Blocking Stat4 activation mimics Stat4-deficient mice, which are unable to elicit a Th1 immune response.[60] Glucocorticoids also profoundly suppress secretion of the Th1 cytokines IFNγ and TNFα, lessening NK and T cytotoxic effector functions. Such massive inhibition of the Th1 immune response by glucocorticoids may lead to severe cellular immunodeficiency and impaired defense to intracellular and opportunistic infections.

HUMORAL IMMUNE RESPONSE

While glucocorticoids profoundly decrease the secretion of Th2 cytokines during primary stimulation, they promote Th2 differentiation during secondary stimulation (FIG. 1). Independent of monocytes/macrophages and dendritic cells, glucocorticoids prime naïve T cells to Th2 commitment upon secondary immune response.[31,55–57] Glucocorticoids present during primary stimulation promote IL-10 secretion during secondary stimulation in both naïve and memory T cells. Sequential exposure to glucocorticoids during primary and/or secondary immune response may influence the pattern of Th2 cytokines. This context-dependent action of glucocorticoids may explain some discrepancies reported in the literature. Very interestingly, by upregulating IL-10 secretion by macrophage/DCs and Th2 cells, glucocorticoids may participate to the emergence of regulatory T cells, high IL-10–producing T cells with major *in vivo* immunoregulatory properties, on experimental allergic encephalomyelitis.[61]

Thus, following exposure to endogenous or exogenous glucocorticoids, a progressive shift takes place from a cellular Th1 to a humoral Th2 immune response. A major question remains—how do glucocorticoids induce Th2 lymphocyte differentiation and humoral immune response? First, it is well recognized that the lack of commitment to develop a Th1 immune response, namely the absence of IL-12, is associated with Th2 development. Therefore, the profound suppression of Th1 differentiation by glucocorticoids may undoubtedly participate to Th2 immune response development. Second, glucocorticoids also have differential action on IL-12 and IL-4 signaling. Interleukin 12 and IL-4 activates Stat4 and Stat6, respectively. Deletion of Stat4 or Stat6 in mice is associated with poor Th1 and Th2 immune responses, respectively. In fact, glucocorticoids block IL-12–mediated Stat4 activation without altering IL-4–induced Stat6 phosphorylation and, hence, help a Th2 immune response to develop. Finally, a growing number of transcription factors have been shown to play a critical role in the hierarchical control of Th1 and Th2 lymphocyte differentiation. Some important Th2 transcription factors could be regulated by glucocorticoids. Thus, by directly acting either on lymphocytes or dendritic cells, endogenous hypersecretion or administration of excessive amounts of glucocorticoids may cause a progressive shift from a Th1-directed cellular toward a Th2-driven humoral immune response.

Glucocorticoids definitely favor humoral immune response and antibody production. *In vivo* administration of glucocorticoids raise IgE serum levels in asthma or atopic patients.[62–64] Despite the paucity of information, glucocorticoids modulate B cell development. They restrain B cell proliferation and early steps of B cell development, but promote generation of antibody secreting–plasma cells and secretion of IgE and IgG4. Similarly to dendritic cells and effector T cells, B cells become resistant to the inhibitory actions of glucocorticoids as they proceed along the different stages of differentiation and maturation.[65,66] Surprisingly, how these hormones influence B cell receptor (BCR) signaling has never been explored, underscoring the lack of interest on glucocorticoid-mediated actions on B cells. Interleukin-4 is the critical cytokine not only for Th2 differentiation but also for B cell differentiation and IgE isotype switching (FIG. 1). Interestingly, glucocorticoids act in synergy with IL-4 on B cell differentiation and isotype switching, leading to immunoglobin (Ig) E–secreting B cells.[67–69] Such IgE isotype switching is dependent on the CD40/CD40 ligand complex since it is not observed in X-linked hyperIgM (CD40L-deficient) patients. In fact, both IL-4 and glucocorticoids upregulate CD40L expression on B cells and explain their synergistic actions on IgE isotype switching.[70]

ALLERGY-MEDIATED IMMUNE RESPONSE AND INFLAMMATION

That glucocorticoids are used in the treatment of atopy and asthma and promote humoral Th2 immune response and IgE secretion represents a disturbing and challenging paradox. In truth, glucocorticoids may enhance IgE secretion but strongly suppress allergic inflammation and chemokine–driven tissue infiltration of eosinophils.[71] Furthermore, it has been suggested that glucocorticoids would inhibit antigen-specific IgE production while raising total IgE level.[62] This may clarify why IgE serum levels remained high in asthma patients while in clinical remission under steroids. While they prime DCs and T cells for Th2 development upon secondary stimulation, glucocorticoids inhibit Th2 cytokine secretion during primary antigen exposure. Despite an ongoing Th2 immune response, glucocorticoids may still prevent the potentially deleterious IgE-induced allergic immune response by mast cells. Indeed, they interfere with IgE R–mediated release of inflammatory mediators and deplete bronchial mucosa from resident mast cells.[72] *In vivo* studies suggest that glucocorticoids deplete tissue mast cells by inhibiting essential survival factors, such as IL-4 and the fibroblast-derived stem cell factor (SCF).[73] Mucosal mast cells are the predominant effector cells orchestrating allergic inflammation. Crosslinking of their FcεRI with specific IgE results in degranulation of preformed inflammatory mediators, such as histamine, proteases, cytokines, and lipid mediators, responsible for the early phase of type I hypersensitivity manifestations. Interestingly, glucocorticoids block IgE-triggered degranulation of human mast cells. They also prevent late phase of type 1 hypersensitivity characterized by mast cell activation with *de novo* synthesis of inflammatory mediators and secondary infiltration of Th2 cells, basophils, and eosinophils. Glucocorticoid-induced downregulation of high-affinity FcεRI and low-affinity FcεRI (CD23) expression may only account for the poor response of IgE-triggered mast cell activation in the late phase.[74] In fact, glucocorticoids interfere with FcεRI signaling through the disruption of raf-1/hsp90 and the subsequent MAP kinase activation and phospholipase A2 responsible for the *de novo* synthesis of arachidonic acid–derived metabolites.[75] In the late phase, glucocorticoids inhibit cytokine synthesis of pro-inflammatory cytokines and chemokines by mast cells necessary for their own survival and the chemotaxis and expansion of eosinophils and basophils.[74,76–78]

T CELL DEVELOPMENT AND HOMEOSTASIS

While supra-pharmacologic doses of glucocorticoids induce T cell apoptosis, adrenal or thymus-derived glucocorticoids or physiologic doses of glucocorticoids can induce T cell survival or apoptosis, depending on T cell type and differentiation stage. Moreover, both the degree of T cell activation and the timing of glucocorticoid exposure (before, during, or after activation) render T cells sensitive or resistant to glucocorticoid-induced apoptosis. Several studies have shown that concomitant T cell receptor (TCR) and glucocorticoid receptor (GR) signaling promote T cell survival, whereas either TCR signaling alone or GR signaling alone induces T cell apoptosis.[79] Indeed, thymus-derived glucocorticoids appear to play a role in early thymocyte expansion and in central positive selection (see Ref. 48 for review). Yet, $GR^{dim-/-}$ knockout mice demonstrate normal thymus, suggesting that glucocorticoid actions on TCR signaling could occur through nongenomic actions that are independent of GR transcriptional activity.[80,81] Similarly, there is some evidence suggesting that glucocorticoids could influence peripheral T cell development and selection by simultaneously preventing TCR-induced T cell deletion and enhancing T cell survival.[82] Interestingly, glucocorticoids enhance a key cytokine receptor for T cell development, IL-7Ra, whose deletion in mice and humans is associated with lack of T cells.[83–86] Moreover, IL-7 potently enhances thymic-independent peripheral expansion and restores immunity in athymic T-cell–depleted hosts in mice.[87] The positive regulation of IL-7Rα expression by glucocorticoids thus suggests their influence in the maintenance of peripheral T cell pool homeostasis.

TOLERANCE

The beneficial therapeutic use of glucocorticoids in patients with autoimmune diseases and acute/chronic graft rejection could be partially explained by a pharmacologically induced tolerance. Mice with deficient HPA axis or pharmacologically induced glucocorticoid resistance develop increased susceptibility to experimental autoimmune diseases.[88] While glucocorticoids could contribute to immune tolerance by triggering T cell apoptosis (peripheral negative selection, see above) or by mediating T cell anergy, they could also promote the development of suppressive T cells. Glucocorticoids strongly inhibit antigen presentation, suppress class II MHC and co-stimulatory molecule expression such as B7.2 (and B7.1), and prevent DC maturation necessary to elicit a specific T cell immune response.[22] Yet immature DCs favor not only T cell anergy but also the emergence of suppressive regulatory T cells.[89] Interleukin-10 has been shown to induce a Th1 to Th2 shift and protect against organ-specific autoimmune diseases in mice. In macrophages/immature DCs, glucocorticoids enhance the expression of IL-10, known to induce the differentiation of suppressor/regulatory T cells (Tregs) such as high IL-10–producing T cells.[90] Similarly, glucocorticoids enhance IL-10 secretion by CD4+ T cells.[31,55,56] The immunosuppressive action of glucocorticoids on a specific T cell response disappears in the presence of anti–IL-10, suggesting the critical role of IL-10 in glucocorticoid-induced T cell tolerance and anergy.[91] Glucocorticoids might therefore participate in the differentiation and expansion of these high IL-10–producing Tregs through their combined actions on DC maturation and on DC- and T cell–mediated IL-10 secretion. Importantly, a very recent study has demonstrated that glucocorticoids, together with vitamin D3, could favor *in vitro* the emergence of high IL-10–producing T cells with major *in vivo* immunoregulatory properties on experimental allergic encephalomyelitis (EAE).[61] Whether alone or in synergy with other additional mediators, glucocorticoids appear to influence *in vivo* the differentiation of Tregs, which could in turn participate to the glucocorticoid-mediated tolerance.

CONCLUDING REMARKS

The description of the actions of glucocorticoids on the immune response demonstrates their positive and negative effects on several components of the innate and adaptive immune response. Glucocorticoid-induced immunomodulation requires both immunoenhancing and immunosuppressive actions simultaneously and should be integrated in a dynamic ongoing process. Indeed, pro-inflammatory mediators participate to the same extent as of anti-inflammatory mediators in the so-called immunosuppressive actions of glucocorticoids. Unraveling their mechanisms of action allows a better understanding of their beneficial use in various inflammatory, autoimmune, allergic, and infectious diseases. Favoring innate immune and antibody responses may explain their clinical benefit in infectious diseases, such as in bacterial meningitis. They strongly suppress molecular mediators and deplete the cellular component involved in allergic inflammation, which helps explain their efficacy in asthma. Similarly, that glucocorticoids enhance phagocytosis of neutrophils and macrophages, upregulating the scavenging and opsonization systems, which underscores the clinical benefit and importance of early and continuous administration of glucocorticoids in patients with septic shock. In Th1 autoimmune diseases, they restrain Th1-driven immune responses through a multistep inhibitory process from DC maturation to cytokine expression and signaling. Importantly, their beneficial effects coexist with severe deleterious adverse effects. Because of their suppressive actions on Th1 immunity, they increase susceptibility to intracellular and opportunistic infections, such as tuberculosis or *Pneumocystis carinii* pneumonia. In addition, glucocorticoids may even worsen disease activity due to their selective Th2 immunoenhancing actions. This was sometimes reported in steroid-resistant asthma and ulcerative colitis.

The overall knowledge of the actions of glucocorticoids on adaptive immunity clearly revealed that by decreasing dendritic cell ability to elicit a T cell immune response, by promoting Th2 differentiation, and by helping the development of regulatory T cells, glucocorticoids may induce tolerance. Thus, glucocorticoids could be extremely helpful in cell therapy through their *ex vivo* action not only on dendritic cells but also potentially on regulatory T cells.[92] This *ex vivo* therapeutic use of glucocorticoids would be very beneficial thanks to their immune modifier effect, while avoiding their long-term deleterious adverse effects. Glucocorticoids could represent an adjuvant therapy for cell therapies in autoimmune diseases and organ transplantations.

REFERENCES

1. Cox, G. 1995. Glucocorticoid treatment inhibits apoptosis in human neutrophils. Separation of survival and activation outcomes. J. Immunol. **154:** 4719–4725.
2. Liles, W.C., D.C. Dale & S.J. Klebanoff. 1995. Glucocorticoids inhibit apoptosis of human neutrophils. Blood **86:** 3181–3188.
3. Meagher, L.C., J.M. Cousin, J.R. Seckl & C. Haslett. 1996. Opposing effects of glucocorticoids on the rate of apoptosis in neutrophilic and eosinophilic granulocytes. J. Immunol. **156:** 4422–44428.
4. Schleimer, R.P. & B.S. Bochner. 1994. The effects of glucocorticoids on human eosinophils. J. Allergy Clin. Immunol. **94:** 1202–1213.
5. Yoshimura, C., M. Miyamasu, H. Nagase, *et al.* 2001. Glucocorticoids induce basophil apoptosis. J. Allergy Clin. Immunol. **108:** 215–220.
6. Riddick, C.A., W.L. Ring, J.R. Baker, *et al.* 1997 Dexamethasone increases expression of 5-lipoxygenase and its activating protein in human monocytes and THP-1 cells. Eur. J. Biochem. **246:** 112–118.
7. Stankova, J., S. Turcotte, J. Harris, *et al.* 2002. Modulation of leukotriene B4 receptor-1 expression by dexamethasone: potential mechanism for enhanced neutrophil survival. J. Immunol. **168:** 3570–3576.
8. Cronstein, B.N., S.C. Kimmel, R.I. Levin, *et al.* 1992. A mechanism for the antiinflammatory effects of corticosteroids: the glucocorticoid receptor regulates leukocyte adhesion to endothelial cells and expression of endothelial-leukocyte adhesion molecule 1 and intercellular adhesion molecule 1. Proc. Natl. Acad. Sci. USA **89:** 9991–9995.
9. Nakagawa, M., G.P. Bondy, D. Waisman, *et al.* 1999. The effect of glucocorticoids on the expression of L-selectin on polymorphonuclear leukocyte. Blood **93:** 2730–2737.
10. Filep, J.G., A. Delalandre, Y. Payette & E. Foldes-Filep. 1997. Glucocorticoid receptor regulates expression of L-selectin and CD11/CD18 on human neutrophils. Circulation **96:** 295–301.
11. Jilma, B. & P. Stohlawetz. 1998. Dexamethasone downregulates L-selectin *in vitro* and *in vivo*. Circulation **97:** 2279–2281.

12. PERRETTI, M., J.D. CROXTALL, S.K. WHELLER, *et al.* 1996. Mobilizing lipocortin 1 in adherent human leukocytes downregulates their transmigration. Nat. Med. **2:** 1259–1262.
13. STRAUSBAUGH, H.J. & S.D. ROSEN. 2001. A potential role for annexin 1 as a physiologic mediator of glucocorticoid-induced L-selectin shedding from myeloid cells. J. Immunol. **166:** 6294–6300.
14. STRAUSBAUGH, H.J., P.G. GREEN, E. LO, *et al.* 1999. Painful stimulation suppresses joint inflammation by inducing shedding of L-selectin from neutrophils. Nat. Med. **5:** 1057–1061.
15. GRENIER, A., M. DEHOUX, A. BOUTTEN, *et al.* 1999. Oncostatin M production and regulation by human polymorphonuclear neutrophils. Blood **93:** 1413–1421.
16. WERTHEIM, W.A., S.L. KUNKEL, T.J. STANDIFORD, *et al.* 1993. Regulation of neutrophil-derived IL-8: the role of prostaglandin E2, dexamethasone, and IL-4. J. Immunol. **151:** 2166–2175.
17. STELLATO, C., L.A. BECK, G.A. GORGONE, *et al.* 1995. Expression of the chemokine RANTES by a human bronchial epithelial cell line. Modulation by cytokines and glucocorticoids. J. Immunol. **155:** 410–418.
18. PYPE, J.L., L.J. DUPONT, P. MENTEN, *et al.* 1999. Expression of monocyte chemotactic protein (MCP)-1, MCP-2, and MCP-3 by human airway smooth-muscle cells. Modulation by corticosteroids and T-helper 2 cytokines. Am. J. Respir. Cell Mol. Biol. **21:** 528–536.
19. JAHNSEN, F.L., R. HAYE, E. GRAN, *et al.* 1999. Glucocorticosteroids inhibit mRNA expression for eotaxin, eotaxin-2, and monocyte-chemotactic protein-4 in human airway inflammation with eosinophilia. J. Immunol. **163:** 1545–1551.
20. NAGASE, H., M. MIYAMASU, M. YAMAGUCHI, *et al.* 2000. Glucocorticoids preferentially upregulate functional CXCR4 expression in eosinophils. J. Allergy Clin. Immunol. **106:** 1132–1139.
21. LIU, Y., J.M. COUSIN, J. HUGHES, *et al.* 1999. Glucocorticoids promote nonphlogistic phagocytosis of apoptotic leukocytes. J. Immunol. **162:** 3639–3646.
22. PIEMONTI, L., P. MONTI, P. ALLAVENA, *et al.* 1999. Glucocorticoids affect human dendritic cell differentiation and maturation. J. Immunol. **162:** 6473–6481.
23. VAN DER GOES, A., K. HOEKSTRA, T.K. VAN DEN BERG & C.D. DIJKSTRA. 2000. Dexamethasone promotes phagocytosis and bacterial killing by human monocytes/macrophages *in vitro*. J. Leukoc. Biol. **67:** 801–807.
24. HOGGER, P., J. DREIER, A. DROSTE, *et al.* 1998. Identification of the integral membrane protein RM3/1 on human monocytes as a glucocorticoid-inducible member of the scavenger receptor cysteine-rich family (CD163). J. Immunol. **161:** 1883–1890.
25. PIEMONTI, L., P. MONTI, P. ALLAVENA, *et al.* 1999. Glucocorticoids increase the endocytic activity of human dendritic cells. Int. Immunol. **11:** 1519–1526.
26. PAN, L.Y., D.B. MENDEL, J. ZURLO & P.M. GUYRE. 1990. Regulation of the steady state level of Fc gamma RI mRNA by IFN-gamma and dexamethasone in human monocytes, neutrophils, and U-937 cells. J. Immunol. **145:** 267–275.
27. BATUMAN, O.A., A. FERRERO, C. CUPP, *et al.* 1995. Differential regulation of transforming growth factor beta-1 gene expression by glucocorticoids in human T and glial cells. J. Immunol. **155:** 4397–4405.
28. ELENKOV, I.J., D.A. PAPANICOLAOU, R.L. WILDER & G.P. CHROUSOS. 1996. Modulatory effects of glucocorticoids and catecholamines on human interleukin-12 and interleukin-10 production: clinical implications. Proc. Assoc. Am. Physicians **108:** 374–381.
29. FRANCHIMONT, D., H. MARTENS, M.T. HAGELSTEIN, *et al.* 1999. Tumor necrosis factor alpha decreases, and interleukin-10 increases, the sensitivity of human monocytes to dexamethasone: potential regulation of the glucocorticoid receptor. J. Clin. Endocrinol. Metab. **84:** 2834–2839.
30. GALON, J., D. FRANCHIMONT, N. HIROI, *et al.* 2002. Gene profiling reveals unknown enhancing and suppressive actions of glucocorticoids on immune cells. FASEB J. **16:** 61–71.
31. RAMIERZ, F., D.J. FOWELL, M. PUKLAVEC, *et al.* 1996. Glucocorticoids promote a TH2 cytokine response by CD4+ T cells *in vitro*. J. Immunol. **156:** 2406–2412.

32. RE, F., M. MUZIO, M. DE ROSSI, *et al.* 1994. The type II "receptor" as a decoy target for interleukin 1 in polymorphonuclear leukocytes: characterization of induction by dexamethasone and ligand binding properties of the released decoy receptor. J. Exp. Med. **179:** 739–743.
33. BERNHAGEN, J., T. CALANDRA, R.A. MITCHELL, *et al.* 1993. MIF is a pituitary-derived cytokine that potentiates lethal endotoxaemia. Nature **365:** 756–759.
34. CALANDRA, T., J. BERNHAGEN, C.N. METZ, *et al.* 1995. MIF as a glucocorticoid-induced modulator of cytokine production. Nature **377:** 68–71.
35. CALANDRA, T., B. ECHTENACHER, D.L. ROY, *et al.* 2000. Protection from septic shock by neutralization of macrophage migration inhibitory factor. Nat. Med. **6:** 164–170.
36. DONNELLY, S.C., C. HASLETT, P.T. REID, *et al.* 1997. Regulatory role for macrophage migration inhibitory factor in acute respiratory distress syndrome. Nat. Med. **3:** 320–323.
37. ROGER, T., J. DAVID, M.P. GLAUSER & T. CALANDRA. 2001. MIF regulates innate immune responses through modulation of Toll-like receptor 4. Nature **414:** 920–924.
38. SEBALDT, R.J., J.R. SHELLER, J.A. OATES, *et al.* 1990. Inhibition of eicosanoid biosynthesis by glucocorticoids in humans. Proc. Natl. Acad. Sci. USA **87:** 6974–6978.
39. YOSS, E.B., E.W. SPANNHAKE, J.T. FLYNN, *et al.* 1990. Arachidonic acid metabolism in normal human alveolar macrophages: stimulus specificity for mediator release and phospholipid metabolism, and pharmacologic modulation *in vitro* and *in vivo*. Am. J. Respir. Cell Mol. Biol. **2:** 69–80.
40. NEWTON, R., J. SEYBOLD, L.M. KUITERT, *et al.* 1998. Repression of cyclooxygenase-2 and prostaglandin E2 release by dexamethasone occurs by transcriptional and post-transcriptional mechanisms involving loss of polyadenylated mRNA. J. Biol. Chem. **273:** 32312–32321.
41. COWBURN, A.S., S.T. HOLGATE & A.P. SAMPSON. 1999. AP. IL-5 increases expression of 5-lipoxygenase-activating protein and translocates 5-lipoxygenase to the nucleus in human blood eosinophils. J. Immunol. **163:** 456–465.
42. POULIOT, M., P.P. MCDONALD, P. BORGEAT & S.R. MCCOLL. 1994. Granulocyte/macrophage colony-stimulating factor stimulates the expression of the 5-lipoxygenase-activating protein (FLAP) in human neutrophils. J. Exp. Med. **179:** 1225–1232.
43. UZ, T., Y. DWIVEDI, A. QELI, *et al.* 2001. Glucocorticoid receptors are required for up-regulation of neuronal 5-lipoxygenase (5LOX) expression by dexamethasone. FASEB J. **15:** 1792–1794.
44. AMEZAGA, M.A., F. BAZZONI, C. SORIO, *et al.* 1992. Evidence for the involvement of distinct signal transduction pathways in the regulation of constitutive and interferon gamma-dependent gene expression of NADPH oxidase components (gp91-phox, p47-phox, and p22-phox) and high-affinity receptor for IgG (Fc gamma R-I) in human polymorphonuclear leukocytes. Blood **79:** 735–744.
45. SALVEMINI, D., P.T. MANNING, B.S. ZWEIFEL, *et al.* 1995. Dual inhibition of nitric oxide and prostaglandin production contributes to the antiinflammatory properties of nitric oxide synthase inhibitors. J. Clin. Invest. **96:** 301–308.
46. PALUCKA, K. & J. BANCHEREAU. 1999. J. Linking innate and adaptive immunity. Nat. Med. **5:** 868–870.
47. MELLMAN, I. & R.M. STEINMAN. 2001. Dendritic cells: specialized and regulated antigen processing machines. Cell **106:** 255–258.
48. ADEREM, A. & R.J. ULEVITCH. 2000. Toll-like receptors in the induction of the innate immune response. Nature **406:** 782–787.
49. KADOWAKI, N., S. HO, S. ANTONENKO, *et al.* 2001. Subsets of human dendritic cell precursors express different toll-like receptors and respond to different microbial antigens. J. Exp. Med. **194:** 863–869.
50. PULENDRAN, B., K. PALUCKA & J. BANCHEREAU. 2001. Sensing pathogens and tuning immune responses. Science **293:** 253–256.
51. BANCHEREAU, J. & R.M. STEINMAN. 1998. Dendritic cells and the control of immunity. Nature **392:** 245–252.
52. LANZAVECCHIA, A. & F. SALLUSTO. 2001. Regulation of T cell immunity by dendritic cells. Cell **106:** 263–266.

53. VANDERHEYDE, N., V. VERHASSELT, M. GOLDMAN & F. WILLEMS. 1999. Inhibition of human dendritic cell functions by methylprednisolone. Transplantation **67:** 1342–1347.
54. MATYSZAK, M.K., S. CITTERIO, M. RESCIGNO & P. RICCIARDI-CASTAGNOLI. 2000. Differential effects of corticosteroids during different stages of dendritic cell maturation. Eur. J. Immunol. **30:** 1233–1242.
55. BLOTTA, M.H., R.H. DEKRUYFF & D.T. UMETSU. 1997. Corticosteroids inhibit IL-12 production in human monocytes and enhance their capacity to induce IL-4 synthesis in CD4+ lymphocytes. J. Immunol. **158:** 5589–5595.
56. VIEIRA, P.L., P. KALINSKI, E.A. WIERENGA, *et al.* 1998. Glucocorticoids inhibit bioactive IL-12p70 production by *in vitro*-generated human dendritic cells without affecting their T cell stimulatory potential. J. Immunol. **161:** 5245–5251.
57. DEKRUYFF, R.H., Y. FANG & D.T. UMETSU. 1998. Corticosteroids enhance the capacity of macrophages to induce Th2 cytokine synthesis in CD4+ lymphocytes by inhibiting IL-12 production. J. Immunol. **160:** 2231–2237.
58. WU, C.Y., K. WANG, J.F. MCDYER & R.A. SEDER. 1998. Prostaglandin E2 and dexamethasone inhibit IL-12 receptor expression and IL-12 responsiveness. J. Immunol. **161:** 2723–2730.
59. FRANCHIMONT, D., J. GALON, M. GADINA, *et al.* 2000. Inhibition of Th1 immune response by glucocorticoids: dexamethasone selectively inhibits IL-12-induced Stat4 phosphorylation in T lymphocytes. J. Immunol. **164:** 1768–1774.
60. KAPLAN, M.H., Y.L. SUN, T. HOEY & M.J. GRUSBY. 1996. Impaired IL-12 responses and enhanced development of Th2 cells in Stat4-deficient mice. Nature **382:** 174–177.
61. BARRAT, F.J., D.J. CUA, A. BOONSTRA, *et al.* 2002. *In vitro* generation of interleukin 10-producing regulatory CD4(+) T cells is induced by immunosuppressive drugs and inhibited by T helper type 1 (Th1)- and Th2-inducing cytokines. J. Exp. Med. **195:** 603–616.
62. AKDIS, C.A., T. BLESKEN, M. AKDIS, *et al.* 2001. Glucocorticoids inhibit human antigen-specific and enhance total IgE and IgG4 production due to differential effects on T and B cells *in vitro*. Eur. J. Immunol. **27:** 2351–2357.
63. BARNES, P.J. 2001. Corticosteroids, IgE, and atopy. J. Clin. Invest. **107:** 265–266.
64. KIMATA, H., I. LINDLEY & K. FURUSHO. 1995. Effect of hydrocortisone on spontaneous IgE and IgG4 production in atopic patients. J. Immunol. **154:** 3557–3566.
65. CUPPS, T.R., L.C. EDGAR, C.A. THOMAS & A.S. FAUCI. 1984. Multiple mechanisms of B cell immunoregulation in man after administration of *in vivo* corticosteroids. J. Immunol. **132:** 170–175.
66. CUPPS, T.R., T.L. GERRARD, R.J. FALKOFF, *et al.* 1985. Effects of *in vitro* corticosteroids on B cell activation, proliferation, and differentiation. J. Clin. Invest. **75:** 754–761.
67. JABARA, H.H., D.J. AHERN, D. VERCELLI & R.S. GEHA. 1991. Hydrocortisone and IL-4 induce IgE isotype switching in human B cells. J. Immunol. **147:** 1557–1560.
68. JABARA, H.H., R. LOH, N. RAMESH, *et al.* 1993. Sequential switching from mu to epsilon via gamma 4 in human B cells stimulated with IL-4 and hydrocortisone. J. Immunol. **151:** 4528–4533.
69. WU, C.Y., M. SARFATI, C. HEUSSER, *et al.* 2001. Glucocorticoids increase the synthesis of immunoglobulin E by interleukin 4-stimulated human lymphocytes. J. Clin. Invest. **87:** 870–877.
70. JABARA, H.H., S.R. BRODEUR & R.S. GEHA. 2001. Glucocorticoids upregulate CD40 ligand expression and induce CD40L-dependent immunoglobulin isotype switching. J. Clin. Invest. **107:** 371–378.
71. MIYAMASU, M., Y. MISAKI, S. IZUMI, *et al.* 1998. Glucocorticoids inhibit chemokine generation by human eosinophils. J. Allergy Clin. Immunol. **101:** 75–83.
72. MURAKAMI, M., C.O. BINGHAM, 3RD, R. MATSUMOTO, *et al.* 1995. IgE-dependent activation of cytokine-primed mouse cultured mast cells induces a delayed phase of prostaglandin D2 generation via prostaglandin endoperoxide synthase-2. J. Immunol. **155:** 4445–4453.

73. FINOTTO, S., Y.A. MEKORI & D.D. METCALFE. 1997. Glucocorticoids decrease tissue mast cell number by reducing the production of the c-kit ligand, stem cell factor, by resident cells: *in vitro* and *in vivo* evidence in murine systems. J. Clin. Invest. **99:** 1721–1728.
74. EKLUND, K.K., D.E. HUMPHRIES, Z. XIA, *et al.* 1997. Glucocorticoids inhibit the cytokine-induced proliferation of mast cells, the high affinity IgE receptor-mediated expression of TNF-alpha, and the IL-10-induced expression of chymases. J. Immunol. **158:** 4373–4380.
75. CISSEL, D.S. & M.A. BEAVEN. 2000. Disruption of Raf-1/heat shock protein 90 complex and Raf signaling by dexamethasone in mast cells. J. Biol. Chem. **275:** 7066–7070.
76. SCHWIEBERT, L.M., L.A. BECK, C. STELLATO, *et al.* 1996. Glucocorticosteroid inhibition of cytokine production: relevance to antiallergic actions. J. Allergy Clin. Immunol. **97:** 143–152.
77. WERSHIL, B.K., G.T. FURUTA, J.A. LAVIGNE, *et al.* 1995. Dexamethasone or cyclosporin A suppress mast cell–leukocyte cytokine cascades. Multiple mechanisms of inhibition of IgE- and mast cell-dependent cutaneous inflammation in the mouse. J. Immunol. **154:** 1391–1398.
78. YOSHIKAWA, H., Y. NAKAJIMA & K. TASAKA. 1999. Glucocorticoid suppresses autocrine survival of mast cells by inhibiting IL-4 production and ICAM-1 expression. J. Immunol. **162:** 6162–6170.
79. ZACHARCHUK, C.M., M. MERCEP, P.K. CHAKRABORTI, *et al.* 1990. Programmed T lymphocyte death. Cell activation- and steroid-induced pathways are mutually antagonistic. J. Immunol. **145:** 4037–4045.
80. VAN LAETHEM, F., E. BAUS, L.A. SMYTH, *et al.* 2001. Glucocorticoids attenuate T cell receptor signaling. J. Exp. Med. **193:** 803–814.
81. PURTON, J.F., R.L. BOYD, T.J. COLE & D.I. GODFREY. 2000. Intrathymic T cell development and selection proceeds normally in the absence of glucocorticoid receptor signaling. Immunity **13:** 179–186.
82. GONZALO, J.A., A. GONZALEZ-GARCIA, C. MARTINEZ & G. KROEMER. 1993. Glucocorticoid-mediated control of the activation and clonal deletion of peripheral T cells *in vivo*. J. Exp. Med. **177:** 1239–1246.
83. FRANCHIMONT, D., J. GALON, M.S. VACCHIO, *et al.* 2002. Positive effects of glucocorticoids on T cell function by up-regulation of IL-7 receptor alpha. J. Immunol. **168:** 2212–2218.
84. PUEL, A., S.F. ZIEGLER, R.H. BUCKLEY & W.J. LEONARD. 1998. Defective IL7R expression in T(-)B(+)NK(+) severe combined immunodeficiency. Nat. Genet. **20:** 394–397.
85. VON FREEDEN-JEFFRY, U., P. VIEIRA, L.A. LUCIAN, *et al.* 1995. Lymphopenia in interleukin (IL)-7 gene-deleted mice identifies IL-7 as a nonredundant cytokine. J. Exp. Med. **181:** 1519–1526.
86. PESCHON, J.J., P.J. MORRISSEY, K.H. GRABSTEIN, *et al.* 1994. Early lymphocyte expansion is severely impaired in interleukin 7 receptor-deficient mice. J. Exp. Med. **180:** 1955–1960.
87. SCHLUNS, K.S., W.C. KIEPER, J.C. JAMESON & L. LEFRANCOIS. 2000. Interleukin-7 mediates the homeostasis of naive and memory CD8 T cells *in vivo*. Nat. Immunol. **1:** 426–432.
88. CHROUSOS, G.P. 1995. The hypothalamic-pituitary-adrenal axis and immune-mediated inflammation. N. Engl. J. Med. **332:** 1351–1362.
89. JONULEIT, H., E. SCHMITT, G. SCHULER, *et al.* 2000. Induction of interleukin 10-producing, nonproliferating CD4(+) T cells with regulatory properties by repetitive stimulation with allogeneic immature human dendritic cells. J. Exp. Med. **192:** 1213–1222.
90. READ, S. & F. POWRIE. 2001. CD4(+) regulatory T cells. Curr. Opin. Immunol. **13:** 644–649.
91. DOZMOROV, I.M. & R.A. MILLER. 1998. Generation of antigen-specific Th2 cells from unprimed mice *in vitro*: effects of dexamethasone and anti-IL-10 antibody. J. Immunol. **160:** 2700–2705.
92. BANCHEREAU, J., B. SCHULER-THURNER, A.K. PALUCKA & G. SCHULER. 2001. Dendritic cells as vectors for therapy. Cell **106:** 271–274.

Glucocorticoids and the Th1/Th2 Balance

ILIA J. ELENKOV

Clinical Neuroendocrinology Branch, National Institute of Mental Health, National Institutes of Health, Bethesda, Maryland 20892, USA

ABSTRACT: Evidence accumulated over the last 5–10 years indicates that glucocorticoids (GCs) inhibit the production of interleukin (IL)-12, interferon (IFN)-γ, IFN-α, and tumor necrosis factor (TNF)-α by antigen-presenting cells (APCs) and T helper (Th)1 cells, but upregulate the production of IL-4, IL-10, and IL-13 by Th2 cells. Through this mechanism increased levels of GCs may systemically cause a selective suppression of the Th1–cellular immunity axis, and a shift toward Th2-mediated humoral immunity, rather than generalized immunosuppression. During an immune response and inflammation, the activation of the stress system, and thus increased levels of systemic GCs through induction of a Th2 shift, may actually protect the organism from systemic "overshooting" with Th1/pro-inflammatory cytokines and other products of activated macrophages with tissue-damaging potential. However, conditions associated with significant changes of GCs levels, such as acute or chronic stress or cessation of chronic stress, severe exercise, and pregnancy and postpartum, through modulation of the Th1/Th2 balance may affect the susceptibility to or the course of infections as well as autoimmune and atopic/allergic diseases.

KEYWORDS: glucocorticoids; stress; interleukin-12; interleukin-10; Th1 cells; Th2 cells; autoimmunity; allergy; inflammation; rheumatoid arthritis; multiple sclerosis

INTRODUCTION

The T helper (Th)1/Th2 balance is critically skewed, one way or the other, in several common human diseases, such as acute and chronic infections, autoimmunity, and atopy/allergy.[1–4] These diseases frequently develop and progress in settings of hyperactivity or hypoactivity of the hypothalamic-pituitary-adrenal (HPA) axis.[5–9] Studies in the 1970s and the 1980s revealed that glucocorticoids (GCs), the end products of HPA axis activity, inhibit lymphocyte proliferation and cytotoxicity, and the secretion of TNFα, IL-2, and IFN-γ.[10,11] These observations, in the context of the broad clinical use of GCs, initially led to the conclusion that GCs are, in general, immunosuppressive. Recent evidence indicates, however, that systemically GCs cause selective suppression of the Th1–cellular immunity axis and a shift toward Th2-mediated humoral immunity, rather than generalized immunosuppression. This

Address correspondence to: Dr. Ilia J. Elenkov, Clinical Neuroendocrinology Branch, National Institute of Mental Health, National Institutes of Health, Building 10, Room 2D46, 10 Center Drive, Bethesda, MD 20892.
Elenkovi@mail.nih.gov

**Ann. N.Y. Acad. Sci. 1024: 138–146 (2004). © 2004 New York Academy of Sciences.
doi: 10.1196/annals.1321.010**

new concept helps explain some well-known, but often contradictory, effects of GCs on the immune system and on the onset and course of certain infectious, autoimmune, and atopic/allergic diseases. This new understanding is briefly outlined below.

THE TH1/TH2 PARADIGM: ROLE OF TH1 AND TH2 CYTOKINES

The immune system is classified into innate (or non-specific, natural) and adaptive (or specific, acquired) immunity. Innate immunity provides a rapid, non-specific host response against different bacteria, viruses, or tumors that precedes the adaptive immunity. Moreover, innate immunity also has an important role in determining the nature of downstream adaptive immune responses. Thus, immune responses are regulated by antigen-presenting cells (APC)—monocytes/macrophages, dendritic cells (DCs), and by natural killer (NK) cells— which are components of innate immunity, and by the recently described Th lymphocyte subclasses Th1 and Th2, which are components of adaptive (acquired) immunity. Th1 cells primarily secrete IFN-γ, IL-2, and TNFα, which promote cellular immunity, whereas Th2 cells secrete a different set of cytokines, primarily IL-4, IL-10, and IL-13, which promote humoral immunity.[1–3] Naive $CD4^+$ (antigen-inexperienced) Th0 cells are clearly bipotential and serve as precursors of Th1 and Th2 cells. IL-12, produced by APC—monocytes/macrophages and DCs—is the major inducer of Th1 differentiation and hence cellular immunity. IL-12 also synergizes with IL-18 to induce the production of IFN-γ by natural killer (NK) cells. Thus, IL-12 in concert with IL-18, IFN-α, and IFN-γ promote the differentiation of Th0 toward the Th1 phenotype. IL-1, IL-12, TNFα, and IFN-γ stimulate the functional activity of T cytotoxic cells (Tc), NK cells, and activated macrophages, which are the major components of cellular immunity. The type 1 cytokines IL-12, TNFα, and IFN-γ also stimulate the synthesis of nitric oxide (NO) and other inflammatory mediators that drive chronic delayed-type inflammatory responses. Because of these synergistic roles in inflammation, IL-12, TNFα, and IFN-γ are considered the major pro-inflammatory cytokines.[1–3] Th1 and Th2 responses are mutually inhibitory. Thus, IL-12 and IFN-γ inhibit Th2 cells' activities, while IL-4 and IL-10 inhibit Th1 responses. IL-4 and IL-10 promote humoral immunity by stimulating the growth and activation of mast cells and eosinophils (Eo), the differentiation of B cells into antibody-secreting B cells, and B cell immunoglobulin switching to IgE. Importantly, these cytokines also inhibit macrophage activation, T cell proliferation, and the production of pro-inflammatory cytokines.[1–3] Therefore, the Th2 (type 2) cytokines IL-4 and IL-10 are the major anti-inflammatory cytokines.

SYSTEMIC EFFECTS OF GCs

Previous studies have shown that GCs suppress the production of TNFα, IFN-γ, and IL-2 *in vitro* and *in vivo* in animals and humans.[5,11] Recent evidence indicates that GCs systemically suppress the Th1–cellular immunity axis and mediate a Th2 shift by suppressing APC- and Th1- and upregulating Th2-cytokine production. Thus, GCs act through their classic cytoplasmic/nuclear receptors on APCs to suppress the production of the main inducer of Th1 responses, IL-12 *in vitro* and *ex vivo*.[12,13] Because IL-12 is extremely potent in enhancing IFN-γ and inhibiting IL-4 synthesis by T cells, the inhibition of IL-12 production may represent a major mechanism by which GCs affect the Th1/Th2 balance. Thus, GC-treated monocytes/macrophages produce significantly less IL-12, leading to a decreased capacity of these cells to induce IFN-γ production by antigen-primed $CD4^+$ T cells. The same treatment of monocytes/macrophages is also associated with an increased production of IL-4 by T cells, probably resulting from disinhibition from the suppressive effects of IL-12 on Th2 activity.[14] Furthermore, GCs potently downregulate the expression of IL-12 receptors on T and NK cells. This explains why human peripheral blood mononuclear cells stimulated with immobilized anti-CD3 lose their ability to produce IFN-γ in the presence of GCs.[15] Thus, although GCs may have a direct suppressive effect on Th1 cells, the overall inhibition of IFN-γ production by these cells appears to result mainly from the inhibition of IL-12 production by APCs and from the loss of IL-12 responsiveness of NK and Th1 cells. GCs also suppress the production of IL-18 (which synergizes with IL-12 to promote Th1 responses) in LPS/IL-2-stimulated peripheral blood mononuclear cells (PBMCs), although the inhibition seems to be incomplete even at high concentrations.[16] This is consistent, however, with an observation that in GC-responsive patients with Graves' ophthalmopathy the GCs treatment results in decreased IL-18 serum concentrations as compared to the pretreatment values.[17] GCs also inhibit the production of IFN-γ, another cytokine that synergizes with IL-12 to promote Th1 responses—in PBMCs of healthy adult volunteers stimulated with Newcastle disease virus (NDV), GCs reduce the IFN-α release by 50 to 60%.[18]

GCs also have a direct effect on Th2 cells by upregulating their IL-4, IL-10, and IL-13 production.[13,19] GCs have no effect on the production of the potent anti-inflammatory cytokine IL-10 by monocytes[12,20]; yet, lymphocyte-derived IL-10 production appears to be upregulated by GCs. Thus, rat $CD4^+$ T cells pretreated with dexamethasone exhibit increased levels of mRNA for IL-10.[21] Similarly, during experimental endotoxemia or cardiopulmonary bypass, or in multiple sclerosis patients having an acute relapse, the treatment with GCs is associated with an increased plasma IL-10 secretion.[20,22,23] This might have resulted from a direct stimulatory effect of GCs on T cell IL-10 production and/or from the disinhibition of the restraining inputs of IL-12 and IFN-γ on monocyte/lymphocyte IL-10 production.

LOCAL VERSUS SYSTEMIC EFFECTS

The systemic Th2-inducing properties of GCs may not pertain to certain conditions or local responses in specific compartments of the body. Thus, steroid treatment results in a significant increase in the number of IL-12^+ cells with concurrent reduction in the number of IL-13^+ expressing cells in bronchial biopsy specimens of subjects with asthma. Interestingly, this occurs only in steroid-sensitive but not steroid-resistant asthmatic subjects.[24] This may reflect only a redistribution of cells, but the number of IL-4^+ cells in the bronchial and nasal mucosa is also reduced by prednisone treatment.[25,26] Furthermore, the synthesis of transforming growth factor (TGF)-β, another cytokine with potent anti-inflammatory activities, is enhanced by GCs in human T cells but suppressed in glial cells,[27] and low concentrations of dexamethasone can indeed activate alveolar macrophages, leading to increased LPS-induced IL-1β production.[28]

In contrast to the abovementioned inhibitory effect of GCs on the production of IL-18 by LPS/IL-2-stimulated PBMCs,[16] in cultured unstimulated PBMC and promonocytic cell lines, prednisolone upregulates IL-18 transcription in parallel with increasing the IL-18 protein release into cell culture supernatants.[29] It is highly likely that GCs downregulate IL-18 production in activated macrophages but upregulate IL-18 production in resting, non-activated cells. In this context it is an interesting observation that Cushing's patients that have high cortisol levels have also markedly elevated levels of IL-18.[30]

CRH/SP–Mast Cell–Histamine Interactions

As the primary regulator of corticotropin (ACTH) secretion from the pituitary and thus, glucocorticoid secretion from the adrenal gland, corticotropin-releasing hormone (CRH) modulates immune/inflammatory reactions through receptor-mediated actions of GCs on their target immune tissues, effects that are in general anti-inflammatory. CRH, however, is also secreted peripherally at inflammatory sites (peripheral or immune CRH), where it exerts mostly pro-inflammatory activities.[31] Immunoreactive CRH is identified locally in tissues from patients with rheumatoid arthritis (RA), autoimmune thyroid disease (ATD), and ulcerative colitis. CRH in early inflammation is of peripheral nerve rather than immune cell origin.[31,32] Peripheral CRH has vascular permeability–enhancing and vasodilatory actions. An intradermal CRH injection induces a marked increase of vascular permeability and mast cell degranulation, mediated through CRH type 1 receptors.[33] It appears that the mast cell is a major target of immune CRH. Substance P (SP) and peripheral CRH, which are released from sensory peptidergic neurons, are two of the most potent mast cell secretagogues.[33–36] Thus, peripheral CRH and SP activate mast cells via a CRH type 1 and NK1 receptor–dependent mechanism leading to release of histamine and other contents of the mast cell granules that subsequently may cause vasodilatation, increased vascular permeability, and other manifestations of inflammation.

CLINICAL IMPLICATIONS

Intracellular Infections

A major factor governing the outcome of infectious diseases is the selection of Th1- versus Th2-predominant adaptive responses during and after the initial invasion of the host. Since the major catecholamines (CAs), epinephrine (EPI) and norepinephrine (NE), also systemically induce a Th2 shift,[4,37] stress-induced—and hence a GC- and CA-induced—Th2 shift may have a profound effect on the susceptibility of the organism to an infection, and/or may influence its course. Thus, stress has a substantial effect on the defense system where cellular immunity mechanisms have a primary rule.

Cellular immunity, and particularly IL-12 and IL-12-dependent IFN-γ secretion in humans, seems essential in the control of mycobacterial infections.[38] In the 1950s, Thomas Holmes (cf. Ref. 7) reported that individuals who had experienced stressful life events were more likely to develop tuberculosis and less likely to recover from it. Although it is still a matter of some speculation, stress hormone–induced inhibition of IL-12 and IFN-γ production, and the consequent suppression of cellular immunity, may amply explain the pathophysiologic mechanisms of these observations.

Helicobacter pylori infection is the most common cause of chronic gastritis that in some cases progresses to peptic ulcer disease. The role of stress in promoting peptic ulcers has been recognized for many years.[39,40] Thus, increased systemic stress hormone levels, in concert with an increased local concentration of histamine, induced by inflammatory or stress-related mediators, may skew the local responses toward Th2 and thus may allow the onset or progression of a *Helicobacter pylori* infection.

HIV^+ patients have IL-12 deficiency, while disease progression has been correlated with a Th2 shift. Progression of HIV infection is also characterized by increased cortisol secretion in both the early and late stages of the disease. Thus, increased glucocorticoid production, probably triggered by the chronic infection, was recently proposed to contribute to HIV progression.[41] In another recent study, Kino and colleagues found that one of the HIV-1 accessory proteins, Vpr, acts as a potent coactivator of the host glucocorticoid receptor rendering lymphoid cells hyperresponsive to GCs.[42] The extracellular Vpr also enhances the suppressive actions of the ligand-activated glucocorticoid receptor on IL-12 secretion by human monocytes/macrophages.[43] Through this effect, Vpr may contribute to the suppression of innate and cellular immunities of HIV-1–infected individuals and AIDS patients. Thus, on the one hand, stress hormones suppress Th1 and cellular immunity responses, while, on the other hand, retroviruses may increase the sensitivity of lymphoid cells to the suppressive effects of GCs.

In a recent study, an association was demonstrated between stress and the susceptibility to common cold among 394 persons who had been intentionally exposed to five different upper respiratory viruses. Psychological stress was found to be associated in a dose-dependent manner with an increased risk of acute infectious respiratory illness, and this risk was attributed to increased rates of infection rather than to an increased frequency of symptoms after infection.[44] Thus, stress hormones, through their selective inhibition of cellular immunity, may play substantial roles in the increased risk of an individual to acute respiratory infections caused by common cold viruses. In addition, stress hormones–induced Th2 can compromise the host's cellular immune response and trigger herpes simplex viral reactivation.[45]

Atopy/Allergy

Allergic diseases, such as asthma, seasonal and perennial allergic rhinitis, eczema, and IgE-mediated food allergy, are characterized by dominant Th2 responses—an overproduction of IL-4, IL-5, IL-9, and IL-13, histamine and a shift to IgE production.[46,47] Interestingly, elevations in IL-13 appear to be more associated with asthma than with atopy. An impaired IL-12 production coupled to an overproduction of IL-13 by alveolar macrophages may underlie to a great extent the Th2-biased response in asthma.[24,47] The effects of stress and GCs on atopic/allergic reactions are complex, at multiple levels, and can be in either direction.[4,48,49] Stress episodes preceding the development of the disease through induction of the Th2-potential may increase the susceptibility of the individual.[49] When the disease is already established, stress may induce a Th2 shift and also can activate the CRH–mast cell–histamine axis (see above) and, thus may facilitate or sustain atopic reactions; however, these effects can be antagonized by the effects of stress hormones on the mast cell.[4] GCs and CAs (through β_2-adrenergic receptors [ARs]) suppress the release of histamine by mast cells, thus abolishing its pro-inflammatory, allergic, and bronchoconstrictor effects. Consequently, reduced levels of epinephrine and cortisol in the very early morning could contribute to nocturnal wheezing and have been linked to high circulating histamine levels in asthmatics.[50] This may also explain the beneficial effect of GCs and β_2-agonists in asthma. It is noteworthy that infusion of high doses of adrenaline, however, causes a rise in circulating histamine levels that may be due to an α-adrenergic-mediated increase in mediator release (cf. Ref. 50). Thus, severe acute stress associated with high EPI concentrations and/or high local secretion of CRH could lead to mast cell degranulation. As a result, a substantial amount of histamine could be released, which consequently would not antagonize, but rather amplify the Th2 shift through H2 receptors, while in parallel, by acting on H1 receptors, it could initiate a new episode or exacerbate a chronic allergic condition.

GCs alone or in combination with β_2-AR-agonists are broadly used in the treatment of atopic reactions, and particularly asthma. *In vivo*, *ex vivo*, and *in vitro* exposure to GCs and β_2-agonists result in a reduction of IL-12 production, which persists at least several days.[12,14,51] Thus, glucocorticoid and/or β_2-AR-agonist therapy is likely to reduce the capacity of APC to produce IL-12, to greatly suppress type 2 cytokine synthesis in activated, but not resting T cells, and to abolish eosinophilia.[14] If, however, resting, (cytokine-uncommitted) T cells are subsequently activated by APCs preexposed to GCs and/or β_2-AR agonists, enhanced IL-4 production, but limited IFN-γ synthesis, could be induced.[14] Thus, while in the short term, the effect of GCs and β_2-AR agonists may be beneficial, their long-term effects might be to sustain the increased vulnerability of the patient to the allergic condition. This is further substantiated by the observations that both GCs and β_2-AR agonists potentiate the IgE production *in vitro* and *in vivo*.[52,53]

Th1-Related Autoimmunity

Several autoimmune diseases are characterized by common alterations of the Th1 versus Th2 and pro- versus anti-inflammatory cytokine balance. In rheumatoid arthritis (RA), multiple sclerosis (MS), type 1 diabetes mellitus, and autoimmune thyroid disease (ATD) the balance is skewed toward Th1 and an excess of IL-12 and TNFα production, whereas Th2 activity and the production of IL-10 appear to be deficient. This appears to be a critical factor that determines the proliferation and differentiation of Th1-related autoreactive cellular immune responses in these disorders.[54] A hypoactive stress system may facilitate or sustain the Th1 shift in Th1-mediated autoimmunity.[4,55,56] Animal studies and certain clinical observations support this hypothesis. Thus, Fischer rats, which have a hyperactive stress system, are extremely resistant to experimental induction of Th1-mediated autoimmune states, including collagen- and adjuvant-induced arthritis and experimental allergic encephalomyelitis (EAE). Conversely, Lewis rats, which exhibit a hypoactive stress system, are extremely prone to develop the abovementioned experimentally induced Th1-mediated disease models.[57,58] Recent studies suggest that suboptimal production of cortisol is involved in the onset and/or progression of RA.[6,59,60] Patients with RA have "inappropriately normal" or low cortisol and CA levels in the setting of severe, chronic inflammation, characterized by increased production of TNFα, IL-1, and IL-6. This may actually facilitate or sustain the pro-inflammatory shift in this disease. Whether this abnormality is primary or secondary has not been established.[59] Clinical observations also indicate that RA and MS frequently remit during pregnancy but exacerbate, or have their onset, in the postpartum period. Recent evidence suggests that a cortisol-, NE-, and 1,25-dihydroxyvitamin D3–induced inhibition and subsequent rebound of IL-12 and TNFα production may represent a major mechanism by which pregnancy and postpartum alter the course of or susceptibility to RA and MS.[9]

REFERENCES

1. MOSMANN, T.R. & S. SAD. 1996. The expanding universe of T-cell subsets: Th1, Th2 and more. Immunol. Today **17:** 138–146.
2. FEARON, D.T. & R.M. LOCKSLEY. 1996. The instructive role of innate immunity in the acquired immune response. Science **272:** 50–53.
3. TRINCHIERI, G. 2003. Interleukin-12 and the regulation of innate resistance and adaptive immunity. Nat. Rev. Immunol. **3:** 133–146.
4. ELENKOV, I.J. & G.P. CHROUSOS. 1999. Stress hormones, Th1/Th2 patterns, pro/anti-inflammatory cytokines and susceptibility to disease. Trends Endocrinol. Metab. **10:** 359–368.
5. CHROUSOS, G.P. 1995. The hypothalamic-pituitary-adrenal axis and immune-mediated inflammation [see comments]. N. Engl. J. Med. **332:** 1351–1362.
6. WILDER, R.L. 1995. Neuroendocrine-immune system interactions and autoimmunity. Annu. Rev. Immunol. **13:** 307–338.
7. LERNER, B.H. 1996. Can stress cause disease? Revisiting the tuberculosis research of Thomas Holmes, 1949–1961. Ann. Intern. Med. **124:** 673–680.
8. VON HERTZEN, L.C. 2002. Maternal stress and T-cell differentiation of the developing immune system: possible implications for the development of asthma and atopy. J. Allergy Clin. Immunol. **109:** 923–928.
9. ELENKOV, I.J., R.L. WILDER, V.K. BAKALOV, *et al.* 2001. IL-12, TNF-alpha, and hormonal changes during late pregnancy and early postpartum: implications for autoimmune disease activity during these times. J. Clin. Endocrinol. Metab. **86:** 4933–4938.
10. BEUTLER, B., N. KROCHIN, I.W. MILSARK, *et al.* 1986. Control of cachectin (tumor necrosis factor) synthesis: mechanisms of endotoxin resistance. Science **232:** 977–980.
11. BOUMPAS, D.T., G.P. CHROUSOS, R.L. WILDER, *et al.* 1993. Glucocorticoid therapy for immune-mediated diseases: basic and clinical correlates. Ann. Intern. Med. **119:** 1198–1208.
12. ELENKOV, I.J., D.A. PAPANICOLAOU, R.L. WILDER & G.P. CHROUSOS. 1996. Modulatory effects of glucocorticoids and catecholamines on human interleukin-12 and interleukin-10 production: clinical implications. Proc. Assoc. Am. Physicians. **108:** 374–381.
13. BLOTTA, M.H., R.H. DEKRUYFF & D.T. UMETSU. 1997. Corticosteroids inhibit IL-12 production in human monocytes and enhance their capacity to induce IL-4 synthesis in CD4+ lymphocytes. J. Immunol. **158:** 5589–5595.
14. DEKRUYFF, R.H., Y. FANG & D.T. UMETSU. 1998. Corticosteroids enhance the capacity of macrophages to induce Th2 cytokine synthesis in CD4+ lymphocytes by inhibiting IL-12 production. J. Immunol. **160:** 2231–2237.
15. WU, C.Y., K. WANG, J.F. MCDYER & R.A. SEDER. 1998. Prostaglandin E2 and dexamethasone inhibit IL-12 receptor expression and IL-12 responsiveness. J. Immunol. **161:** 2723–2730.
16. KODAMA, M., H.K. TAKAHASHI, H. IWAGAKI, *et al.* 2002. Effect of steroids on lipopolysaccharide/interleukin 2-induced interleukin 18 production in peripheral blood mononuclear cells. J. Int. Med. Res. **30:** 144–160.

17. MYSLIWIEC, J., A. KRETOWSKI, A. STEPIEN, *et al.* 2003. Interleukin 18 and transforming growth factor beta1 in the serum of patients with Graves' ophthalmopathy treated with corticosteroids. Int. Immunopharmacol. **3:** 549–552.
18. REISSLAND, P. & K.P. WANDINGER. 1999. Increased cortisol levels in human umbilical cord blood inhibit interferon alpha production of neonates. Immunobiology **200:** 227–233.
19. RAMIERZ, F., D.J. FOWELL, M. PUKLAVEC, *et al.* 1996. Glucocorticoids promote a TH2 cytokine response by CD4+ T cells in vitro. J. Immunol. **156:** 2406–2412.
20. VAN DER POLL, T., A.E. BARBER, S.M. COYLE & S.F. LOWRY. 1996. Hypercortisolemia increases plasma interleukin-10 concentrations during human endotoxemia—a clinical research center study. J. Clin. Endocrinol. Metab. **81:** 3604–3606.
21. SHADIACK, A.M., C.D. CARLSON, M. DING, *et al.* Lipopolysaccharide induces substance P in sympathetic ganglia via ganglionic interleukin-1 production. J. Neuroimmunol. **49:** 51–58.
22. TABARDEL, Y., J. DUCHATEAU, D. SCHMARTZ, *et al.* 1996. Corticosteroids increase blood interleukin-10 levels during cardiopulmonary bypass in men. Surgery **119:** 76–80.
23. GAYO, A., L. MOZO, A. SUAREZ, *et al.* 1998. Glucocorticoids increase IL-10 expression in multiple sclerosis patients with acute relapse. J. Neuroimmunol. **85:** 122–130.
24. NASEER, T., E.M. MINSHALL, D.Y. LEUNG, *et al.* 1997. Expression of IL-12 and IL-13 mRNA in asthma and their modulation in response to steroid therapy. Am. J. Respir. Crit. Care Med. **155:** 845–851.
25. BENTLEY, A.M., Q. HAMID, D.S. ROBINSON, *et al.* 1996. Prednisolone treatment in asthma. Reduction in the numbers of eosinophils, T cells, tryptase-only positive mast cells, and modulation of IL-4, IL-5, and interferon-gamma cytokine gene expression within the bronchial mucosa. Am. J. Respir. Crit. Care Med. **153:** 551–556.
26. BRADDING, P., I.H. FEATHER, S. WILSON, *et al.* 1995. Cytokine immunoreactivity in seasonal rhinitis: regulation by a topical corticosteroid. Am. J. Respir. Crit. Care Med. **151:** 1900–1906.
27. BATUMAN, O.A., A. FERRERO, C. CUPP, *et al.* 1995. Differential regulation of transforming growth factor beta-1 gene expression by glucocorticoids in human T and glial cells. J. Immunol. **155:** 4397–4405.
28. BROUG-HOLUB, E. & G. KRAAL. 1996. Dose- and time-dependent activation of rat alveolar macrophages by glucocorticoids. Clin. Exp. Immunol. **104:** 332–336.
29. MOLLER, B., N. KUKOC-ZIVOJNOV, N. KOYAMA, *et al.* 2002. Prednisolone induces interleukin-18 expression in mononuclear blood and myeloid progenitor cells. Inflamm. Res. **51:** 457–463.
30. KRISTO, C., K. GODANG, T. UELAND, *et al.* 2002. Raised serum levels of interleukin-8 and interleukin-18 in relation to bone metabolism in endogenous Cushing's syndrome. Eur. J. Endocrinol. **146:** 389–395.
31. KARALIS, K., H. SANO, J. REDWINE, *et al.* 1991. Autocrine or paracrine inflammatory actions of corticotropin-releasing hormone in vivo. Science **254:** 421–423.
32. ELENKOV, I.J., E.L. WEBSTER, D.J. TORPY & G.P. CHROUSOS. 1999. Stress, corticotropin-releasing hormone, glucocorticoids, and the immune/inflammatory response: acute and chronic effects. Ann. N.Y. Acad. Sci. **876:** 1–11.
33. THEOHARIDES, T.C., L.K. SINGH, W. BOUCHER, *et al.* 1998. Corticotropin-releasing hormone induces skin mast cell degranulation and increased vascular permeability, a possible explanation for its proinflammatory effects. Endocrinology **139:** 403–413.
34. FOREMAN, J.C. 1987. Substance P and calcitonin gene-related peptide: effects on mast cells and in human skin. Int. Arch. Allergy Appl. Immunol. **82:** 366–371.
35. CHURCH, M.K., M.A. LOWMAN, C. ROBINSON, *et al.* 1989. Interaction of neuropeptides with human mast cells. Int. Arch. Allergy Appl. Immunol. **88:** 70–78.
36. ROZNIECKI, J.J., E. WEBSTER & G.P. CHROUSOS. 1995. Stress-induced intracranial mast cell degranulation: a corticotropin-releasing hormone-mediated effect. Endocrinology **136:** 5745–5750.
37. ELENKOV, I.J., R.L. WILDER, G.P. CHROUSOS & E.S. VIZI. 2000. The sympathetic nerve—an integrative interface between two supersystems: the brain and the immune system. Pharmacol. Rev. **52:** 595–638.
38. ALTARE, F., A. DURANDY, D. LAMMAS, *et al.* 1998. Impairment of mycobacterial immunity in human interleukin-12 receptor deficiency. Science **280:** 1432–1435.

39. LEVENSTEIN, S. 1998. Stress and peptic ulcer: life beyond Helicobacter. Brit. Med. J. **316:** 538–541.
40. LEVENSTEIN, S., S. ACKERMAN, J.K. KIECOLT-GLASER & A. DUBOIS. 1999. Stress and peptic ulcer disease. J. Am. Med. Assoc. **281:** 10–11.
41. CLERICI, M., M. BEVILACQUA, T. VAGO, *et al.* 1994. An immunoendocrinological hypothesis of HIV infection [see comments]. Lancet **343:** 1552–1553.
42. KINO, T., A. GRAGEROV, J.B. KOPP, *et al.* 1999. The HIV-1 virion-associated protein vpr is a coactivator of the human glucocorticoid receptor. [In Process Citation] J. Exp. Med. **189:** 51–62.
43. MIRANI, M., I. ELENKOV, S. VOLPI, *et al.* 2002. HIV-1 protein Vpr suppresses IL-12 production from human monocytes by enhancing glucocorticoid action: potential implications of Vpr coactivator activity for the innate and cellular immunity deficits observed in HIV-1 infection. J. Immunol. **169:** 6361–6368.
44. COHEN, S., D.A. TYRRELL & A.P. SMITH. 1991. Psychological stress and susceptibility to the common cold [see comments]. N. Engl. J. Med. **325:** 606–612.
45. SAINZ, B., J.M. LOUTSCH, M.E. MARQUART & J.M. HILL. 2001. Stress-associated immunomodulation and herpes simplex virus infections. Med. Hypotheses **56:** 348–356.
46. HUMBERT, M., G. MENZ, S. YING, *et al.* 1999. The immunopathology of extrinsic (atopic) and intrinsic (non-atopic) asthma: more similarities than differences. Immunol. Today **20:** 528–533.
47. WILLS-KARP, M. 2001. IL-12/IL-13 axis in allergic asthma. J. Allergy Clin. Immunol. **107:** 9–18.
48. MARSHALL, G.D., JR. & S.K. AGARWAL. 2000. Stress, immune regulation, and immunity: applications for asthma. Allergy Asthma Proc. **21:** 241–246.
49. VON HERTZEN, L.C. 2002. Maternal stress and T-cell differentiation of the developing immune system: possible implications for the development of asthma and atopy. J. Allergy Clin. Immunol. **109:** 923–928.
50. BARNES, P., G. FITZGERALD, M. BROWN & C. DOLLERY. 1980. Nocturnal asthma and changes in circulating epinephrine, histamine, and cortisol. N. Engl. J. Med. **303:** 263–267.
51. PANINA-BORDIGNON, P., D. MAZZEO, P.D. LUCIA, *et al.* 1997. Beta2-agonists prevent Th1 development by selective inhibition of interleukin 12. J. Clin. Invest. **100:** 1513–1519.
52. ZIEG, G., G. LACK, R.J. HARBECK, *et al.* 1994. In vivo effects of glucocorticoids on IgE production [see comments]. J. Allergy Clin. Immunol. **94:** 222–230.
53. COQUERET, O., V. LAGENTE, C.P. FRERE, *et al.* 1994. Regulation of IgE production by beta 2-adrenoceptor agonists. Ann. N.Y. Acad. Sci. **725:** 44–49.
54. SEGAL, B.M., B.K. DWYER & E.M. SHEVACH. 1998. An interleukin (IL)-10/IL-12 immunoregulatory circuit controls susceptibility to autoimmune disease. J. Exp. Med. **187:** 537–546.
55. STERNBERG, E.M. 2001. Neuroendocrine regulation of autoimmune/inflammatory disease. J. Endocrinol. **169:** 429–435.
56. ELENKOV, I.J. & G.P. CHROUSOS. 2002. Stress hormones, proinflammatory and antiinflammatory cytokines, and autoimmunity. Ann. N.Y. Acad. Sci. **966:** 290–303.
57. STERNBERG, E.M., W.S. YOUNG, R. BERNARDINI, *et al.* 1989. A central nervous system defect in biosynthesis of corticotropin-releasing hormone is associated with susceptibility to streptococcal cell wall-induced arthritis in Lewis rats. Proc. Natl. Acad. Sci. USA **86:** 4771–4775.
58. STERNBERG, E.M., J.M. HILL, G.P. CHROUSOS, *et al.* 1989. Inflammatory mediator-induced hypothalamic-pituitary-adrenal axis activation is defective in streptococcal cell wall arthritis-susceptible Lewis rats. Proc. Natl. Acad. Sci. USA **86:** 2374–2378.
59. WILDER, R.L. & I.J. ELENKOV. 1999. Hormonal regulation of tumor necrosis factor-alpha, interleukin-12 and interleukin-10 production by activated macrophages. A disease-modifying mechanism in rheumatoid arthritis and systemic lupus erythematosus? Ann. N.Y. Acad. Sci. **876:** 14–31.
60. STRAUB, R.H. & M. CUTOLO. 2001. Involvement of the hypothalamic-pituitary-adrenal/gonadal axis and the peripheral nervous system in rheumatoid arthritis: viewpoint based on a systemic pathogenetic role. Arthritis Rheum. **44:** 493–507.

L-Carnitine Is a Modulator of the Glucocorticoid Receptor Alpha

SALVATORE ALESCI,[a,b] MASSIMO U. DE MARTINO,[b] TOMOSHIGE KINO,[b] AND IOANNIS ILIAS[b,c]

[a]*Clinical Neuroendocrinology Branch, National Institute of Mental Health, National Institutes of Health, Bethesda, Maryland, USA*

[b]*Pediatric and Reproductive Endocrinology Branch, National Institute of Child Health and Human Development, National Institutes of Health, Bethesda, Maryland, USA*

[c]*Department of Pharmacology, Faculty of Medicine, University of Patras, Rion-Patras, Greece*

ABSTRACT: L-Carnitine (LC) is a nutrient with an essential role in cellular energy production. At high doses, LC can mimic some of the biological activities of glucocorticoids, particularly immunomodulation. To explore the molecular bases of this property, we tested the influence of LC on glucocorticoid receptor-α (GRα) functions. LC reduced the binding capacity of GRα, induced its nuclear translocation, and stimulated its transcriptional activity. Moreover, LC suppressed TNFα and IL-12 release from human monocytes in glucocorticoid-like fashion. We conclude that pharmacologic doses of LC can activate GRα and, via this mechanism, regulate glucocorticoid-responsive genes, potentially sharing some of the biological and therapeutic properties of glucocorticoids.

KEYWORDS: glucocorticoid receptor alpha; immunity; cytokines

BACKGROUND

L-Carnitine (LC) is a quaternary amine, synthesized from the essential amino acids lysine and methionine (FIG. 1).[1] In humans, LC is produced in small quantities by the brain, liver, and kidney, but is predominatly introduced through dietary consumption of meat and dairy products. Absorbed by simple diffusion in the jejunum, LC circulates at levels of ~0.05 mM, as opposed to tissue concentrations of 1–100 mM, depending on the active uptake of LC from the extracellular space.[2,3] The metabolism of LC is very limited, and its normal excretion (approximately 5%) is renal and proportional to its plasma concentration.[4] The main recognized physiologic role of LC is to transport short-, medium-, and long-chain free fatty acids through cell compartments as acyl-LC esters. Transfer of acyl groups in this form from peroxisomes to mitochondria, followed by their oxidation in the Krebs cycle, is essen-

Address for correspondence: Salvatore Alesci, M.D., Ph.D., CNE/NIMH/NIH, 10 Center Drive, Bldg. 10, Rm. 2D46, MSC 1284, Bethesda, MD 20892-1284. Voice: 301-496-6886; fax: 301-402-1561.
alescisa@mail.nih.gov

Ann. N.Y. Acad. Sci. 1024: 147–152 (2004).
doi: 10.1196/annals.1321.012

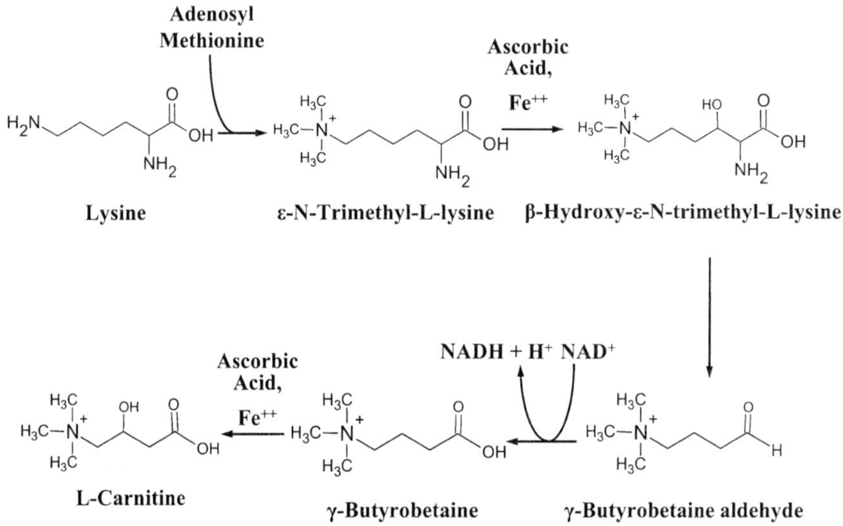

FIGURE 1. Simplified L-carnitine biosynthetic pathway. L-Carnitine biosynthesis begins with the methylation of lysine residues in proteins, such as myosin, actin, and histones, through the action of histone-lysine methyltransferase. Although the biosynthetic steps shown take place in most tissues, in humans the last step occurs only in the liver, brain, and kidney, where the necessary enzyme (γ-butyrobetaine, 2-oxoglutarate dioxygenase) is present.

tial for cell energy production and storage and for membrane synthesis and repair.[5,6] Currently, the only FDA-approved indication to administering pharmacologic doses of LC (up to 600 mg/kg body weight/day) is the treatment of primary and secondary LC deficiencies.[7–10] However, in the United States, LC is also widely available over the counter as a nutritional supplement in different doses and formulations.[11]

Glucocorticoids (GCs) exert their actions by interacting with intracellular receptors expressed in almost every tissue. In the absence of cognate ligands, the glucocorticoid receptor-α (GRα) is cytoplasmic and transcriptionally inactive, bound to receptor-associated proteins (RAPs), such as chaperones (hsp90, hsp70), co-chaperones (hip, hop), immunophilins (FKBP59, Cyp40), and others.[12,13] Binding of GCs to GRα induces dissociation of these RAPs, leading to its activation and translocation into the cell nucleus.[14] In the nucleus, homodimers of the activated receptor modulate the transcription of many responsive genes by binding to specific DNA-associated GC-responsive elements (GREs) in the promoters of these genes.[15] Both steroidal and nonsteroidal compounds can influence the transcription of GC-responsive genes by binding to GREs and/or interacting directly with either GRα or the RAPs.[15–17]

EVIDENCE FOR GC-LIKE PROPERTIES OF LC

Animal and human studies suggest that, at pharmacologic doses, LC may mimic some actions of GCs, including their well-established immunomodulatory effect. In rodents, LC (50–100 mg/kg body weight) markedly suppressed lipopolysaccharide (LPS)-induced cytokine production and improved their survival during cachexia and septic shock.[18,19] Similarly, LC reduced the *ex vivo* release of tumor necrosis factor-α (TNFα) by *S. aureus*–stimulated human polymorphonuclear white blood cells.[20] Moreover, profoundly decreased serum TNFα levels have been reported after LC supplementation in surgical patients (8 g i.v. at the end of surgery and 24 h afterward) and HIV+ patients (6 g/day for 2 weeks).[21,22] In pregnant rats, the administration of LC (100 mg/kg) was equally effective as betamethasone (at doses of 0.1–0.2 mg/kg) in increasing the dipalmitoyl-phosphatidylcholine content of the fetal lung.[23–25] In pregnant women, the combination of low-dose betamethasone (given at 2 mg/1 day) and high-dose LC (given at 4 g/5 days) decreased impressively the incidence and mortality of respiratory distress syndrome in their preterm newborns.[26,27]

MODULATORY EFFECTS OF LC ON GRα FUNCTIONS

To evaluate whether the proposed GC-like effects of LC could be mediated through activation of GRα, we assessed the influence of LC on the various properties of GRα, including binding capacity, cell trafficking, and transcriptional and biological activities.[28] Millimolar concentrations of LC, which were not cytotoxic *in vitro*, significantly reduced the whole-cell binding of [^{3}H]dexamethasone (Dex) to GRα in HeLa cells, with a significant increase in the K_d values of GRα for Dex, and no change in B_{max} values, indicating a decrease in the affinity of this receptor for its steroid ligand.[28] Additionally, at the same concentrations, LC was able to trigger nuclear translocation of green fluorescent protein (GFP)-fused human GRα and transactivate the GC-responsive mouse mammary tumor virus (MMTV) and TAT3 promoters in a dose-dependent fashion. These effects were dependent on the presence of GREs on the promoter, and on the expression of functional GRα by the cell.[28]

TNFα and the "immunomodulatory" cytokine IL-12, along with other type-1 cytokines, stimulate the synthesis of nitric oxide and other inflammatory mediators that drive acute and chronic inflammatory responses.[29,30] The secretion of TNFα and IL-12 is consistently reduced by GCs, both *in vitro* and *ex vivo*.[31–33] Similar to GCs, LC at concentrations that had maximally stimulated the transcription of GC-responsive promoters also suppressed the *ex vivo* release of TNFα and IL-12 by IFN-γ primed and/or LPS stimulated human primary monocytes, mimicking Dex.[28] The transactivation of GRα, as well as the suppression of cytokine release by LC, was completely abrogated in the presence of the GRα-antagonist RU 486.[28]

Thus, a common signal transduction pathway for LC and Dex may be suggested. LC may function as an allosteric regulator of the GRα. The decreased affinity of GRα for Dex in the presence of LC might be explained by the ability of this nutrient to interact with a segment of the receptor outside the GC-binding pocket, and modify the allosteric structure of the latter. This structural modification would, at the same time, reduce the affinity of the binding pocket for Dex and create conformational changes similar to those induced by the native ligand, ultimately resulting in GRα

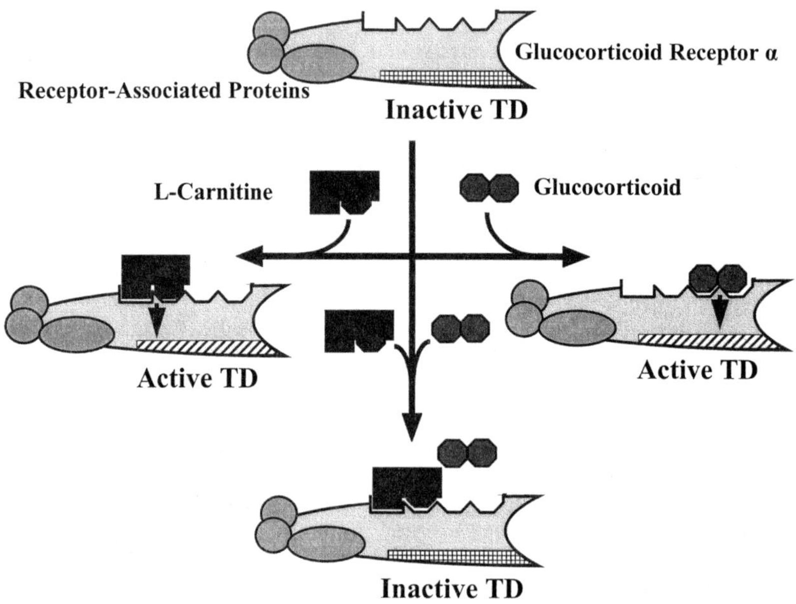

FIGURE 2. Proposed model of interrelation between L-carnitine and the glucocorticoid receptor alpha (GRa). L-Carnitine might interact with a segment of the receptor outside the glucocorticoid binding pocket, modifying its allosteric structure. This structural modification would cause conformational changes similar to those induced by glucocorticoids, resulting in GRa transactivation in the absence of native ligand. At the same time, it would decrease the affinity of the binding pocket for glucocorticoids in their presence, reducing their transactivational capacity. TD: transactivation domain.

activation. The ability of LC to reduce GRα affinity for Dex, combined with its weaker transactivating effect compared with Dex, lend support to the possibility of this compound to act as a partial GC agonist/antagonist, able to both transactivate GRα in the absence of the native ligand, and compete for GRα transactivation with the native ligand in its presence (FIG. 2).

Interestingly, millimolar concentrations of LC were also shown to increase alkaline phosphatase activity and collagen type I levels in primary porcine osteoblast-like cell cultures.[34] Moreover, LC supplementation in women with benign nodular thyroid disease treated with TSH-suppressive doses of L-thyroxine, resulted in positive effects on bone mineralization when compared to placebo.[35] Together with our findings, this evidence indicates that LC might share the beneficial immunomodulatory properties of GCs, but not their deleterious effects on the bone.

CONCLUSIONS

High doses of the nutrient LC can directly influence the various activities of the GRα *in vitro*, which can provide a molecular mechanism for previously reported

GC-like properties of LC supplementation. The GRα-mediated suppression of cytokine secretion by LC, coupled with its recently described positive effects on parameters of bone metabolism, lead us to suggest that GRα modulation by this naturally occurring compound may be tissue/gene-specific and/or selective. Potential clinical and therapeutic applications of our findings remain to be assessed.

REFERENCES

1. REBOUCHE, C.J. & H. SEIM. 1998. Carnitine metabolism and its regulation in microorganisms and mammals. Annu. Rev. Nutr. **18:** 39–61.
2. BREMER, J. 1997. The role of carnitine in cell metabolism. *In* Carnitine Today. C. De Simone & G. Famularo, Eds.: 1–37. R.G. Landes Company. Austin, TX.
3. REBOUCHE, C.J. 1992. Carnitine function and requirements during the life cycle. FASEB J. **6:** 3379–3386.
4. FURLONG, J.H. 1996. Acetyl-L-carnitine: metabolism and applications in clinical practice. Altern. Med. Rev. **1**.
5. ARDUINI, A. *et al.* 1992. Role of carnitine and carnitine palmitoyltransferase as integral components of the pathway for membrane phospholipid fatty acid turnover in intact human erythrocytes. J. Biol. Chem. **267:** 12673–12681.
6. FRITZ, I.B. & N.R. MARQUIS. 1965. The role of acylcarnitine esters and carnitine palmityltransferase in the transport of fatty acyl groups across mitochondrial membranes. Proc. Natl. Acad. Sci. USA **54:** 1226–1233.
7. ANGELINI, C., S. LUCKE & F. CANTARUTTI. 1976. Carnitine deficiency of skeletal muscle: report of a treated case. Neurology **26:** 633–637.
8. BOHLES, H. & W. LEHNERT. 1984. The effect of intravenous L-carnitine on propionic acid excretion in acute propionic acidaemia. Eur. J. Pediatr. **143:** 61–63.
9. ROE, C.R. *et al.* 1984. L-Carnitine enhances excretion of propionyl coenzyme A as propionylcarnitine in propionic acidemia. J. Clin. Invest. **73:** 1785–1788.
10. WALTER, P. & A.O. SCHAFFHAUSER. 2000. L-Carnitine, a "vitamin-like substance" for functional food. Proceedings of the Symposium on L-Carnitine, April 28–May 1, 2000, Zermatt, Switzerland. Ann. Nutr. Metab. **44:** 75–96.
11. BORUM, P.R. 2000. Supplements: questions to ask to reduce confusion. Am. J. Clin. Nutr. **72:** 538S–40S.
12. PRATT, W.B. 1992. Control of steroid receptor function and cytoplasmic-nuclear transport by heat shock proteins. Bioessays **14:** 841–848.
13. PRATT, W.B., U. GEHRING & D.O. TOFT. 1996. Molecular chaperoning of steroid hormone receptors. Experientia Suppl. **77:** 79–95.
14. BAMBERGER, C.M., H.M. SCHULTE & G.P. CHROUSOS. 1996. Molecular determinants of glucocorticoid receptor function and tissue sensitivity to glucocorticoids. Endocr. Rev. **17:** 245–261.
15. LEFEBVRE, P. *et al.* 1988. Association of the glucocorticoid receptor binding subunit with the 90K nonsteroid-binding component is stabilized by both steroidal and nonsteroidal antiglucocorticoids in intact cells. Biochemistry **27:** 9186–9194.
16. DAO-PHAN, H.P., P. FORMSTECHER & P. LEFEBVRE. 1997. Disruption of the glucocorticoid receptor assembly with heat shock protein 90 by a peptidic antiglucocorticoid. Mol. Endocrinol. **11:** 962–972.
17. BAMBERGER, C.M. *et al.* 1997. Inhibition of mineralocorticoid and glucocorticoid receptor function by the heat shock protein 90-binding agent geldanamycin. Mol. Cell Endocrinol. **131:** 233–240.
18. WINTER, B.K., G. FISKUM & L.L. GALLO. 1995. Effects of L-carnitine on serum triglyceride and cytokine levels in rat models of cachexia and septic shock. Br. J. Cancer **72:** 1173–1179.
19. RUGGIERO, V. *et al.* 1993. LPS-induced serum TNF production and lethality in mice: effect of L-carnitine and some acyl-derivatives. Med. Inflamm. **2:** S43–S50.

20. FATTOROSSI, A. *et al.* 1993. Regulation of normal human polymorphonuclear leucocytes by carnitine. Med. Inflamm. **2:** S37–S41.
21. DELOGU, G. *et al.* 1993. Anaesthetics modulate tumour necrosis factor alpha: effects of L-carnitine supplementation in surgical patients. Preliminary results. Med. Inflamm. **2:** S33–S36.
22. DE SIMONE, C. *et al.* 1993. High dose L-carnitine improves immunologic and metabolic parameters in AIDS patients. Immunopharmacol. Immunotoxicol. **15:** 1–12.
23. LOHNINGER, A. *et al.* 1996. Prenatal effects of betamethasone-L-carnitine combinations on fetal rat lung. J. Perinat. Med. **24:** 591–599.
24. LOHNINGER, A. *et al.* 1984. Comparison of the effects of betamethasone and L-carnitine on dipalmitoylphosphatidylcholine content and phosphatidylcholine species composition in fetal rat lungs. Pediatr. Res. **18:** 1246–1252.
25. LOHNINGER, A. *et al.* 1986. Studies on the effects of betamethasone, L-carnitine, and betamethasone-L-carnitine combinations on the dipalmitoyl phosphatidylcholine content and phosphatidylcholine species composition in foetal rat lungs. J. Clin. Chem. Clin. Biochem. **24:** 361–368.
26. SALZER, H. *et al.* 1983. 1st report: alternatives to cortisone therapy. 1st clinical experiences with a carnitine-betamethasone combination for the stimulation of fetal lung maturity. Wien Klin. Wochenschr. **95:** 724–728.
27. KURZ, C. *et al.* 1993. L-Carnitine-betamethasone combination therapy versus betamethasone therapy alone in prevention of respiratory distress syndrome. Z. Geburtshilfe Perinatol. **197:** 215–219.
28. ALESCI, S. *et al.* 2003. L-carnitine: a nutritional modulator of glucocorticoid receptor functions. FASEB J. **17:** 1553–1555.
29. FEARON, D.T. & R.M. LOCKSLEY. 1996. The instructive role of innate immunity in the acquired immune response. Science **272:** 50–53.
30. TRINCHIERI, G. 1995. Interleukin-12: a proinflammatory cytokine with immunoregulatory functions that bridge innate resistance and antigen-specific adaptive immunity. Annu. Rev. Immunol. **13:** 251–276.
31. ELENKOV, I.J. *et al.* 1996. Modulatory effects of glucocorticoids and catecholamines on human interleukin-12 and interleukin-10 production: clinical implications. Proc. Assoc. Am. Physicians **108:** 374–381.
32. FOSTER, S.J. *et al.* 1993. Production of TNF alpha by LPS-stimulated murine, rat and human blood and its pharmacological modulation. Agents Actions **38:** C77–79.
33. KATAKAMI, N., Y. NAKAO & T. FUJITA. 1991. Suppressive effect of 1,25(OH)2D3 and glucocorticoids on production of tumor necrosis factor alpha by human peripheral blood adherent cells. Kobe J. Med. Sci. **37:** 179–188.
34. CHIU, K.M. *et al.* 1999. Carnitine and dehydroepiandrosterone sulfate induce protein synthesis in porcine primary osteoblast-like cells. Calcif. Tissue Int. **64:** 527–533.
35. BENVENGA, S. *et al.* 2001. Usefulness of L-carnitine, a naturally occurring peripheral antagonist of thyroid hormone action, in iatrogenic hyperthyroidism: a randomized, double-blind, placebo-controlled clinical trial. J. Clin. Endocrinol. Metab. **86:** 3579–3594.

Human Immunodeficiency Virus Type-1 Accessory Protein Vpr

A Causative Agent of the AIDS-Related Insulin Resistance/Lipodystrophy Syndrome?

TOMOSHIGE KINO AND GEORGE P. CHROUSOS

Pediatric and Reproductive Endocrinology Branch, National Institute of Child Health and Human Development, National Institutes of Health, Bethesda, Maryland 20892, USA

ABSTRACT: Recent advances in the development of three different types of antiviral drugs, the nucleotide and non-nucleotide analogues acting as reverse transcriptase inhibitors (NRTIs) and the nonpeptidic viral protease inhibitors (PI), and their introduction in the management of patients with AIDS, either alone or in combination, have dramatically improved the clinical course of the disease and prolonged life expectancy in patients with AIDS. The increase in life expectancy in association with the long-term use of the above antiviral agents, however, have generated novel morbidities and complications. Central among them is the quite common AIDS-related insulin resistance and lipodystrophy syndrome, which is characterized by a striking phenotype and marked metabolic disturbances. To look for the pathologic causes of this particular syndrome, we focused on one of the HIV-1 accessory proteins, Vpr, which has multiple functions, such as virion incorporation, nuclear translocation of the HIV-1 preintegration complex, nucleo-cytoplasmic shuttling, transcriptional activation, and induction of apoptosis. Vpr may also act like a hormone, which is secreted into the extracellular space and affects the function of distant organs. Vpr functions as a coactivator of the glucocorticoid receptor and potentiates the action of glucocorticoid hormones, thereby inducing tissue glucocorticoid hypersensitivity. Vpr also arrests host cells at the G2/M phase of the cell cycle by interacting with novel 14-3-3 proteins. Vpr facilitates the interaction of 14-3-3 and its partner protein Cdc25C, which is critical for the transition of G2/M checkpoint in the cell cycle, and suppresses its activity by segregating it into the cytoplasm. The same Vpr protein also suppresses the association of 14-3-3 with other partner molecules, the Foxo transcription factors. Since the Foxo proteins function as negative transcription factors for insulin, Vpr may cause resistance of tissues to insulin. Through these two newly identified functions of Vpr, namely, coactivation of glucocorticoid receptor activity and inhibition of insulin effects on Foxo proteins, Vpr may participate in the development of AIDS-related insulin resistance/lipodystrophy syndrome.

KEYWORDS: Vpr; HIV-1 accessory protein; insulin resistance; lipodystrophy syndrome

Address for correspondence: Tomoshige Kino, M.D.,Ph.D., Pediatric and Reproductive Endocrinology Branch, National Institute of Child Health and Human Development, National Institutes of Health, Bldg. 10, Rm. 9D42, 10 Center Drive MSC 1583, Bethesda, MD 20892-1583. Voice: 301-496-6417; fax: 301-402-0884.

kinot@mail.nih.gov

Ann. N.Y. Acad. Sci. 1024: 153–167 (2004). © 2004 New York Academy of Sciences. doi: 10.1196/annals.1321.013

INTRODUCTION

Patients with the acquired immunodeficiency syndrome (AIDS), caused by the human immunodeficiency virus type-1 (HIV-1), develop profound immunosuppression, particularly of their innate and T-helper 1–directed, cellular immunity.[1] The same patients may also develop dysfunction of many organ systems, including the liver, adipose tissue, skeletal muscle, and central nervous system, mediated by as yet unclear mechanisms. Recent advances in the development and clinical use of three different types of antiviral drugs, the nucleotide and non-nucleotide analogues acting as reverse transcriptase inhibitors (NRTIs) and the nonpeptidic viral protease inhibitors (PI) (especially the combination therapy using any three of the above drugs — termed highly active antiretroviral therapy or HAART) have dramatically improved the clinical course of AIDS patients and prolonged their lives.[2–6] However, the prolongation of life expectancy and/or the long-term use of the above antiviral agents have generated novel morbidities and complications, which influence the patients' quality of life and add new risk factors for premature death. Central among them is the quite common AIDS-related insulin resistance and lipodystrophy syndrome, which is characterized by a striking phenotype and marked metabolic disturbances that increase the risk for cardiovascular disease.[7–10] The patients with this syndrome

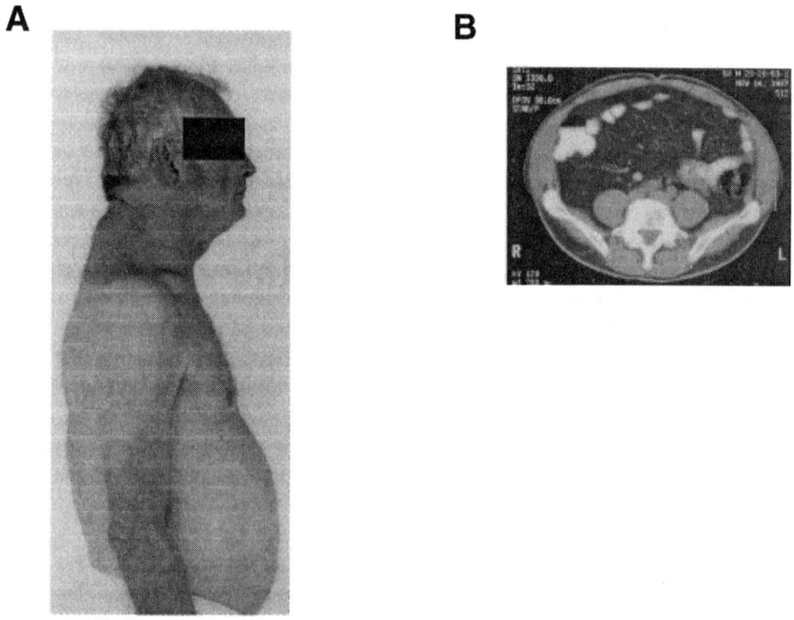

FIGURE 1. (**A**) Clinical features of a patient with AIDS-related insulin resistance/lipodystrophy syndrome. The patient demonstrates loss of facial fat, accumulation of dorsocervical tissue, and abdominal distention. (**B**) An abdominal computed tomography scan shows abundant visceral abdominal adipose tissue causing abdominal distention. (Adapted from Yanovski *et al.*,[74] with permission.)

have a combination of regional lipodystrophy with characteristic redistribution of their adipose tissue. They have an enlargement of their dorsocervical fat pad ("buffalo hump"), axial fat pads (bilateral symmetric lipomatosis), lipomastia, and expansion in abdominal girth ("Crix-belly" or "protease paunch"), as well as thinning of the extremities and muscle wasting (FIG. 1). Since all these manifestations are reminiscent of the typical phenotype of chronic glucocorticoid excess or Cushing's syndrome, this condition was initially referred to as a pseudo-Cushing's state, a term reserved for obese, depressive, or alcoholic patients with biochemical hypercortisolism, who are frequently hard to differentiate from true Cushing's syndrome.[8] In addition, the AIDS-related lipodystrophy/insulin resistance syndrome is accompanied by profound dyslipidemia and carbohydrate intolerance or overt diabetes mellitus, which are also recognized in true Cushing's syndrome and some of the congenital lipodystrophy syndromes.[7,9,11–18]

One of the most likely causative agents of this syndrome is the use of antiviral drugs, especially protease inhibitors (PIs), based on the evidence that the majority of AIDS patients develop this syndrome after taking such compounds.[15,19] Mitochondrial dysfunction caused by NRTIs has been also hypothesized to be a potential cause of this syndrome, due to its phenotypic similarity to multiple symmetric lipomatosis (MSL).[20,21] Although adverse effects of the antiviral drugs may be the most likely candidates, some patients still develop the characteristic features of the syndrome prior to treatment with these compounds, indicating that the HIV-1 infection itself and/or infection-related pathologic changes could also induce this pathologic status or could increase the vulnerability of patients to these antiviral drugs.[8,9,22] In this context, the use of antiviral drugs might just exacerbate the already existing lipodystrophy.

After the HIV-1 infects host cells and enters into their cytoplasm, its 9.8 kb genomic information is integrated into the host genome and produces three precursor proteins (Gag, RNA polymerase, and Envelope), whose processed products are reverse transcriptase, protease, integrase, matrix, capsid, as well as six accessory proteins, Tat, Rev, Nef, Vif, Vpr, and Vpu[23] (FIG. 2). Some of these polypeptides are virion-associated proteins incorporated into the viral particle and others are expressed in host cells where they direct viral replication and several host cell functions. Since infection with HIV-1 has a dramatic impact on host target cells, it is quite possible that some of these viral proteins modulate host cell glucose and lipid metabolism, participating in the development of AIDS-related insulin resistance/lipodystrophy syndrome.

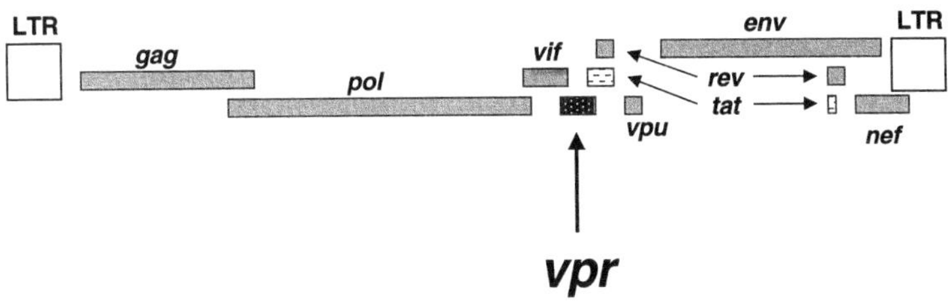

FIGURE 2. Linearized structure of the HIV-1 genome and localization of *Vpr*-coding region. LTR, long terminal repeat.

Therefore, to explore the pathologic causes of this syndrome, we recently focused on one of the HIV-1 accessory proteins, Vpr, which is known to act as a modulator of the host cell activity.[23,24] This viral protein is secreted into extracellular spaces, such as sera and the cerebrospinal fluid, and exerts its biologic activities from outside of the cells by penetrating the cytoplasmic membrane in a manner similar to classic hormones.[25,26] In the following sections, we will describe the biologic activities of HIV-1 Vpr and its potential contribution to the pathogenesis of the AIDS-related lipodystrophy/insulin resistance syndrome.

HIV-1 VPR

The human immunodeficiency virus (HIV) type-1 accessory protein Vpr (viral protein R) is a 96-amino-acid virion-associated protein shown to be important for virus replication/propagation *in vivo*.[27–29] It is a small basic protein conserved in HIV-1, HIV-2, and the simian immunodeficiency virus (SIV). The Vpr molecule exhibits three α-helices, which are folded around a hydrophobic core and has flexible amino- and carboxyl-termini, in the three-dimensional structure observed in the nuclear magnetic resonance (NMR) analysis.[30] Vpr is packaged in significant quantities into viral particles[31,32] and is imported into the nucleus early after infection. Vpr plays a role in the nuclear translocation of the HIV-1–preintegration complex, has the ability to shuttle between the cytoplasm and the nucleus, and induces apoptosis[23,24,33–39] (FIG. 3).

In addition to these activities, Vpr was the first molecule to be reported to act as a weak activator of the HIV-1 long terminal repeat (LTR) and several heterologous viral promoters.[40] Vpr can stimulate the basal transcriptional activity of the HIV-1 long terminal repeat (LTR) promoter by associating with the transcription factor SP1.[41] Vpr has been also shown to interact with one of the general transcription factors (TF) TFIIB.[42] Subsequent analyses also indicated that Vpr functions as a potent enhancer of Tat-induced activation of the HIV-1-LTR.[43,44]

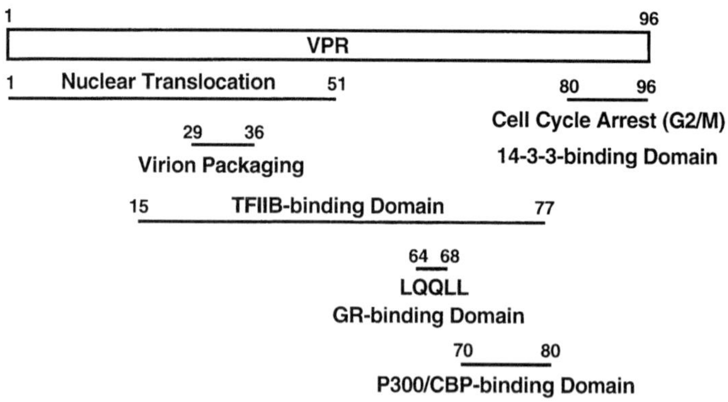

FIGURE 3. Linearized structure of Vpr and distribution of its known activities and binding domains.[42,45–47,62,82,83,98–100] TFIIB, transcription factor IIB; GR, glucocorticoid receptor; CBP, cAMP-responsive element-binding protein (CREB)-binding protein.

Furthermore, Vpr has a strong activity to arrest host cells, such as peripheral monocytes and lymphocytes, at the G2/M boundary of the cell cycle.[45–47] Because of this activity, Vpr has been proposed to facilitate viral propagation.[29] Transition through the G2/M checkpoint in mammalian cells is controlled by activation of a protein complex formed by a catalytic subunit, the cyclin-dependent kinase Cdc2, and its regulatory partner cyclinB1, through coordinated phosphorylation/dephosphorylation events.[48,49] The protein kinases Wee1 and Myt1 inactivate this complex by phosphorylating threonine residues at amino acids 14 and 15 of Cdc2, while the phosphatase Cdc25C activates it by dephosphorylating the same threonine residues.[48–50] Threonine at amino acid 161 of Cdc2 is phosphorylated by Cdk-activating kinases. The upstream kinases, such as Chk1 or Cdk2, which are stimulated by several signals such as DNA damage, stimulate Wee1, but suppress Cdc25C by phosphorylating serine residues at amino acid 549 of Wee1 and 216 of Cdc25C, respectively, and induce cell cycle arrest at G2/M phase. Protein phosphatase 2A dephosphorylates serine residue at amino acid 549 of Wee1 and inactivates this kinase. Stimulated Cdc2/cyclinB1 complexes then phosphorylate Wee1 and Cdc25C, thus creating a positive feedback loop.

In this cascade, it is known that Vpr inactivates the Cdc2/cyclinB1 complex by keeping Cdc2 at a hyperphosphorylated state.[45–47,51–53] This is possible by modulating the function of host protein(s), which act upstream of Cdc2/cyclinB1, such as PP2A, Wee1, Myt1, and Cdc25C. Vpr is reported to modulate the activity and/or protein levels of PP2A and Wee1 with yet unknown mechanisms.[54–56] Using genetic analysis in fission yeast, the cell cycle–arresting activity of Vpr is associated with the presence of *pp2a*, *wee1*, and *rad24*.[51] *rad24* encodes the 14-3-3 family proteins in humans.[57]

The 14-3-3 family of proteins consists of nine isotypes produced from at least seven distinct genes in vertebrates. The 14-3-3 proteins bind phosphorylated serine/threonine residues at specific positions of their partner proteins and regulate their activities by changing their subcellular localization and/or stability. They contain nine α-helical structures and form homo- and heterodimers through their N-terminal portion.[57–61] The central third-to-fifth α-helices create a binding pocket for a phosphorylated serine/threonine residue and the C-terminal seventh-to-ninth helices determine the specificity to target peptide motifs.[60,61] Finally, 14-3-3 contains a nuclear export signal (NES) in the ninth helix.[61,62]

The 14-3-3 proteins play a significant role in cell cycle progression at several different stages. First, they regulate Cdc25C activity.[3,61,63–68] Second, they bind Wee1 kinase and increase the stability and activity of this protein.[69,70] Third, they bind and activate the Chk1 and Cdk2 kinases, by appropriately sequestering these molecules inside the nucleus.[71] Activation of Chk1 causes phosphorylation of Cdc25C, producing a binding site for 14-3-3 proteins, which leads to inactivation of the activity of its phosphatase.[66,67] Finally, the 14-3-3 proteins bind to the phosphorylated Cdc2 and cyclinB1, and inactivate their complex by exporting it into the cytoplasm.[72,73]

To explore host molecules that support this Vpr activity, we have recently performed extensive yeast two-hybrid screening assays using a panel of wild-type and mutant Vprs and found that 14-3-3 is a specific partner of Vpr for its cell cycle–arresting activity.[62] Vpr bound to the C-terminal portion of 14-3-3, which is located outside of the phosphopeptide-binding pocket but determines specificity of its binding activity to phosphopeptides.[61] Through direct binding to 14-3-3, Vpr facilitated the association of 14-3-3 to its partner protein Cdc25C and abrogated the ability of the latter to stimulate cell cycle progression by retaining it in the nucleus. Since Vpr binds 14-3-3 proteins at their C-terminal part, binding of Vpr may alter the binding affinity of 14-3-3 to its partner proteins. Therefore, it is highly possible that Vpr also modulates the activities of Wee1, Chk1, Cdk2, and, possibly, Cdc2/cyclinB1 complex by changing their binding specificity to 14-3-3 proteins. Indeed, a recent report indicates that Vpr downregulates the protein levels of Weel.[54] Since 14-3-3 increases the stability of this kinase,[70] this Vpr effect may be possible by its potentiation of 14-3-3/Wee1 association.

IMPLICATIONS OF VPR TO THE DEVELOPMENT OF AIDS-RELATED LIPODYSTROPHY/INSULIN RESISTANCE SYNDROME

Vpr: A Viral Coactivator of GR

Since the clinical picture of the AIDS-related insulin resistance/lipodystrophy syndrome shares many features with those observed in Cushing's syndrome, hypercortisolism was originally hypothesized as a potential factor leading to AIDS-related lipodystrophy syndrome. We examined the adrenal function of patients with lipodystrophy syndrome and showed that this syndrome is distinct from the glucocorticoid-induced condition.[74] Thus, patients with this syndrome had normal plasma concentrations of basal and CRH-stimulated ACTH and cortisol. Moreover, their GRs were in normal concentrations and their affinity to dexamethasone was similar to that of controls. Therefore, biochemical hypercortisolism is not likely to be a major cause of AIDS-related lipodystrophy. Rather, it is still possible that localized or tissue-specific hypersensitivity to glucocorticoids may be involved.

Action of the Glucocorticoid Receptor

Glucocorticoids exert their effects on their target cells through the glucocorticoid receptor (GR), a ligand-specific and -dependent transcription factor, ubiquitously expressed in almost all tissues.[75] The GR shuttles between the cytoplasm and the nucleus. Binding of glucocorticoids to the GR causes it to dissociate from a cytoplasmic hetero-oligomer, containing heat-shock proteins, and to translocate into the nucleus via the nuclear pore. There, ligand-bound GR molecules bind as dimers to specific DNA enhancer sequences, the glucocorticoid-responsive elements (GREs), in the promoters of glucocorticoid-responsive genes, to modulate the transcription of these genes.

The GRE-bound GR interacts with newly described "coactivator complexes," which possess histone acetyltransferase (HAT) activity, as well as other chromatin modulatory protein complexes, such as SWI/SNF, SMCC, and TRAP/DRIP.[76,77] One family of the coactivator molecules consisting of the homologous p300 and cAMP-responsive element binding protein (CBP) may, in addition to nuclear receptors, serve as macromolecular docking "platforms" for many other transcription factors from different signal transduction cascades. Another coactivator, p/CAF, originally reported as a human homologue of yeast GCN5 that interacts with p300/CBP, is also a broad coactivator with HAT activity.[76,78] Coactivator molecules interacting preferentially with nuclear receptors have also been described.[76,77,79] They include members of the p160 family of proteins: steroid receptor coactivator-1 (SRC-1); TIF-II or glucocorticoid receptor interacting polypeptide-1 (GRIP-1), also called SRC-2; the p300/CBP/co-integrator–associated protein (p/CIP), ACTR or RAC3, also called SRC-3; and the recently reported riboprotein steroid receptor coactivator (SRA).[76]

These different classes of coactivator proteins form complexes by binding to each other as well as to the ligand-activated nuclear receptors, which interact with components of the transcription machinery on the promoter regions of responsive genes. p300/CBP and the members of the p160 family of coactivators contain one or more copies of the coactivator signature motif sequence LXXLL, which is essential for the interaction with nuclear receptors.[80] The receptor-coactivator complexes not only help transduce the hormonal signal to the transcription initiation complex but also loosen chromatin structure by acetylating histones through their intrinsic histone acetyltransferase activity and facilitate the binding of the transcription machinery components to DNA.[76]

The complex system of glucocorticoid receptor signaling suggests that the glucocorticoid activity is modulated by numerous factors at the level of the peripheral tissues. This is referred to as "sensitivity of tissues to glucocorticoids," which determines effectiveness of glucocorticoids in peripheral tissues. Depending on its direction, decreased or increased, it is divided into two subgroups: resistance and hypersensitivity. Both states may be generalized or tissue-specific, as well as congenital or acquired.

Mechanism of Vpr Coactivator Activity on GR Transactivation

Since Vpr has been shown to interact with the GR via cellular 41 kDa protein,[81] we recently investigated in detail the action of Vpr on the GR-mediated transcriptional activity in order to address a possible implication of Vpr to the development of AIDS-related insulin resistance/lipodystrophy syndrome. We found that, in contrast to the previous report,[81] Vpr binds directly to the GR via its conserved LXXLL motif located at amino acids 64 to 68, and markedly potentiates the action of glucocorticoid receptor on its responsive promoters, acting as a nuclear receptor coactivator in cooperation with the host cell coactivator p300/CBP.[82] We also found that Vpr is a general coregulator of nuclear receptors that influences not only the glucocorticoid receptor but also the progesterone and estrogen receptors. We showed that Vpr acts as an adaptor molecule bridging promoter-bound transcription factors and the transcriptional coactivator p300/cAMP-responsive element-binding protein (CREB)-binding protein (CBP).[44] Vpr, via its third α-helix, binds directly to amino acids 2,045–2,191 of human p300, which are also known to associate with the host coactivators p160 family proteins.[76,83] Furthermore, we found that extracellularly administered Vpr suppressed interleukin (IL)-12 production from peripheral monocytes by potentiating GR activity, possibly contributing to the suppression of innate and cellular immunity of HIV-1–infected individuals and AIDS patients.[26]

HIV-1 Tat: Another Enhancer of the GR Transactivation

We have recently studied another HIV-1 accessory protein, Tat. This protein is the most potent transactivator of the HIV-1-LTR, important for the expression of HIV-1–encoded proteins.[23] Tat binds to a stem loop structure of a short transcribed

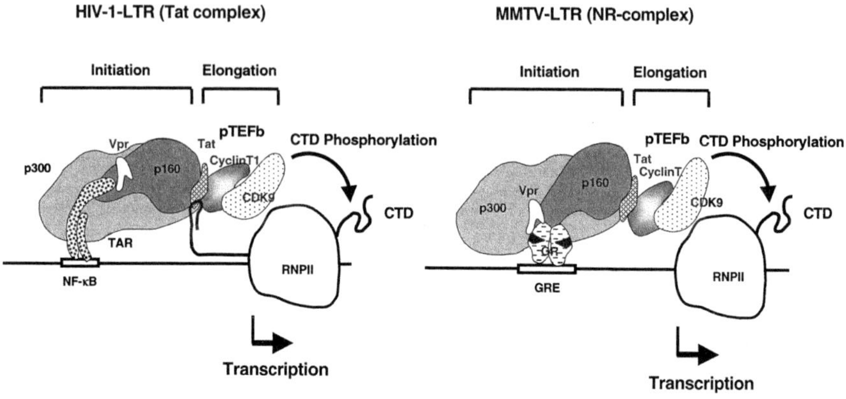

FIGURE 4. Simplified model of Vpr and Tat actions on HIV-1 LTR and glucocorticoid-responsive genes. Both promoters use the same set of coactivator molecules.[83,85] CDK9, cyclin-dependent kinase 9; RNPII, RNA polymerase Iip; GRE, glucocorticoid-responsive element.

mRNA called TAR, through which it is tethered to the HIV-1-LTR. Tat helps accumulate the positive transcription elongation factor-b (pTEF-b) on the HIV-1-LTR by binding to one of its components, cyclinT1, that is also important for its binding to TAR.[84] We found that activation of HIV-1-LTR by Tat uses components of nuclear hormone receptor coactivator system by directly binding not only to p300/CBP but also to p160 coactivators where Vpr functions as an enhancer.[83,85] We also found that Tat moderately potentiated GR activity, possibly through accumulation of the pTEF-b complex on glucocorticoid-responsive promoters.[44] Since Tat also circulates in blood and exerts its actions as an auto/paracrine or endocrine factor by penetrating the cell membrane,[86] it is possible that, like Vpr, it modulates tissue sensitivity to glucocorticoids irrespective of a cell's infection by HIV-1. Concomitantly with Vpr, Tat may induce tissue hypersensitivity to glucocorticoids that might contribute to viral proliferation indirectly by suppressing the host immune system activity and by altering the host's metabolic balance, both functions governed by glucocorticoids (FIG. 4).

Vpr: A Viral Inhibitor of the Insulin Signal Pathway

Insulin Actions

Insulin regulates diverse physiologic functions of cells and tissues, such as carbohydrate and lipid metabolism, protein synthesis, DNA replication, cell growth and differentiation, and inhibition of apoptosis.[87] Binding of insulin to its receptor stimulates many signaling cascades via phosphorylation-mediated reactions and activates several transcription factors, which, finally, regulate expression of target molecules.[87] Insulin-responsive genes have at least eight distinct consensus insulin-responsive sequences (IRSs) in their promoter regions, which positively or negatively respond to insulin stimuli.[88] Consensus sequences, such as those of activator protein 1 (AP-1), Ets, E-box, and thyroid transcription factor 2 (TTF-2), mediate positive transcriptional effect of insulin, while an element with the consensus sequence T(G/A)TTT(T/G)-(G/T), also referred to as the phosphoenolpyruvate carboxykinase (PEPCK)-like motif, mediates the inhibitory effect of insulin on several insulin-responsive genes.[88] The forkhead in human rhabdomyosarcoma (FKHR or Foxo1a), one of the FOXO subfamily of the forkhead transcription factors that share the forkhead DNA-binding domain and play diverse roles in developmental and metabolic functions, has recently been shown to bind this PEPCK-like IRS in response to insulin stimuli and to mediate negative effect of this hormone.[89]

Insulin stimuli regulate the activity of several FOXO proteins, such as Foxo3a (FKHR-L1) and Foxo4 (AFX), in addition to Foxo1a (FKHR).[87,90,91] In the absence of insulin, they are located in the nucleus, bind to their responsive promoters, and activate the transcription rate of their target genes, including the key gluconeogenesis enzyme phosphoenolpyruvate carboxykinase (PEPCK), the insulin-like growth factor–binding protein 1 and the key glycolysis enzyme glucose 6-phosphatase (G6Pase).[88,92–94] Once insulin induces the phosphorylation of specific serine and threonine residues of these FOXO proteins via activation of Akt or protein kinase B, these phosphorylated amino acids create binding sites for 14-3-3.[95,96] Upon binding to protein 14-3-3, FOXO proteins translocate from the nucleus into the cytoplasm, leading to inactivation of their transcriptional activity. Thus, FOXOs function as negative transcription factors of insulin, and their binding to 14-3-3 is a crucial step in insulin's ability to exert its actions.

Vpr: An Inhibitor of Insulin Action on FOXO Proteins

Since Vpr binds 14-3-3 and changes its binding specificity to its partner protein Cdc25C, we hypothesized that Vpr might also modulate the binding activity of 14-3-3 to other partner proteins, the FOXO subfamily of the forkhead proteins. Indeed, Vpr moderately inhibited insulin or its downstream Akt-induced translocation of Foxo3a (FKHR-L1) into the cytoplasm, and interfered with insulin-induced coprecipitation of 14-3-3 and Foxo3a *in vivo*. Wild-type Vpr antagonized the negative effect of insulin on Foxo3a-induced transactivation of a FOXO-responsive promoter. Moreover, Vpr antagonized insulin-induced suppression of glucose 6-phosphatase mRNA, an endogenous FOXO-responsive gene, in HepG2 cells. These findings indicate that Vpr interferes with the negative effects of insulin on FOXO-mediated inhibition of target genes by inhibiting the association between these transcription factors and 14-3-3. These results may also indicate that Vpr appears to modulate the binding specificity of 14-3-3 positively or negatively, depending on partner molecules that bind to 14-3-3 proteins. These *in vitro* findings from our laboratory and a recent report indicating the involvement of FOXOs in the adipocyte development[97] suggest that Vpr may be a key viral factor that induces insulin resistance as well as lipodystrophy and hyperlipidemia by interfering with and/or modulating cellular activities, such as transactivation of nuclear receptors or insulin.

SUMMARY

The AIDS-related insulin resistance/lipodystrophy syndrome is a newly recognized severe pathologic condition, which may compromise the quality and expect-

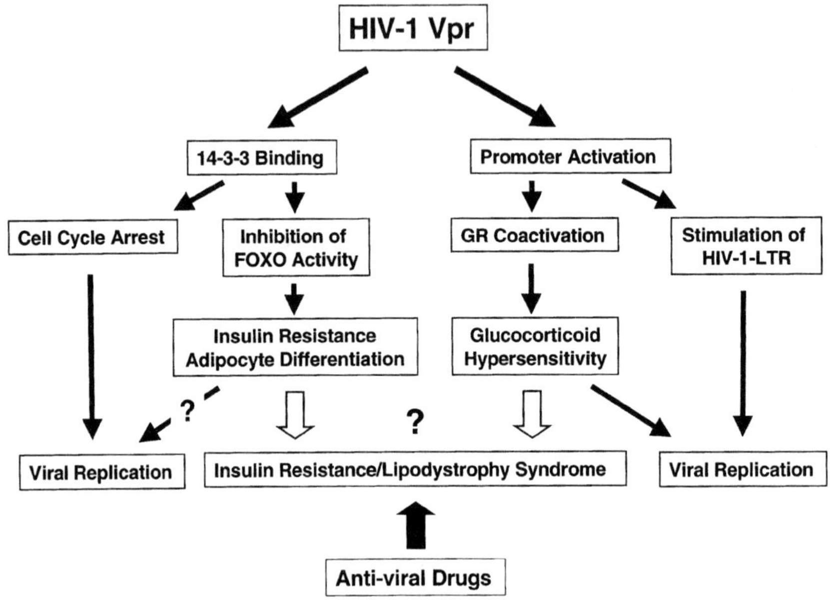

FIGURE 5. Possible contribution of Vpr to the pathogenesis of AIDS-related insulin resistance/lipodystrophy syndrome via its two distinct activities. Vpr may play a role in the development of this syndrome by modulating the activity of GR and FOXO proteins.

ancy of life in affected patients. To explore potential viral factors that cause this syndrome, we focused on the HIV-1 accessory protein Vpr and examined its effect on the glucocorticoid and insulin signal pathways. We demonstrated that this viral molecule upregulates the GR-induced transcriptional activity at the level of coactivators and potentially induces the glucocorticoid hypersensitivity in HIV-1–infected patients. In addition, Vpr inhibits the effects of insulin on FOXOs through interacting with the novel 14-3-3 proteins, thereby inducing insulin resistance (FIG. 5). Although further clinical evidence is required to prove a direct involvement of Vpr in the AIDS-related insulin resistance/lipodystrophy syndrome, neutralization of these molecules could be a potential target for the development of new therapeutic interventions for this syndrome.

ACKNOWLEDGMENT

We appreciate Dr. Evangelia Charmandari for critically reading the manuscript.

REFERENCES

1. PANTALEO, G., C. GRAZIOSI & A.S. FAUCI. 1993. New concepts in the immunopathogenesis of human immunodeficiency virus infection. N. Engl. J. Med. **328:** 327–335.

2. TASHIMA, K.T. & T.P. FLANAGAN. 2000. Antiretroviral therapy in the year 2000. Infect. Dis. Clin. North Am. **14:** 827–849.
3. MORRIS, A.B., S. CU-UVIN, J.I. HARWELL, *et al.* 2000. Multicenter review of protease inhibitors in 89 pregnancies. J. Acquir. Immune Defic. Syndr. **25:** 306–311.
4. KAUFMANN, G.R. & D.A. COOPER. 2000. Antiretroviral therapy of HIV-1 infection: established treatment strategies and new therapeutic options. Curr. Opin. Microbiol. **3:** 508–514.
5. HARRINGTON, M. & C.C. CARPENTER. 2000. Hit HIV-1 hard, but only when necessary. Lancet **355:** 2147–2152.
6. VELLA, S. & L. PALMISANO. 2000. Antiretroviral therapy: state of the HAART. Antiviral Res. **45:** 1–7.
7. LO, J.C., K. MULLIGAN, V.W. TAI, *et al.* 1998. "Buffalo hump" in men with HIV-1 infection. Lancet **351:** 867–870.
8. MILLER, K.K., P.A. DALY, D. SENTOCHNIK, *et al.* 1998. Pseudo-Cushing's syndrome in human immunodeficiency virus-infected patients. Clin. Infect. Dis. **27:** 68–72.
9. LO, J.C., K. MULLIGAN, V.W. TAI, *et al.* 1998. "Buffalo hump" in men with HIV-1 infection. Lancet **351:** 867–870.
10. FRIIS-MOLLER, N., C.A. SABIN, R. WEBER, *et al.* 2003. Combination antiretroviral therapy and the risk of myocardial infarction. N. Engl. J. Med. **349:** 1993–2003.
11. STOCKER, D.N., P.J. MEIER, R. STOLLER & K.E. FATTINGER. 1998. "Buffalo hump" in HIV-1 infection. Lancet **352:** 320–321.
12. SAINT-MARC, T. & J.L. TOURAINE. 1998. "Buffalo hump" in HIV-1 infection. Lancet **352:** 319–320.
13. CHRISTEFF, N., J.C. MELCHIOR, P. DE TRUCHIS, *et al.* 1999. Lipodystrophy defined by a clinical score in HIV-infected men on highly active antiretroviral therapy: correlation between dyslipidaemia and steroid hormone alterations. AIDS **13:** 2251–2260.
14. DORNIER, C., M. POSTH, F. GRANEL, *et al.* 1999. "Buffalo hump" related to indinavir treatment. Ann. Dermatol. Venereol. **126:** 720–722.
15. GRAHAM, N.M. 2000. Metabolic disorders among HIV-infected patients treated with protease inhibitors: a review. J. Acquir. Immune Defic. Syndr. **25:** S4–11.
16. CARR, A. & D.A. COOPER. 2000. Adverse effects of antiretroviral therapy. Lancet **356:** 1423–1430.
17. BEHRENS, G.M., M. STOLL & R.E. SCHMIDT. 2000. Lipodystrophy syndrome in HIV infection: what is it, what causes it and how can it be managed? Drug Saf. **23:** 57–76.
18. QAQISH, R.B., E. FISHER, J. RUBLEIN & D.A. WOHL. 2000. HIV-associated lipodystrophy syndrome. Pharmacotherapy **20:** 13–22.
19. HADIGAN, C., J.B. MEIGS, C. CORCORAN, *et al.* 2001. Metabolic abnormalities and cardiovascular disease risk factors in adults with human immunodeficiency virus infection and lipodystrophy. Clin. Infect. Dis. **32:** 130–139.
20. BRINKMAN, K., J.A. SMEITINK, J.A. ROMIJN & P. REISS. 1999. Mitochondrial toxicity induced by nucleoside-analogue reverse-transcriptase inhibitors is a key factor in the pathogenesis of antiretroviral-therapy-related lipodystrophy. Lancet **354:** 1112–1115.
21. COSSARIZZA, A., C. MUSSINI & A. VIGANO. 2001. Mitochondria in the pathogenesis of lipodystrophy induced by anti-HIV antiretroviral drugs: actors or bystanders? BioEssays **23:** 1070–1080.
22. KINO, T., M. MIRANI, S. ALESCI & G.P. CHROUSOS. 2003. AIDS-related lipodystrophy/insulin resistance syndrome. Horm. Metab. Res. **35:** 129–136.
23. PAVLAKIS G.N. 1996. The molecular biology of HIV-1. *In* AIDS: Diagnosis, Treatment and Prevention. V.T. DeVita, S. Hellman & S.A. Rosenberg, Eds.: 45–74. Lippincott Raven. Philadelphia.
24. EMERMAN, M. 1996. HIV-1, Vpr and the cell cycle. Curr. Biol. **6:** 1096–1103.
25. HENKLEIN, P., K. BRUNS, M.P. SHERMAN, *et al.* 2000. Functional and structural characterization of synthetic HIV-1 vpr that transduces cells, localizes to the nucleus, and induces G2 cell cycle arrest. J. Biol. Chem. **275:** 32016–32026.
26. MIRANI, M., I. ELENKOV, S. VOLPI, *et al.* 2002. HIV-1 protein Vpr suppresses IL-12 production from human monocytes by enhancing glucocorticoid action: potential implications of Vpr coactivator activity for the innate and cellular immunity deficits observed in HIV-1 infection. J. Immunol. **169:** 6361–6368.

27. Connor, R.I., B.K. Chen, S. Choe & N.R. Landau. 1995. Vpr is required for efficient replication of human immunodeficiency virus type-1 in mononuclear phagocytes. Virology **206:** 935–944.
28. Gibbs, J.S., A.A. Lackner, S.M. Lang, *et al.* 1995. Progression to AIDS in the absence of a gene for vpr or vpx. J. Virol. **69:** 2378–2383.
29. Goh, W.C., M.E. Rogel, C.M. Kinsey, *et al.* 1998. HIV-1 Vpr increases viral expression by manipulation of the cell cycle: a mechanism for selection of Vpr *in vivo*. Nat. Med. **4:** 65–71.
30. Morellet, N., S. Bouaziz, P. Petitjean & B.P. Roques. 2003. NMR structure of the HIV-1 regulatory protein VPR. J. Mol. Biol. **327:** 215–227.
31. Cohen, E.A., G. Dehni, J.G. Sodroski & W.A. Haseltine. 1990. Human immunodeficiency virus vpr product is a virion-associated regulatory protein. J. Virol. **64:** 3097–3099.
32. Paxton, W., R.I. Connor & N.R. Landau. 1993. Incorporation of Vpr into human immunodeficiency virus type 1 virions: requirement for the p6 region of gag and mutational analysis. J. Virol. **67:** 7229–7237.
33. Vodicka, M.A., D.M. Koepp, P.A. Silver & M. Emerman. 1998. HIV-1 Vpr interacts with the nuclear transport pathway to promote macrophage infection. Genes Dev. **12:** 175–185.
34. Popov, S., M. Rexach, L. Ratner, *et al.* 1998. Viral protein R regulates docking of the HIV-1 preintegration complex to the nuclear pore complex. J. Biol. Chem. **273:** 13347–13352.
35. Heinzinger, N.K., M.I. Bukinsky, S.A. Haggerty, *et al.* 1994. The Vpr protein of human immunodeficiency virus type 1 influences nuclear localization of viral nucleic acids in nondividing host cells. Proc. Natl. Acad. Sci. USA **91:** 7311–7315.
36. Cohen, E.A., R.A. Subramanian & H.G. Gottlinger. 1996. Role of auxiliary proteins in retroviral morphogenesis. Curr. Top. Microbiol. Immunol. **214:** 219–235.
37. Sherman, M.P., C.M. de Noronha, M.I. Heusch, *et al.* 2001. Nucleocytoplasmic shuttling by human immunodeficiency virus type 1 Vpr. J. Virol. **75:** 1522–1532.
38. Jacotot, E., L. Ravagnan, M. Loeffler, *et al.* 2000. The HIV-1 viral protein R induces apoptosis via a direct effect on the mitochondrial permeability transition pore. J. Exp. Med. **191:** 33–46.
39. Stewart, S.A, B. Poon, J.B. Jowett & I.S. Chen. 1997. Human immunodeficiency virus type 1 Vpr induces apoptosis following cell cycle arrest. J. Virol. **71:** 5579–5592.
40. Cohen, E.A., E.F. Terwilliger, Y. Jalinoos, *et al.* 1990. Identification of HIV-1 vpr product and function. J. Acquir. Immune Defic. Syndr. **3:** 11–18.
41. Wang, L., S. Mukherjee, F. Jia, *et al.* 1995. Interaction of virion protein Vpr of human immunodeficiency virus type 1 with cellular transcription factor Sp1 and trans-activation of viral long terminal repeat. J. Biol. Chem. **270:** 25564–25569.
42. Agostini, I., J.M. Navarro, F. Rey, *et al.* 1996. The human immunodeficiency virus type 1 Vpr transactivator: cooperation with promoter-bound activator domains and binding to TFIIB. J. Mol. Biol. **261:** 599–606.
43. Forget, J., X.J. Yao, J. Mercier & E.A. Cohen. 1998. Human immunodeficiency virus type 1 vpr protein transactivation function: mechanism and identification of domains involved. J. Mol. Biol. **284:** 915–923.
44. Kino, T., A. Gragerov, O. Slobodskaya, *et al.* 2002. Human immunodeficiency virus type-1 (HIV-1) accessory protein Vpr induces transcription of the HIV-1 and glucocorticoid-responsive promoters by binding directly to p300/CBP coactivators. J. Virol. **76:** 9724–9734.
45. Jowett, J.B., V. Planelles, B. Poon, *et al.* 1995. The human immunodeficiency virus type 1 vpr gene arrests infected T cells in the G2+M phase of the cell cycle. J. Virol. **69:** 6304–6313.
46. He, J., S. Choe, R. Walker, *et al.* 1995. Human immunodeficiency virus type 1 viral protein R (Vpr) arrests cells in the G2 phase of the cell cycle by inhibiting p34cdc2 activity. J. Virol. **69:** 6705–6711.
47. Re, F., D. Braaten, E.K. Franke & J. Luban. 1995. Human immunodeficiency virus type 1 Vpr arrests the cell cycle in G2 by inhibiting the activation of p34cdc2-cyclin B. J. Virol. **69:** 6859–6864.

48. OHI, R. & K.L. GOULD. 1999. Regulating the onset of mitosis. Curr. Opin. Cell Biol. **11:** 267–273.
49. JACKMAN, M.R. & J.N. PINES. 1997. Cyclins and the G2/M transition. Cancer Surv. **29:** 47–73.
50. NEBREDA, A.R. & I. FERBY. 2000. Regulation of the meiotic cell cycle in oocytes. Curr. Opin. Cell Biol. **12:** 666–675.
51. MASUDA, M., Y. NAGAI, N. OSHIMA, *et al.* 2000. Genetic studies with the fission yeast *Schizosaccharomyces pombe* suggest involvement of wee1, ppa2, and rad24 in induction of cell cycle arrest by human immunodeficiency virus type 1 Vpr. J. Virol. **74:** 2636–2646.
52. ROGEL, M.E., L.I. WU & M. EMERMAN. 1995. The human immunodeficiency virus type 1 vpr gene prevents cell proliferation during chronic infection. J. Virol. **69:** 882–888.
53. ELDER, R.T., M. YU, M. CHEN, *et al.* 2000. Cell cycle G2 arrest induced by HIV-1 Vpr in fission yeast (*Schizosaccharomyces pombe*) is independent of cell death and early genes in the DNA damage checkpoint. Virus Res. **68:** 161–173.
54. YUAN, H., Y.M. XIE & I.S. CHEN. 2003. Depletion of Wee-1 kinase is necessary for both human immunodeficiency virus type 1 Vpr- and gamma irradiation-induced apoptosis. J. Virol. **77:** 2063–2070.
55. TUNG, H.Y., H. DE ROCQUIGNY, L.J. ZHAO, *et al.* 1997. Direct activation of protein phosphatase-2A0 by HIV-1 encoded protein complex NCp7:vpr. FEBS Lett. **401:** 197–201.
56. ELDER, R.T., M. YU, M. CHEN, *et al.* 2001. HIV-1 Vpr induces cell cycle G2 arrest in fission yeast (*Schizosaccharomyces pombe*) through a pathway involving regulatory and catalytic subunits of PP2A and acting on both Wee1 and Cdc25. Virology **287:** 359–370.
57. ROSENQUIST, M., P. SEHNKE, R.J. FERL, *et al.* 2000. Evolution of the 14-3-3 protein family: does the large number of isoforms in multicellular organisms reflect functional specificity? J. Mol. Evol. **51:** 446–458.
58. MUSLIN, A.J. & H. XING. 2000. 14-3-3 proteins: regulation of subcellular localization by molecular interference. Cell Signal. **12:** 703–709.
59. FU, H., R.R. SUBRAMANIAN & S.C. MASTERS. 2000. 14-3-3 proteins: structure, function, and regulation. Annu. Rev. Pharmacol. Toxicol. **40:** 617–647.
60. YAFFE, M.B., K. RITTINGER, S. VOLINIA, *et al.* 1997. The structural basis for 14-3-3:phosphopeptide binding specificity. Cell **91:** 961–971.
61. RITTINGER, K., J. BUDMAN, J. XU, *et al.* 1999. Structural analysis of 14-3-3 phosphopeptide complexes identifies a dual role for the nuclear export signal of 14-3-3 in ligand binding. Mol. Cell. **4:** 153–166.
62. TSOPANOMICHALOU, M., T. KINO, A. GRAGEROV, *et al.* 2002. Involvement of HIV Vpr protein in G2/M arrest through interaction with 14-3-3 and Cdc25C. Retrovirus Meeting. Cold Spring Harbor, NY, May 21–26, p235.
63. LOPEZ-GIRONA, A., B. FURNARI, O. MONDESERT & P. RUSSELL. 1999. Nuclear localization of Cdc25 is regulated by DNA damage and a 14-3-3 protein. Nature **397:** 172–175.
64. GRAVES, P.R., C.M. LOVLY, G.L. UY & H. PIWNICA-WORMS. 2001. Localization of human Cdc25C is regulated both by nuclear export and 14-3-3 protein binding. Oncogene **20:** 1839–1851.
65. DALAL, S.N., C.M. SCHWEITZER, J. GAN & J.A. DECAPRIO. 1999. Cytoplasmic localization of human cdc25C during interphase requires an intact 14-3-3 binding site. Mol. Cell. Biol. **19:** 4465–4479.
66. SANCHEZ, Y., C. WONG, R.S. THOMA, *et al.* 1997. Conservation of the Chk1 checkpoint pathway in mammals: linkage of DNA damage to Cdk regulation through Cdc25. Science **277:** 1497–1501.
67. ZENG, Y., K.C. FORBES, Z. WU, *et al.* 1998. Replication checkpoint requires phosphorylation of the phosphatase Cdc25 by Cds1 or Chk1. Nature **395:** 507–510.
68. PENG, C.Y., P.R. GRAVES, R.S. THOMA, *et al.* 1997. Mitotic and G2 checkpoint control: regulation of 14-3-3 protein binding by phosphorylation of Cdc25C on serine-216. Science **277:** 1501–1505.
69. WANG, Y., C. JACOBS, K.E. HOOK, *et al.* 2000. Binding of 14-3-3beta to the carboxyl terminus of Wee1 increases Wee1 stability, kinase activity, and G2-M cell population. Cell Growth Differ. **11:** 211–219.

70. LEE, J., A. KUMAGAI & W.G. DUNPHY. 2001. Positive regulation of Wee1 by Chk1 and 14-3-3 proteins. Mol. Biol. Cell **12:** 551–563.
71. CHEN, L., T.H. LIU & N.C. WALWORTH. 1999. Association of Chk1 with 14-3-3 proteins is stimulated by DNA damage. Genes Dev. **13:** 675–685.
72. CHAN, T.A., H. HERMEKING, C. LENGAUER, *et al.* 1999. 14-3-3Sigma is required to prevent mitotic catastrophe after DNA damage. Nature **401:** 616–620.
73. LARONGA, C., H.Y. YANG, C. NEAL & M.H. LEE. 2000. Association of the cyclin-dependent kinases and 14-3-3 sigma negatively regulates cell cycle progression. J. Biol. Chem. **275:** 23106–23112.
74. YANOVSKI, J.A., K.D. MILLER, T. KINO, *et al.* 1999. Endocrine and metabolic evaluation of human immunodeficiency virus-infected patients with evidence of protease inhibitor-associated lipodystrophy. J. Clin. Endocrinol. Metab. **84:** 1925–1931.
75. BAMBERGER, C.M., H.M. SCHULTE & G.P. CHROUSOS. 1996. Molecular determinants of glucocorticoid receptor function and tissue sensitivity to glucocorticoids. Endocr. Rev. **17:** 245–261.
76. MCKENNA, N.J., R.B. LANZ & B.W. O'MALLEY. 1999. Nuclear receptor coregulators: cellular and molecular biology. Endocr. Rev. **20:** 321–344.
77. GLASS, C.K. & M.G. ROSENFELD. 2000. The coregulator exchange in transcriptional functions of nuclear receptors. Genes Dev. **14:** 121–141.
78. YANG, X.J., V.V. OGRYZKO, J. NISHIKAWA, *et al.* 1996. A p300/CBP-associated factor that competes with the adenoviral oncoprotein E1A. Nature **382:** 319–324.
79. LEO, C. & J.D. CHEN. 2000. The SRC family of nuclear receptor coactivators. Gene **245:** 1–11.
80. HEERY, D.M., E. KALKHOVEN, S. HOARE & M.G. PARKER. 1997. A signature motif in transcriptional co-activators mediates binding to nuclear receptors. Nature **387:** 733–736.
81. REFAELI, Y., D.N. LEVY & D.B. WEINER. 1995. The glucocorticoid receptor type II complex is a target of the HIV-1 vpr gene product. Proc. Natl. Acad. Sci. USA **92:** 3621–3625.
82. KINO, T., A. GRAGEROV & J.B. KOPP, *et al.* 1999. The HIV-1 virion-associated protein vpr is a coactivator of the human glucocorticoid receptor. J. Exp. Med. **189:** 51–62.
83. KINO, T., A. GRAGEROV, O. SLOBODSKAYA, *et al.* 2002. Human immunodeficiency virus type-1 accessory protein Vpr induces transcription of the HIV-1 and glucocorticoid-responsive promoters by binding directly to p300/CBP coactivators. J. Virol. **76:** 9724–9734.
84. WEI, P., M.E. GARBER, S.M. FANG, *et al.* 1998. A novel CDK9-associated C-type cyclin interacts directly with HIV-1 Tat and mediates its high-affinity, loop-specific binding to TAR RNA. Cell **92:** 451–462.
85. KINO, T., O. SLOBODSKAYA, G.N. PAVLAKIS & G.P. CHROUSOS. 2002. Nuclear receptor coactivator p160 proteins enhance the HIV-1 long terminal repeat promoter by bridging promoter-bound factors and the Tat-P-TEFb complex. J. Biol. Chem. **277:** 2396–2405.
86. FAWELL, S., J. SEERY, Y. DAIKH, *et al.* 1994. Tat-mediated delivery of heterologous proteins into cells. Proc. Natl. Acad. Sci. USA **91:** 664–668.
87. NAKAE, J. & D. ACCILI. 1999. The mechanism of insulin action. J. Pediatr. Endocrinol. Metab. **12:** 721–731.
88. O'BRIEN, R.M., R.S. STREEPER, J.E. AYALA, *et al.* 2001. Insulin-regulated gene expression. Biochem. Soc. Trans. **29:** 552–558.
89. CARLSSON, P. & M. MAHLAPUU. 2002. Forkhead transcription factors: key players in development and metabolism. Dev. Biol. **250:** 1–23.
90. WANG, J.C., J.M. STAFFORD, D.K. SCOTT, *et al.* 2000. The molecular physiology of hepatic nuclear factor 3 in the regulation of gluconeogenesis. J. Biol. Chem. **275:** 14717–14721.
91. EISENBERGER, C.L., H. NECHUSHTAN, H. COHEN, *et al.* 1992. Differential regulation of the rat phosphoenolpyruvate carboxykinase gene expression in several tissues of transgenic mice. Mol. Cell. Biol. **12:** 1396–1403.
92. GUO, S., G. RENA, S. CICHY, *et al.* 1999. Phosphorylation of serine 256 by protein kinase B disrupts transactivation by FKHR and mediates effects of insulin on insulin-

like growth factor-binding protein-1 promoter activity through a conserved insulin response sequence. J. Biol. Chem. **274:** 17184–17192.

93. TOMIZAWA, M., A. KUMAR, V. PERROT, *et al.* 2000. Insulin inhibits the activation of transcription by a C-terminal fragment of the forkhead transcription factor FKHR. A mechanism for insulin inhibition of insulin-like growth factor-binding protein-1 transcription. J. Biol. Chem. **275:** 7289–7295.
94. NAKAE, J., T. KITAMURA, D.L. SILVER & D. ACCILI. 2001. The forkhead transcription factor Foxo1 (Fkhr) confers insulin sensitivity onto glucose-6-phosphatase expression. J. Clin. Invest. **108:** 1359–1367.
95. BRUNET, A., A. BONNI, M.J. ZIGMOND, *et al.* 1999. Akt promotes cell survival by phosphorylating and inhibiting a Forkhead transcription factor. Cell **96:** 857–868.
96. NAKAE, J., V. BARR & D. ACCILI. 2000. Differential regulation of gene expression by insulin and IGF-1 receptors correlates with phosphorylation of a single amino acid residue in the forkhead transcription factor FKHR. EMBO J. **19:** 989–996.
97. NAKAE, J., T. KITAMURA, K. KITAMURA, *et al.* 2003. The forkhead transcription factor Foxo1 regulates adipocyte differentiation. Dev. Cell **4:** 119–129.
98. JENKINS, Y., O. PORNILLOS, R.L. RICH, *et al.* 2001. Biochemical analyses of the interactions between human immunodeficiency virus type 1 Vpr and p6(Gag). J. Virol. **75:** 10537–10542.
99. MAHALINGAM, S., V. AYYAVOO, M. PATEL, *et al.* 1997. Nuclear import, virion incorporation, and cell cycle arrest/differentiation are mediated by distinct functional domains of human immunodeficiency virus type 1 Vpr. J. Virol. **71:** 6339–6347.
100. YAO, X.J., R.A. SUBRAMANIAN, N. ROUGEAU, *et al.* 1995. Mutagenic analysis of human immunodeficiency virus type 1 Vpr: role of a predicted N-terminal alpha-helical structure in Vpr nuclear localization and virion incorporation. J. Virol. **69:** 7032–7044.

Familial/Sporadic Glucocorticoid Resistance

Clinical Phenotype and Molecular Mechanisms

EVANGELIA CHARMANDARI, TOMOSHIGE KINO, AND GEORGE P. CHROUSOS

Pediatric and Reproductive Endocrinology Branch, National Institute of Child Health and Human Development, National Institutes of Health, Bethesda, Maryland 20892, USA

ABSTRACT: Glucocorticoids regulate a variety of biologic processes and exert profound influences on many physiologic functions. Their actions are mediated by the glucocorticoid receptor (GR), which belongs to the nuclear receptor family of ligand-dependent transcription factors. Alterations in tissue sensitivity to glucocorticoids may manifest as states of resistance or hypersensitivity. Glucocorticoid resistance is a rare, familial or sporadic, condition characterized by generalized, partial target-tissue resistance to glucocorticoids. Compensatory elevations in circulating adrenocorticotropic hormone (ACTH) concentrations lead to increased production of adrenal steroids with mineralocorticoid and/or androgenic activity and their corresponding clinical manifestations, as well as increased urinary free-cortisol excretion in the absence of symptomatology suggestive of hypercortisolism. The molecular basis of the condition has been ascribed to mutations in the GR gene, which impair normal glucocorticoid signal transduction, altering tissue sensitivity to glucocorticoids. The present review focuses on the mechanisms of GR action and the clinical manifestations and molecular mechanisms of familial/sporadic glucocorticoid resistance.

KEYWORDS: glucocorticoid receptor; glucocorticoid resistance; tissue sensitivity to glucocorticoids; mutations in the GR gene

INTRODUCTION

Glucocorticoids regulate a variety of biologic processes and exert profound influences on many physiologic functions, including basal and stress-related homeostasis, growth and development, reproduction, metabolism, and thyroid and immune functions.[1–3] At the cellular level, the actions of glucocorticoids are mediated by an intracellular receptor, the glucocorticoid receptor (GR), which belongs to the nuclear receptor family of ligand-dependent transcription factors.

The physiologic response and sensitivity to glucocorticoids vary among species, individuals, tissues, cell types, and during the cell cycle.[4,5] Abnormalities in glucocorticoid sensitivity can be divided into two major groups: resistance and hypersen-

Address for correspondence: Evangelia Charmandari, M.D., National Institutes of Health, Building 10, Room 9D42, 10 Center Drive MSC 1583, Bethesda, MD 20892-1583. Voice: 301-496-5800; fax: 301-402-0884.

charmane@mail.nih.gov

Ann. N.Y. Acad. Sci. 1024: 168–181 (2004).
doi: 10.1196/annals.1321.014

sitivity. Target tissue resistance to glucocorticoids is characterized by inability of target tissues to respond to these hormones and may be generalized or tissue-specific, transient or permanent, partial or complete, compensated or non compensated.[6,7] Complete glucocorticoid resistance is not compatible with life. Absence of functional GR in $GR^{-/-}$ knockout mice leads to severe neonatal respiratory distress syndrome and death within a few hours after birth.[8]

In the present review, we will provide a brief overview of the mechanisms of action of GR and discuss the clinical phenotype and molecular mechanisms of familial/ sporadic glucocorticoid resistance.

MOLECULAR MECHANISMS OF GLUCOCORTICOID ACTION

The Human Glucocorticoid Receptor Gene, mRNA, and Protein

The human glucocorticoid receptor (hGR) gene is located on chromosome 5 and consists of nine exons (FIG. 1A). Exon 1 and the first part of exon 2 contain the 5′UTR, exons 2–9 the coding sequence, and exon 9 the 3′UTR.[9] The activity of hGR gene is regulated by at least three different promoters, which facilitate tissue-specific expression.[10,11] Two of these promoters (promoter 1B and 1C) are located in the 3-kb region

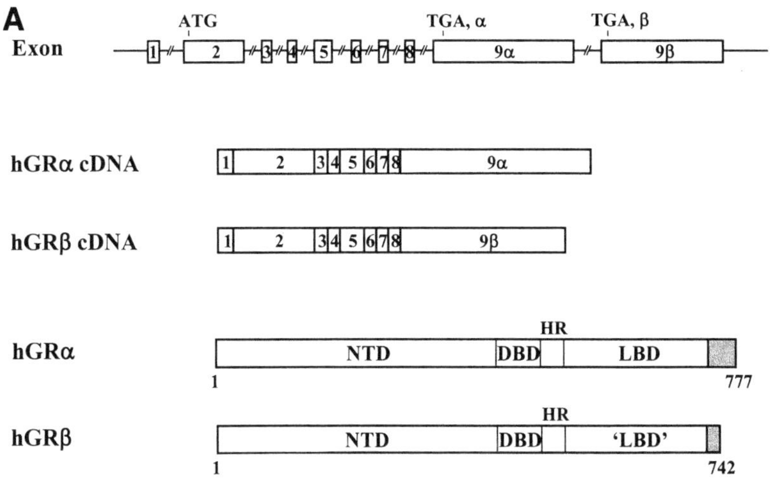

FIGURE 1. (**A**) Schematic representation of the structure of the human glucocorticoid receptor (hGR) gene. Alternative splicing of the primary transcript gives rise to the two mRNA and protein isoforms, hGRα and hGRβ. The functional domains and subdomains are indicated beneath the linearized protein structures. AF, activation function; DBD, DNA-binding domain; LBD, ligand-binding domain; NLS, nuclear localization signal. (**B**) Schematic representation of the interaction of coactivators with the AF-1 and AF-2 domains of the glucocorticoid receptor and their role in transcriptional regulation. AF, activation function; GR, glucocorticoid receptor; GREs, glucocorticoid-response elements. (**C**) Linearized GRIP1 molecule and distribution of its functional domains. AD1, activation domain 1; AD2, activation domain 2; HLH, helix-loop-helix; NIDaux, auxiliary nuclear receptor interacting domain; NRB, nuclear receptor binding; PAS, period aryl hydrogen receptor and single-minded. (From Vottero and colleagues.[54])

upstream of exon 1C and contain several GC boxes, which are binding sites for the transcription factor SP-1. Promoter 1B contains a putative nuclear factor (NF)-κB–binding site, while promoter 1C contains a putative activator protein (AP)-2 site. No obvious TATA or CAAT elements have been found in these promoters.[11] The third promoter, promoter 1A, has been located further upstream, approximately 31 kb upstream of the coding sequence, and contains a putative interferon regulatory factor–binding element and a sequence resembling a GRE.[10]

The hGR cDNA was first cloned in 1985 and two forms were initially described: hGRα and hGRβ cDNA.[12,13] The hGRα and hGRβ mRNA transcripts result from alternative splicing of the hGR gene in exon 9 and are highly homologous. Both transcripts contain exons 1–8 but different versions of exon 9 (9α and 9β, respectively), and are translated into hGRα and hGRβ protein, respectively (FIG. 1A). Recently, a third hGR mRNA has been described, which has a higher expression level in most tissues than the other two previously discovered messengers. This messenger, which is 7.0 kb in length, contains exons 1–8 and the entire exon 9 (exon 9α, the "J region" and exon 9β) and is thought to be translated into hGRα, although potentially it may be translated into hGRβ.[11]

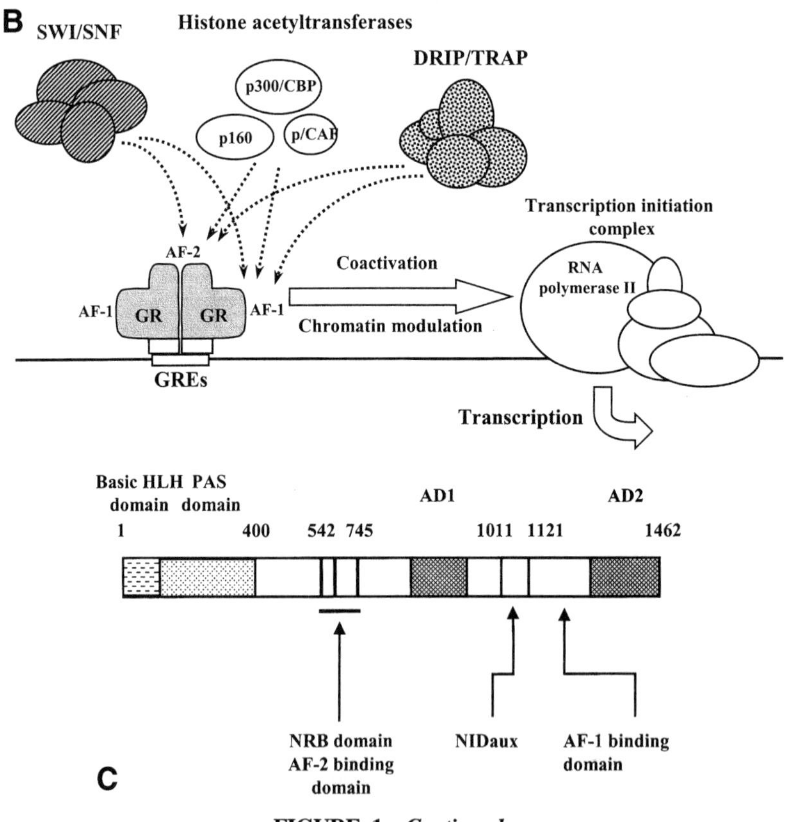

FIGURE 1. *Continued.*

The molecular structure of hGR is similar to that of other steroid receptors and comprises three functional domains: (1) a poorly conserved amino-terminal domain (NTD), which consists of amino acids 1–420 and contains a major transactivation domain, termed activation function (AF)-1 or τ1; (2) a central, highly conserved DNA-binding domain (DBD), which spans over amino acids 421–488, and contains two zinc-finger motifs through which it binds to specific DNA sequences in the promoter region of target genes, the glucocorticoid response elements (GREs). The DBD also contains a dimerization and a nuclear localization domain (NLS1); and (3) a carboxyl-terminal, ligand-binding domain (LBD), which is formed by amino acids 527–777 and contains a second transactivation domain, termed AF-2 or τ2, a second nuclear localization signal (NLS2), and sequences important for interaction with heat-shock proteins (hsps) and nuclear transcription factors, nuclear translocation and receptor dimerization.[14–17]

The two isoforms of hGR are identical through amino acid 727, but then diverge, with hGRα having an additional 50 amino acids and hGRβ having an additional, nonhomologous 15 amino acids (FIG. 1A).[18] hGRα is ubiquitously expressed in almost all human tissues and cells, and represents the classic hGR that functions as a ligand-dependent transcription factor. hGRβ is also ubiquitously expressed in tissues, but resides primarily in the nucleus of cells, does not bind glucocorticoids, and is transcriptionally inactive.[19,20] hGRβ functions as a dominant negative inhibitor of hGRα activity and inhibits hGRα-mediated transactivation of target genes in a dose-dependent manner.[21] The ability of hGRβ to antagonize the function of hGRα indicates that hGRβ may play a role in regulating target tissue sensitivity to glucocorticoids.[22–26] Increased expression of hGRβ has been documented in generalized and tissue-specific glucocorticoid resistance and leads to a reduction in the ability of hGRα to bind to GREs.[22,23] Therefore, an imbalance in hGRα and hGRβ expression may underlie the pathogenesis of several clinical conditions associated with glucocorticoid resistance, such as rheumatoid arthritis, systemic lupus erythematosus, or ulcerative colitis.[27]

Mechanisms of Glucocorticoid Receptor Action

In the absence of ligand, hGRα resides mostly in the cytoplasm of cells as part of a large multiprotein complex, which consists of the receptor polypeptide, two molecules of hsp90, and several other proteins.[28] The hsp90 molecules are thought to sequester hGRα in the cytoplasm of cells by maintaining the receptor in a conformation that masks/inactivates its nuclear localization signals (NLSs), but empowers it to interact with glucocorticoids. Upon hormone binding, the receptor undergoes an allosteric change, which results in dissociation from hsp90 and other proteins, unmasking/activation of the NLSs and phosphorylation at five serine phosporylation sites (positions 113, 141, 203, 211, and 226). In its new conformation, the phosphorylated, ligand-bound glucocorticoid receptor translocates into the nucleus, where it binds as homodimer to GREs located in the promoter region of target genes. hGRα then communicates with the basal transcription machinery and regulates the expression of glucocorticoid-responsive genes positively or negatively, depending on the GRE sequence and promoter context.[28] The receptor can also modulate gene expression independently of GRE-binding, by physically interacting with other transcription factors, such as AP-1 and NF-κB.[29,30]

It is not fully understood how binding of hGRα to GREs leads to initiation of transcription. The basal transcription machinery, which consists of RNA-polymerase II and general transcription factors, such as TATA box-binding protein (TBP) and a host of TBP-associated proteins (TAF_{II}s), must be recruited to the promoter.[31] The transcriptional activity of hGRα also depends on coactivators that facilitate recruitment of the basal transcription machinery or remodel chromatin and are attracted to the promoter region of target genes via the AF-1 and AF-2 of the receptor (FIG. 1B).[32,33] Several families of coactivators have been described, including the p160 coactivators, such as the steroid receptor coactivator 1 (SRC1) and the glucocorticoid receptor-interacting protein 1 (GRIP1), the p300/CREB-binding protein (CBP) cointegrators, the p300/CBP-associated protein (p/CAF), the switching/sucrose nonfermenting (SWI/SNF) complex, and the newly described vitamin D receptor–interacting protein/thyroid hormone associated protein (DRIP/TRAP) complex.[34–36] Both the p160 and CBP/p300 coactivators have multiple amphipathic LXXLL motifs, which, upon ligand binding, serve as interfaces between these coactivators and a hydrophobic cleft formed by the α-helix 12 of the nuclear receptors.[37] The GRIP1 coactivator contains two sites that bind to steroid receptors: one site, the nuclear receptor binding (NRB) site, is located between amino acids 542 and 745, and contains three LXXLL signature motifs, through which it interacts with the AF-2 of hGRα in a ligand-dependent fashion. The other site is located between amino acids 1121 and 1250, and binds to the AF-1 of hGRα in a ligand-independent fashion (FIG. 1C).[38–40] The p160, CBP/p300, and p/CAF proteins have histone acetyltransferase activity, which enables them to acetylate histones in the promoter, a process that is known to enhance transcription.

FAMILIAL/SPORADIC GLUCOCORTICOID RESISTANCE

Clinical Manifestations

Glucocorticoid resistance is a rare, familial or sporadic condition characterized by generalized, partial end-organ insensitivity to physiologic glucocorticoid concentrations.[6,7,41–45] Affected subjects have compensatory elevations in circulating cortisol and adrenocorticotropic hormone (ACTH) concentrations, but no clinical evidence of hypo- or hypercortisolism. Although adequate compensation is achieved by the elevated cortisol concentrations in the majority of patients with familial or sporadic generalized glucocorticoid resistance, the excess ACTH secretion also results in increased production of adrenal steroids with mineralocorticoid activity, such as deoxycorticosterone and corticosterone, and/or androgenic activity, such as androstenedione, dehydroepiandrosterone (DHEA) and DHEA-sulfate (DHEAS).[6,7,41–45] The former accounts for symptoms and signs of mineralocorticoid excess, such as hypertension and hypokalemic alkalosis. The latter accounts for manifestations of androgen excess, such as acne, hirsutism, and infertility in both sexes, male-pattern hair-loss, menstrual irregularities (oligo-amenorrhea), and oligo-anovulation in females, and oligospermia and infertility in males. In adult males, the oligospermia and infertility are most likely the result of feedback inhibition of follicle-stimulating hormone (FSH) secretion by the excessive adrenal androgens and/or ACTH-induced growth of intratesticular adrenal rests. In children, the early and excessive prepuber-

TABLE 1. Mutations of the human glucocorticoid receptor gene causing glucocorticoid resistance

Author	Mutation Position		Biochemical				
	cDNA	Amino Acid	Phenotype	Present Study	Genotype	Transmission	Phenotype
Chrousos *et al.*[6] Hurley *et al.*[48]	2054 (A→T)	641 (D→V)	Affinity for ligand ↓ Transactivation ↓	Transcriptional Activity of LBD ↓ Transdominance (–) Nuclear translocation ↓ DNA binding (+) Abnormal interaction with GRIP1	Homozygous	Autosomal recessive	Hypertension Hypokalemic alkalosis
Karl *et al.*[49]	4 bp deletion in exon-intron 6		hGRα number: 50% of control Inactivation of the affected allele		Heterozygous	Autosomal dominant	Hirsutism Male-pattern baldness Menstrual irregularities
Malchoff *et al.*[50]	2317 (G→A)	729 (V→I)	Affinity for ligand ↓ Transactivation ↓	Transcriptional Activity of LBD ↓ Transdominance (–) Nuclear translocation ↓ DNA binding (+) Abnormal interaction with GRIP1	Homozygous	Autosomal recessive	Precocious puberty Hyperandrogenism

TABLE 1. (*continued*) **Mutations of the human glucocorticoid receptor gene causing glucocorticoid resistance**

Author	Mutation Position		Biochemical				
	cDNA	Amino Acid	Phenotype	Present Study	Genotype	Transmission	Phenotype
Karl *et al.*[46] Kino *et al.*[52]	1808 (T→A)	559 (I→N)	Affinity for ligand ↓ Transactivation ↓ Transdominance (+)	Transcriptional Activity of LBD ↓ DNA binding (+) Abnormal interaction with GRIP1	Heterozygous	Sporadic	Hypertension Oligospermia, Infertility
Mendonca *et al.*[53]	1844 (C→T)	571 (V→A)	Affinity for ligand ↓ Transactivation ↓	Transcriptional Activity of LBD↓ Transdominance (–) Nuclear translocation ↓ DNA binding (+) Abnormal interaction GRIP1	Homozygous	Autosomal recessive	Ambiguous genitalia at birth Hypertension Hyperandrogenism Hypokalemia
Vottero *et al.*[54]	2373 (T→G)	747 (I →M)	Affinity for ligand ↓ Transactivation↓ Transdominance (+) Nuclear translocation ↓	Transcriptional Activity of LBD ↓ DNA binding (+)	Heterozygous	Autosomal dominant	Cystic acne, Hirsutism Oligoamenorrhea
Ruiz *et al.*[51]	1430 (G→A)	477 (R→H)	Transactivation ↓		Heterozygous	Sporadic	Hirsutism, Fatigue
	2035 (G→A)	679 (G→S)	Affinity for ligand ↓ Transactivation ↓		Heterozygous	Sporadic	Hypertension

tal adrenal androgen secretion has been associated with ambiguous genitalia at birth and precocious puberty. The clinical spectrum of glucocorticoid resistance is quite broad, ranging from completely asymptomatic to severe cases of hyperandrogenism, fatigue, and/or mineralocorticoid excess.[41–45]

Diagnosis

Elevated serum cortisol concentrations and 24-h urinary free-cortisol excretion in the absence of clinical features of hypercortisolism are highly suggestive of the condition. The plasma concentrations of ACTH may be normal or high. The circadian pattern of ACTH and cortisol secretion and their responsiveness to stressors are preserved, albeit at higher concentrations, and there is resistance of the HPA axis to dexamethasone suppression. Thymidine incorporation and dexamethasone binding assays on peripheral blood mononuclear cells or cultured skin fibroblasts, as well as sequencing of genomic DNA or complementary DNA may be necessary to confirm the diagnosis.

The differential diagnosis includes: (1) mild forms of Cushing's syndrome, in which hypercortisolism is accompanied by normal or mildly elevated ACTH concentrations, preserved circadian pattern of ACTH and cortisol secretion, and lack of cortisol suppression by dexamethasone; (2) pseudo-Cushing's states; (3) conditions associated with elevated serum concentrations of cortisol-binding globulin; (4) other causes of mineralocorticoid-induced hypertension; and (5) other causes of hyperandrogenism, such as idiopathic hirsutism, polycystic ovarian syndrome, and nonclassic congenital adrenal hyperplasia.

Treatment

The aim of treatment in glucocorticoid resistance is to suppress the excessive secretion of ACTH, and hence the increased production of mineralocorticoids and androgens from the adrenal cortex. Treatment involves administration of high doses of mineralocorticoid-sparing synthetic glucocorticoids, such as dexamethasone (1–3 mg/day), which activate the mutated and/or wild-type hGR and suppress the endogenous secretion of ACTH.[41–45] Adequate suppression of the HPA axis is of particular importance in cases of severe impairment of hGR function, given that longstanding corticotroph hyperstimulation in association with decreased glucocorticoid negative feedback may lead to the development of an ACTH-secreting adenoma.[46]

Despite the pharmacologic doses of dexamethasone employed in the treatment of generalized glucocorticoid resistance, patients do not manifest any side effects or Cushingoid features. Long-term dexamethasone treatment should be carefully titrated based on the clinical manifestations and biochemical profile.[6] Asymptomatic, normotensive subjects with primary glucocorticoid resistance do not require any treatment.

Molecular Mechanisms of Glucocorticoid Resistance

The molecular basis of glucocorticoid resistance in several families and sporadic cases has been ascribed to mutations in the hGRα gene, which impair one or more of the molecular mechanisms of glucocorticoid receptor function, thus altering tissue sensitivity to glucocorticoids.[6,47] Abnormalities of several hGRα characteristics,

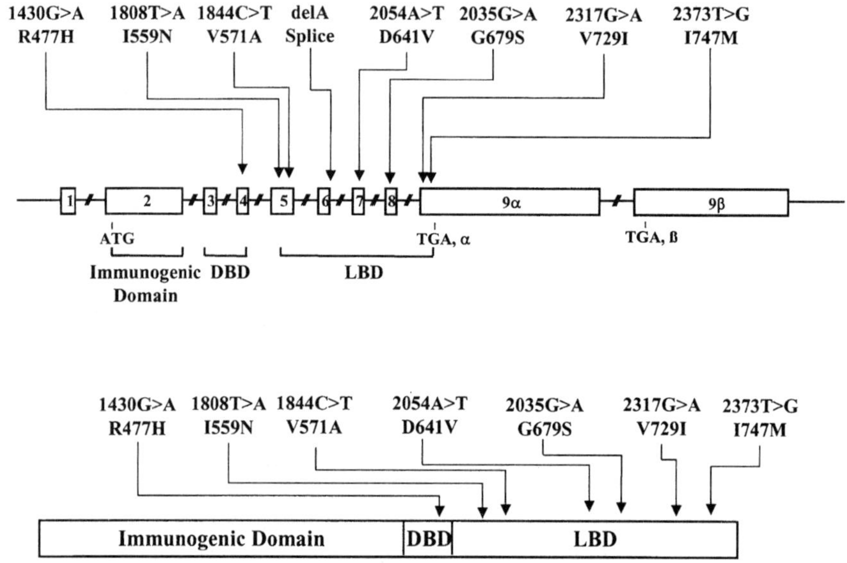

FIGURE 2. Location of the known mutations of the human glucocorticoid receptor gene (*upper panel*) and protein (*lower panel*).

such as cell concentration, affinity for ligand, and translocation into the nucleus have been associated with this condition.[48–55] The molecular defects elucidated in the reported cases are summarized in TABLE 1, while the corresponding mutations in the hGRα gene are shown in both TABLE 1 and FIGURE 2.

The propositus of the original kindred was homozygous for a single point mutation at nucleotide position 2054, which results in a nonconservative amino acid substitution at position 641, replacing aspartic acid with valine.[48] Compared to the wild-type receptor, this mutant receptor results in decreased transactivation of the glucocorticoid-responsive mouse mammary tumor virus (MMTV) promoter and a threefold reduction in the affinity for dexamethasone.[48] In the absence of ligand, the mutant receptor is primarily localized in the cytoplasm of cells. Exposure to dexamethasone (10^{-6} M) induces a slow translocation into the nucleus, which takes 22 min as opposed to 12 min required for nuclear translocation of the wild-type receptor.[55] Finally, the mutant receptor interacts with the amino-terminal but not with the carboxyl-terminal fragment or full-length GRIP1 *in vitro.*[55]

The propositus of the second family was heterozygous for a four-base deletion at 3′-boundary of exon and intron 6. The deletion removed a donor splice site in one allele, affecting the last two bases of the exon 6 and the first two bases of the intron 6. This resulted in complete ablation of the expression of one of the hGRα alleles and a decrease in GRα protein by 50% in affected members of the family.[49]

The propositus of the third kindred was homozygous for a point mutation at nucleotide position 2317, which results in substitution of valine for isoleucine at amino acid 729 of the ligand-binding domain of hGRα.[50] This mutation results in decreased transcriptional activity of the receptor and a fourfold reduction in the affinity for dexamethasone.[50] The mutant receptor is localized primarily in the nucleus of cells in the absence of ligand, while further translocation from the cytoplasm into the nucleus requires longer (120 min) exposure to dexamethasone (10^{-6} M), and demonstrates a weak, ligand-dependent interaction with the full-length and carboxyl-terminal fragment but not with the amino-terminal fragment of GRIP1 *in vitro*.[55]

The first sporadic case of glucocorticoid resistance was due to a *de novo*, germline, heterozygous mutation at nucleotide position 1808, resulting in substitution of isoleucine for asparagine at amino acid 559 in the hormone-binding domain of hGRα. This mutation reduces the transcriptional activity of hGRα significantly and has been associated with the development of an ACTH-secreting adenoma.[46] Although the affinity for ligand was preserved in the patient studied, there was a 50% decrease in the hGR binding sites.[46] Furthermore, the mutant receptor has a markedly delayed nuclear translocation (180 min) and a dominant negative activity upon the wild-type receptor, that is, it decreases the transcriptional activity of hGRα in a dose-dependent manner.[52] The latter may account for manifestation of the disease at the heterozygotic state. There is no interaction between the mutant receptor and GRIP1.[55]

The fifth and sixth cases of glucocorticoid resistance were due to heterozygous mutations at nucleotide positions 1430 and 2035, resulting, respectively, in substitution of arginine for histidine at amino acid 477 and glycine for serine at amino acid 679.[51] The former mutation is located in the second zinc finger of the DNA-binding domain. This mutant receptor has no transactivation activity due to impaired binding to GREs but has the same affinity for ligand as the wild-type receptor. The latter mutation is located in the ligand-binding domain, outside the ligand-binding pocket, and results in a 50% reduction both in the transcriptional activity and the ligand-binding affinity of the receptor.[51]

The propositus of the seventh case was homozygous for a point mutation at nucleotide position 1844, which results in a valine to alanine substitution at amino acid 571 in the ligand-binding domain of hGRα.[53] This mutation causes up to 50-fold decrease in the transcriptional activity of the receptor and a sixfold reduction in the affinity for ligand.[53] The nuclear translocation of the mutant receptor is delayed (25 min), while its interaction with the GRIP1 coactivator occurs mostly via its AF-1 domain.[55]

The eighth case of glucocorticoid resistance was due to a heterozygous mutation at nucleotide position 2373, which causes a isoleucine to methionine substitution at amino acid 747 in the ligand-binding domain of the receptor.[54] This mutation is located at the carboxyl-terminus of the ligand-binding domain, close to helix 12, which plays a pivotal role in the formation of AF-2, a domain that interacts with p160 and other coactivators. The mutant receptor has a 20- to 30-fold decrease in the transactivation of the MMTV promoter, a twofold reduction in the affinity for dexamethasone, and delayed nuclear translocation. It also exerts a dominant negative effect upon the wild-type hGRα and interacts with the GRIP1 coactivator *in vitro* only through its intact AF-1 domain. Overexpression of GRIP1 restores the transcriptional activity and reverses the dominant negative activity of the mutant upon the wild-type receptor.[54]

Further to the above mechanisms, we have demonstrated that (1) the ligand-binding domains of the mutant receptors hGRαI559N, hGRαV571A, hGRαD641V, hGRαV729I, and hGRαI747M have decreased intrinsic transcriptional activity; (2) hGRαV571A, hGRαD641V, and hGRαV729I do not exert a dominant negative effect on the transcriptional activity of hGRα; and (3) all five mutant receptors hGRαI559N, hGRαV571A, hGRαD641V, hGRαV729I, and hGRαI747M preserve their ability to bind to DNA.[55] Therefore, the process through which the above hGRα mutant receptors impair the physiologic mechanisms of glucocorticoid action at the molecular level is multifactorial and involves impaired ability to bind ligand, aberrant nucleocytoplasmic trafficking, and abnormal interaction with the p160 coactivators. These variable functional defects of the mutant receptors upon the glucocorticoid signaling pathway may explain the genetic transmission and the variable clinical phenotype of glucocorticoid resistance.

In addition to mutations in the hGRα gene, steroid receptor coactivator defects may also account for generalized glucocorticoid resistance and/or resistance to other steroid hormones. New and colleagues described two sisters with multiple, partial steroid resistance, whose clinical and biochemical findings were largely consistent with isolated glucocorticoid resistance.[56,57] However, the paradoxical absence of hyperandrogenic manifestations necessitated further evaluation, which also revealed resistance to mineralocorticoids, estrogen, and androgens, but not to vitamin D or thyroid hormones.[56]

SUMMARY

Mutations in the human glucocorticoid receptor gene impair one or more of the molecular mechanisms of glucocorticoid action, affecting normal glucocorticoid signal transduction and altering tissue sensitivity to these hormones. A subsequent increase in the activity of the HPA axis compensates for the reduced sensitivity of peripheral tissues to glucocorticoids, however, at the expense of ACTH hypersecretion-related pathology. The study of the functional defects of natural hGR mutants sheds light on the mechanisms of hGR action, including hGR-mediated transactivation of target genes, ligand-binding, nuclear translocation, DNA binding, and interaction with coactivators, and highlights the importance of integrated cellular and molecular signaling mechanisms for maintaining homeostasis and preserving normal physiology.

REFERENCES

1. TSAI, M.J. & B.W. O'MALLEY. 1994. Molecular mechanisms of action of steroid/thyroid receptor superfamily members. Annu. Rev. Biochem. **63:** 451–486.
2. SIMPSON, E.R. & M.R. WATERMAN. 1995. Steroid biosynthesis in the adrenal cortex and its regulation by adrenocorticotropin. *In* Endocrinology. L.J. DeGroote *et al.*, Eds: 1630–1641. W.B. Saunders. Philadelphia.
3. LAZAR, M.A. 2003. Mechanism of action of hormones that act on nuclear receptors. *In* Williams Textbook of Endocrinology. P.R. Larsen *et al.*, Eds.: 35–44. W.B. Saunders. Philadelphia.
4. HSU, S.C. & D.B. DEFRANCO. 1995. Selectivity of cell cycle regulation of glucocorticoid receptor function. J. Biol. Chem. **270:** 3359–3364.
5. LIM-TIO, S.S., M.C. KEIGHTLEY & P.J. FULLER. 1997. Determinants of specificity of transactivation by the mineralocorticoid or glucocorticoid receptor. Endocrinology **138:** 2537–2543.

6. CHROUSOS, G.P., A. VINGERHOEDS, D. BRANDON, *et al.* 1982. Primary cortisol resistance in man. A glucocorticoid receptor-mediated disease. J. Clin. Invest. **69:** 1261–1269.
7. CHROUSOS, G.P., S.D. DETERA-WADLEIGH & M. KARL. 1993. Syndromes of glucocorticoid resistance. Ann. Intern. Med. **119:** 1113–1124.
8. COLE, T.J., J.A. BLENDY, A.P. MONAGHAN, *et al.* 1995. Targeted disruption of the glucocorticoid receptor gene blocks adrenergic chromaffin cell development and severely retards lung maturation. Genes Dev. **9:** 1608–1621.
9. ENCIO, I.J. & S.D. DETERA-WADLEIGH. 1991. The genomic structure of the human glucocorticoid receptor. J. Biol. Chem. **266:** 7182–7188.
10. BRESLIN, M.B., C.D. GENG & W.V. VEDECKIS. 2001. Multiple promoters exist in the human GR gene, one of which is activated by glucocorticoids. Mol. Endocrinol. **15:** 1381–1395.
11. SCHAAF, M.J. & J.A. CIDLOWSKI. 2002. Molecular mechanisms of glucocorticoid action and resistance. J. Steroid Biochem. Mol. Biol. **83:** 37–48.
12. HOLLENBERG, S.M., C. WEINBERGER, E.S. ONG, *et al.* 1985. Primary structure and expression of a functional human glucocorticoid receptor cDNA. Nature **318:** 635–641.
13. WEINBERGER, C., S.M. HOLLENBERG, E.S. ONG, *et al.* 1985. Identification of human glucocorticoid receptor complementary DNA clones by epitope selection. Science **228:** 740–742.
14. CARSON-JURICA, M.A., W.T. SCHRADER & B.W. O'MALLEY. 1990. Steroid receptor family: structure and functions. Endocr. Rev. **11:** 201–220.
15. PICARD, D. & K.R. YAMAMOTO. 1987. Two signals mediate hormone-dependent nuclear localization of the glucocorticoid receptor. EMBO J. **6:** 3333–3340.
16. HOLLENBERG, S.M. & R.M. EVANS. 1988. Multiple and cooperative trans-activation domains of the human glucocorticoid receptor. Cell **55:** 899–906.
17. DALMAN, F.C., L.C. SCHERRER, L.P. TAYLOR, *et al.* 1991. Localization of the 90-kDa heat shock protein-binding site within the hormone-binding domain of the glucocorticoid receptor by peptide competition. J. Biol. Chem. **266:** 3482–3490.
18. OAKLEY, R.H., M. SAR & J.A. CIDLOWSKI. 1996. The human glucocorticoid receptor beta isoform. Expression, biochemical properties, and putative function. J. Biol. Chem. **271:** 9550–9559.
19. DE CASTRO, M., S. ELLIOT, T. KINO, *et al.* 1996. The non-ligand binding beta-isoform of the human glucocorticoid receptor (hGR beta): tissue levels, mechanism of action, and potential physiologic role. Mol. Med. **2:** 597–607.
20. OAKLEY, R.H., J.C. WEBSTER, M. SAR, *et al.* 1997. Expression and subcellular distribution of the beta-isoform of the human glucocorticoid receptor. Endocrinology **138:** 5028–5038.
21. OAKLEY, R.H., C.M. JEWELL, M.R. YUDT, *et al.* 1999. The dominant negative activity of the human glucocorticoid receptor beta isoform. Specificity and mechanisms of action. J. Biol. Chem. **274:** 27857–27866.
22. LEUNG, D.Y., Q. HAMID, A. VOTTERO, *et al.* 1997. Association of glucocorticoid insensitivity with increased expression of glucocorticoid receptor beta. J. Exp. Med. **186:** 1567–1574.
23. SHAHIDI, H., A. VOTTERO, C.A. STRATAKIS, *et al.* 1999. Imbalanced expression of the glucocorticoid receptor isoforms in cultured lymphocytes from a patient with systemic glucocorticoid resistance and chronic lymphocytic leukemia. Biochem. Biophys. Res. Commun. **254:** 559–565.
24. LONGUI, C.A., A. VOTTERO, P.C. ADAMSON, *et al.* 2000. Low glucocorticoid receptor alpha/beta ratio in T-cell lymphoblastic leukemia. Horm. Metab. Res. **32:** 401–406.
25. HAUK, P.J., Q.A. HAMID, G.P. CHROUSOS & D.Y. LEUNG. 2000. Induction of corticosteroid insensitivity in human PBMCs by microbial superantigens. J. Allergy Clin. Immunol. **105:** 782–787.
26. HAUK, P.J., E. GOLEVA, I. STRICKLAND, *et al.* 2002. Increased glucocorticoid receptor Beta expression converts mouse hybridoma cells to a corticosteroid-insensitive phenotype. Am. J. Respir. Cell Mol. Biol. **27:** 361–367.
27. CHROUSOS, G.P. 1995. The hypothalamic-pituitary-adrenal axis and immune-mediated inflammation. N. Engl. J. Med. **332:** 1351–1362.

28. BAMBERGER, C.M., H.M. SCHULTE & G.P. CHROUSOS. 1996. Molecular determinants of glucocorticoid receptor function and tissue sensitivity to glucocorticoids. Endocr. Rev. **17:** 245–261.
29. JONAT, C., H.J. RAHMSDORF, K.K. PARK, *et al.* 1990. Antitumor promotion and antiinflammation: down-modulation of AP-1 (Fos/Jun) activity by glucocorticoid hormone. Cell **62:** 1189–1204.
30. SCHEINMAN, R.I., A. GUALBERTO, C.M. JEWELL, *et al.* 1995. Characterization of mechanisms involved in transrepression of NF-kappa B by activated glucocorticoid receptors. Mol. Cell Biol. **15:** 943–953.
31. BEATO, M. & A. SANCHEZ-PACHECO. 1996. Interaction of steroid hormone receptors with the transcription initiation complex. Endocr. Rev. **17:** 587–609.
32. MCKENNA, N.J., J. XU, Z. NAWAZ, *et al.* 1999. Nuclear receptor coactivators: multiple enzymes, multiple complexes, multiple functions. J. Steroid Biochem. Mol. Biol. **69:** 3–12.
33. MCKENNA, N.J. & B.W. O'MALLEY. 2002. Combinatorial control of gene expression by nuclear receptors and coregulators. Cell **108:** 465–474.
34. HITTELMAN, A.B., D. BURAKOV, J.A. INIGUEZ-LLUHI, *et al.* 1999. Differential regulation of glucocorticoid receptor transcriptional activation via AF-1-associated proteins. EMBO J. **18:** 5380–5388.
35. AUBOEUF, D., A. HONIG, S.M. BERGET & B.W. O'MALLEY. 2002. Coordinate regulation of transcription and splicing by steroid receptor coregulators. Science **298:** 416–419.
36. TORCHIA, J., D.W. ROSE, J. INOSTROZA, *et al.* 1997. The transcriptional co-activator p/CIP binds CBP and mediates nuclear-receptor function. Nature **387:** 677–684.
37. HEERY, D.M., E. KALKHOVEN, S. HOARE & M.G. PARKER. 1997. A signature motif in transcriptional co-activators mediates binding to nuclear receptors. Nature **387:** 733–736.
38. HONG, H., K. KOHLI, A. TRIVEDI, *et al.* 1996. GRIP1, a novel mouse protein that serves as a transcriptional coactivator in yeast for the hormone binding domains of steroid receptors. Proc. Natl. Acad. Sci. USA **93:** 4948–4952.
39. HONG, H., K. KOHLI, M.J. GARABEDIAN & M.R. STALLCUP. 1997. GRIP1, a transcriptional coactivator for the AF-2 transactivation domain of steroid, thyroid, retinoid, and vitamin D receptors. Mol. Cell Biol. **17:** 2735–2744.
40. DING, X.F., C.M. ANDERSON, H. MA, *et al.* 1998. Nuclear receptor-binding sites of coactivators glucocorticoid receptor interacting protein 1 (GRIP1) and steroid receptor coactivator 1 (SRC-1): multiple motifs with different binding specificities. Mol. Endocrinol. **12:** 302–313.
41. DE CASTRO, M. & G.P. CHROUSOS. 1997. Glucocorticoid resistance. Curr. Ther. Endocrinol. Metab. **6:** 188–189.
42. KINO, T. & G.P. CHROUSOS. 2001. Glucocorticoid and mineralocorticoid resistance/hypersensitivity syndromes. J. Endocrinol. **169:** 437–445.
43. KINO, T. & G.P. CHROUSOS. 2002. Tissue-specific glucocorticoid resistance-hypersensitivity syndromes: multifactorial states of clinical importance. J. Allergy Clin. Immunol. **109:** 609–613.
44. KINO, T., A. VOTTERO, E. CHARMANDARI & G.P. CHROUSOS. 2002. Familial/sporadic glucocorticoid resistance syndrome and hypertension. Ann. N.Y. Acad. Sci. **970:** 101–111.
45. KINO, T., M.U. DE MARTINO, E. CHARMANDARI, *et al.* 2003. Tissue glucocorticoid resistance/hypersensitivity syndromes. J. Steroid Biochem. Mol. Biol. **85:** 457–467.
46. KARL, M., S.W. LAMBERTS, J.W. KOPER, *et al.* 1996. Cushing's disease preceded by generalized glucocorticoid resistance: clinical consequences of a novel, dominant-negative glucocorticoid receptor mutation. Proc. Assoc. Am. Physicians **108:** 296–307.
47. VINGERHOEDS, A.C., J.H. THIJSSEN & F. SCHWARZ. 1976. Spontaneous hypercortisolism without Cushing's syndrome. J. Clin. Endocrinol. Metab. **43:** 1128–1133.
48. HURLEY, D.M., D. ACCILI, C.A. STRATAKIS, *et al.* 1991. Point mutation causing a single amino acid substitution in the hormone binding domain of the glucocorticoid receptor in familial glucocorticoid resistance. J. Clin. Invest. **87:** 680–686.
49. KARL, M., S.W. LAMBERTS, S.D. DETERA-WADLEIGH, *et al.* 1993. Familial glucocorticoid resistance caused by a splice site deletion in the human glucocorticoid receptor gene. J. Clin. Endocrinol. Metab. **76:** 683–689.

50. MALCHOFF, D.M., A. BRUFSKY, G. REARDON, *et al.* 1993. A mutation of the glucocorticoid receptor in primary cortisol resistance. J. Clin. Invest. **91:** 1918–1925.
51. RUIZ, M., U. LIND, M. GAFVELS, *et al.* 2001. Characterization of two novel mutations in the glucocorticoid receptor gene in patients with primary cortisol resistance. Clin. Endocrinol. (Oxford) **55:** 363–371.
52. KINO, T., R.H. STAUBER, J.H. RESAU, *et al.* 2001. Pathologic human GR mutant has a transdominant negative effect on the wild-type GR by inhibiting its translocation into the nucleus: importance of the ligand-binding domain for intracellular GR trafficking. J. Clin. Endocrinol. Metab. **86:** 5600–5608.
53. MENDONCA, B.B., M.V. LEITE, M. DE CASTRO, *et al.* 2002. Female pseudohermaphroditism caused by a novel homozygous missense mutation of the GR gene. J. Clin. Endocrinol. Metab. **87:** 1805–1809.
54. VOTTERO, A., T. KINO, H. COMBE, *et al.* 2002. A novel, C-terminal dominant negative mutation of the GR causes familial glucocorticoid resistance through abnormal interactions with p160 steroid receptor coactivators. J. Clin. Endocrinol. Metab. **87:** 2658–2667.
55. CHARMANDARI, E., T. KINO, A. VOTTERO, *et al.* 2004. Natural glucocorticoid receptor mutants causing generalized glucocorticoid resistance: Molecular genotype, genetic transmission and clinical phenotype. J. Clin. Endocrinol. Metab. **89:** 1939–1949.
56. NEW, M.I., S. NIMKARN, D.D. BRANDON, *et al.* 1999. Resistance to several steroids in two sisters. J. Clin. Endocrinol. Metab. **84:** 4454–4464.
57. CHROUSOS, G.P. 1999. A new "New" syndrome in the new world: is multiple postreceptor steroid hormone resistance due to a coregulator defect? J. Clin. Endocrinol. Metab. **84:** 4450–4453.

Early Environmental Regulation of Hippocampal Glucocorticoid Receptor Gene Expression

Characterization of Intracellular Mediators and Potential Genomic Target Sites

IAN C.G. WEAVER, JOSIE DIORIO, JONATHAN R. SECKL,[a] MOSHE SZYF, AND MICHAEL J. MEANEY

McGill Program for the Study of Behavior, Genes and Environment, Douglas Hospital Research Centre, Departments of Psychiatry, and Neurology and Neurosurgery, McGill University, Montreal, Canada, H4H 1R3

[a]Molecular Endocrinology Laboratory, Department of Medicine, University of Edinburgh, Edinburgh, Scotland, United Kingdom EH4 2XU

ABSTRACT: Environmental conditions in early life permanently alter the development of glucocorticoid receptor gene expression in the hippocampus and hypothalamic-pituitary-adrenal responses to acute or chronic stress. In part, these effects can involve an activation of ascending serotonergic pathways and subsequent changes in the expression of transcription factors that might drive glucocorticoid receptor expression in the hippocampus. This paper summarizes the evidence in favor of these pathways as well as recent studies describing regulatory targets within the chromatin structure of the promoter region of the rat hippocampal glucocorticoid receptor gene.

KEYWORDS: maternal care; stress; glucocorticoid receptor; methylation; epigenomic programming

Several years ago Levine, Denenberg, Zarrow, and colleagues[1–6] showed that early experience modified the development of adrenal glucocorticoid responses to a wide range of stressors. These findings clearly demonstrated that even rudimentary adaptive responses to stress could be modified by environmental events. More recent studies[7] revealed these environmental effects produce sustained alterations in glucocorticoid receptor gene expression in the hippocampus and frontal cortex, which mediate glucocorticoid negative feedback regulation of the hypothalamic-pituitary-adrenal (HPA) axis and thus result in stable differences in HPA responses to stress. These studies reflect the plasticity within brain regions that regulate the activity of the HPA axis, and provide a model for understanding the processes that contribute

Address for correspondence: Michael J. Meaney, Douglas Hospital Research Center, 6875 LaSalle Boulevard, Montreal (Quebec), Canada H4H 1R3. Voice: 514-761-6131 ext. 3938; fax: 514-762-3034.

michael.meaney@mcgill.ca

Ann. N.Y. Acad. Sci. 1024: 182–212 (2004). © 2004 New York Academy of Sciences.

doi: 10.1196/annals.1321.001

to individual differences in neuroendocrine function. The importance of this topic is underscored by the fact that individual differences in adrenal hormone and sympathetic responses to stress appear to be of considerable importance in determining the vulnerability to multiple forms of pathology.[8–11]

ENVIRONMENTAL REGULATION OF HPA AND BEHAVIORAL RESPONSES TO STRESS

Postnatal Handling Studies

Perhaps the strongest evidence for the environmental regulation of the development of HPA responses to stress comes from the postnatal handling research with rodents. Handling involves a brief (i.e., 3–15 min), daily period of separation of the pup from the mother for the first few weeks of life and results in decreased stress reactivity in adulthood.[2–6,12–15] As adults, neonatally handled rats show decreased fearfulness and more modest pituitary ACTH and adrenal corticosterone responses to stress; such effects are apparent in animals tested as late as 26 months of age.[16,17]

The handling effects on the development of HPA responses to stress have important functional consequences. In the rat, glucocorticoid levels often rise with age and are associated with hippocampal degeneration and the emergence of learning and memory deficits.[18–21] Such age-related increases in basal and stress-induced pituitary-adrenal activity are significantly less apparent in the handled animals, and thus these animals show little evidence of hippocampal aging.[16] Likewise, handled animals also show more modest stress-induced suppression of immune function by comparison to non-handled rats.[22]

Such findings may lead to the conclusion that handled animals are hardier or more resistant than non-handled animals. But this misses the point. Handled animals are not better adapted than non-handled animals, they are simply different. The environmental context then serves to determine the adaptive value of increased or decreased stress reactivity. In the examples cited above, it would appear that the handled animals are at some advantage by virtue of a more modest HPA response to stress. But this condition is not universal, Laban and colleagues[23] found that non-handled animals are more resistant to the induction of experimental allergic encephalomyelitis (EAE) than are handled animals. Glucocorticoids are protective against the development of EAE, which can be fatal.[24] Adrenalectomized animals, for example, rarely survive EAE. Hence, the increased HPA responsivity of the non-handled renders an advantage under these circumstances. The cost of such resistance is an increased vulnerability to glucocorticoid-induced illness, but it is not difficult to imagine a scenario whereby such a cost is an acceptable trade-off. In essence, the handling studies represent a robust example of phenotypic plasticity in the expression of defensive responses to threat. One obvious question concerns the nature of the neurobiological mechanisms that mediate such phenotypic variation.

Considering the importance of the corticotropin-releasing hormone (CRH) systems for both behavioral and HPA responses to stress, it is probably not surprising that these systems are critical targets for the handling effect on stress reactivity. Adult animals exposed to postnatal handling show decreased CRH mRNA expression in the paraventricular nucleus of the hypothalamus (PVNh) and the central nucleus of the amygdala,[14,25,26] decreased CRH content in the locus coeruleus,[26] and decreased CRH receptor levels in the locus coeruleus compared with non-handled rats.[26] The release of CRH and the activation of HPA responses to stress are mediated by stress-induced increases in the release of noradrenaline at the level of the PVNh. Indeed, CRH release from the amygdala activates the release of noradrenaline from the locus coeruleus.[27] Together, these findings suggest that there would be more modest CRH-induced activation of the locus coeruleus during stress in the handled animals. At least two findings are consistent with this idea. By comparison to non-handled rats, acute stress in handled animals produces (1) a smaller stress-induced increase in cFOS immunoreactive neurons in the locus coeruleus[28] and (2) more modest increases in extracellular noradrenaline levels in the PVNh.[29] We propose that postnatal handling can decrease the expression of behavioral responses to stress, in part, by altering the development of the central nucleus of the amygdala–locus coeruleus CRH system.

Postnatal handling affects the development of neural systems that regulate CRH gene expression. Levels of CRH mRNA and protein in PVNh neurons are subject to inhibitory regulation via glucocorticoid negative feedback.[9,30] Handled rats show increased negative feedback sensitivity to glucocorticoids.[13,14] This effect is, in turn, related to the increased glucocorticoid receptor expression in the hippocampus and frontal cortex,[14,31–33] regions known to mediate the inhibitory effects of glucocorticoids over CRH synthesis in PVNh neurons.[19,30,34–36] The alterations in glucocorticoid receptor expression are a critical feature for the effect of the early environment on negative feedback sensitivity and HPA responses to stress; reversing the differences in hippocampal glucocorticoid receptor levels eliminates the differences in HPA responses to stress between handled and non-handled animals.[13]

CRH activity within the amygdala–locus coeruleus pathway is subject to γ-aminobutyric acid-(GABA-)ergic inhibition.[37,38] Interestingly, handled rats also show increased $GABA_A$ and benzodiazepine (BZ) receptor levels in the noradrenergic cell body regions of the locus coeruleus and the nuclei tractus solitarius as well as in the basolateral and central nucleus of the amygdala.[39] These effects are associated with increased expression of the mRNA for the $\alpha 1$ and $\gamma 2$ subunit of the $GABA_A$ receptor, which together encode for proteins that form the BZ site. These findings suggest that the composition of the $GABA_A$ receptor complex in brain regions that regulate stress reactivity is influenced by early life events. Handling increases $\alpha 1$ and $\gamma 2$ subunit expression[39,40] and, importantly, this profile is associated with increased GABA binding.[41,42] Interestingly, in humans, individual differences in BZ receptor sensitivity are associated with vulnerability for anxiety disorders.[43]

Together, the effects of handling on glucocorticoid and $GABA_A$/BZ receptor gene expression could serve to dampen CRH synthesis and release and to decrease the effect of CRH at critical target sites, such as the locus coeruleus. We feel that this model provides a reasonable working hypothesis for the mechanisms underlying the handling effect on endocrine and behavioral responses to stress.[7]

WHAT ARE THE CRITICAL FEATURES OF THESE ENVIRONMENTAL MANIPULATIONS?

Some years ago Levine and colleagues[44] suggested that the effects of handling were actually mediated by alterations in mother–infant interactions. Indeed, postnatal handling increases the frequency of pup licking/grooming by mothers.[45] We[40,46,47] examined this question by attempting to define naturally occurring variations in maternal behavior over the first eight days following birth through the simple, albeit time-consuming, observation of mother–pup interactions in normally reared animals. There was considerable variation in two forms of maternal behavior: licking/grooming of pups and arched-back nursing.[48] Licking/grooming included both body as well as anogenital licking. Arched-back nursing, also referred to as "crouching," is characterized by a dam nursing her pups with her back conspicuously arched and legs splayed outward. While common, it is not the only posture from which dams nurse. A blanket posture represents a more relaxed version of the arched-back position, where the mother is almost lying on the suckling pups. This position provides substantially less opportunity for movements such as nipple switching. Dams also nurse from their sides and often will move from one posture to another over the course of a nursing bout. Interestingly, the frequency of licking/grooming and arched-back nursing was highly correlated ($r = +0.91$) across animals and thus we were able to define mothers according to both behaviors—High or Low licking/grooming–arched-back nursing (LG-ABN) mothers. For the sake of most of the studies described here, High and Low LG-ABN mothers were identified as females whose scores on both measures were ± 1 SD above (High) or below (Low) the mean for their cohort.[49]

The critical question, of course, concerns the potential consequences of these differences in maternal behavior for the development of behavioral and neuroendocrine responses to stress. Indeed, if postnatal handling results in more modest behavioral and HPA responses to stress through effects on maternal behavior, then the adult offspring of animals reared by High LG-ABN mothers should resemble animals handled as neonates. As adults, the offspring of High LG-ABN mothers showed reduced plasma ACTH and corticosterone responses to acute stress by comparison to the adult offspring of Low LG-ABN mothers. The High LG-ABN offspring also showed significantly increased hippocampal glucocorticoid receptor mRNA expression, enhanced glucocorticoid negative feedback sensitivity, and decreased hypothalamic CRH mRNA levels. Moreover, the magnitude of the corticosterone response to acute stress was significantly correlated with the frequency of both maternal licking/grooming ($r = -0.61$) and arched-back nursing ($r = -0.64$) during the first week of life, as was the level of hippocampal glucocorticoid receptor mRNA and hypothalamic CRH mRNA expression (all r's > 0.70).[46]

HOW MIGHT MATERNAL CARE REGULATE GENE EXPRESSION IN THE OFFSPRING?

The handling paradigm provides a model for understanding the mechanisms by which environmental stimuli can regulate neural development and physiology. This model is somewhat unique since most paradigms involving alterations in perinatal environmental conditions focus on changes in either synapse formation or neuron survival[50] that ultimately result in effects at the level of morphology. In contrast, handling affects neurochemical differentiation in the hippocampus, specifically altering the sensitivity of hippocampal cells to corticosterone, via an effect on glucocorticoid receptor gene expression and thus receptor density. Such variations in neuronal differentiation underlie important individual differences in tissue sensitivity to hormonal signals and thus represent a biochemical basis for environmental "programming" of neural systems.

The handling effect on the development of glucocorticoid receptor density in the hippocampus shows the common characteristics of a developmental effect. First, there is a specific "critical period" during which the organism is maximally responsive to the effects of handling. Second, the effects of handling during the first 21 days of life on glucocorticoid receptor density endure throughout the life of the animal. Finally, there is substantial specificity to the handling effect. Handling alters the glucocorticoid, but not mineralocorticoid, receptor gene expression. Interestingly, glucocorticoid and mineralocorticoid receptors are co-expressed in virtually all hippocampal neurons. Thus, the handling effect on gene expression is specific.

Temporal Features of the Handling Effect

Handling during the first week of life is as effective as handling during the entire first three weeks of life in reducing adrenal steroid responses to stress[51] and in increasing hippocampal glucocorticoid receptor density.[52] Handling over the second week of life is less effective, whereas animals handled between days 15 and 21 do not differ from non-handled animals in glucocorticoid receptor binding. Thus, in terms of both HPA activity and glucocorticoid receptor binding, the sensitivity of the system to environmental regulation decreases progressively over the first three weeks of life. Moreover, in comparison to same-aged non-handled animals, handled animals exhibited significantly increased hippocampal glucocorticoid receptor density as early as day 7 of life and the magnitude of the effect did not increase thereafter.[52] Thus, glucocorticoid receptor binding capacity appears to be especially sensitive to environmental regulation during the first week of life. However, please note that these findings do not preclude the possibility of other periods of environmental regulation. Indeed, a so-called critical period must be defined not only in terms of the target outcome, but also by the relevant input stimulus. Hence, we assume that these findings suggest that the critical period for the effect of handling on glucocorticoid receptor gene expression occurs during the first week of life.

The Role of Thyroid Hormones

Handling during the first week of life activates the hypothalamic-pituitary-thyroid axis leading to increased levels of circulating thyroxine (T_4) and increased intracellular levels of the biologically more potent T_4 metabolite, triiodothyronine (T_3). The pituitary-thyroid axis is a major regulator of HPA development (see Ref. 53 for a review). Neonatal treatment with either T_4 or T_3 resulted in significantly increased glucocorticoid receptor binding capacity in the hippocampus in animals examined as adults.[54] Like the handling manipulation, neither T_4 nor T_3 treatment affected hypothalamic or pituitary glucocorticoid receptor density. Moreover, administration of the thyroid hormone synthesis inhibitor, propylthiouracil (PTU), to handled pups for the first two weeks of life completely blocked the effects of handling on hippocampal glucocorticoid receptor binding capacity. These data are consistent with the idea that the thyroid hormones might mediate, in part at least, the effects of neonatal handling on the development of the forebrain glucocorticoid receptor system.

Systemic injections of neonatal rat pups represent a rather crude manipulation, particularly procedures involving thyroid hormones. While these data might implicate the thyroid hormones, there is no indication that the hippocampus is actually the critical site of action. To examine whether thyroid hormones might act directly on hippocampal cells we used an *in vitro* system, involving primary cultures of dissociated hippocampal cells.[55] The hippocampal cells were taken from embryonic rat pups (E20) and beginning on the fifth day following plating the cultures were exposed to 0, 1, 10, or 100 nM T_3. These cells exhibit both mineralocorticoid and glucocorticoid receptor binding.[56] Indeed, both receptors as well as their mRNAs can be detected using material from a 60-mm dish. The results of several experiments have failed to detect any effect of thyroid hormones on glucocorticoid receptor density in cultured hippocampal cells. These *in vitro* data suggest that (1) the effects of the thyroid hormones on the glucocorticoid receptor binding occur at some site distal to the hippocampal cells or (2) thyroid hormones interact at the level of the hippocampus with some other hormonal signal that is obligatory for the expression of the thyroid hormone effect.

The Role of Serotonin

Thyroid hormones have pervasive effects throughout the developing central nervous system (CNS) and one such effect involves the regulation of central serotonergic neurons.[57] Thyroid hormones increase serotonin (5-HT) turnover in the hippocampus of the neonatal rat. [58] Handling also increases hippocampal 5-HT turnover[58,59] and thus both manipulations increase serotonergic stimulation of hippocampal neurons. There is also direct evidence for an effect of 5-HT on glucocorticoid receptor density in the neonatal rat. Lesioning of the raphe 5-HT neurons with 5,7-dihydroxytryptamine (5,7-DHT) dramatically reduces the ascending serotonergic input into the hippocampus. Rat pups administered 5,7-DHT on the second day of life showed reduced hippocampal glucocorticoid receptor density as adults.[58] Interestingly, neonatal administration of 5,7-DHT produces only a transient effect, such that by adulthood 5-HT innervations to the hippocampus are restored. The effect of hippocampal glucocorticoid receptor levels, however, persists into adulthood. This finding suggests that the effect of 5-HT on hippocampal glucocorticoid receptor expression, like handling itself, is unique to the first week of life.

Serotonin significantly increases glucocorticoid receptor density in cultured hippocampal cells.[56,60] In hippocampal cells cultured in the presence of increasing concentrations of 5-HT, there was a twofold increase in glucocorticoid receptor binding. The effect of 5-HT was dose related, with an EC_{50} of 4–5 nM and a maximal effect achieved at 10 nM concentrations that require a four-day treatment period. Shorter periods of exposure were ineffective, suggesting that the effect of 5-HT involves the increased synthesis of receptors. In support of this idea, we found that the effect of 5-HT on glucocorticoid receptor density in cultured hippocampal cells is blocked by either actinomycin-D or cycloheximide and is paralleled by an increase in glucocorticoid receptor mRNA levels.

The effect of 5-HT on glucocorticoid receptor expression occurs uniquely in the neuronal cell population. We found no effect of 5-HT on glucocorticoid receptor binding in hippocampal glial-enriched cell cultures. This finding is not surprising, since our initial studies were performed with cultures composed largely (~85%) of neuron-like cells.[56] Moreover, the composition of the cultures is unaffected by 5-HT treatment. We also examined the potential involvement of the glial cells by using a conditioned-medium experiment in which glial-enriched cultures were treated for five days with 5-HT and the medium was then used to feed neuronal cultures. This procedure had no effect on glucocorticoid receptor density, suggesting that the effect was not due to a 5-HT–induced glial secretory product.

The effects of 10 nM 5-HT on glucocorticoid receptor density in cultured hippocampal cells are completely blocked by the 5-HT_2 receptor antagonists, ketanserin and mianserin.[56,60] Moreover, the 5-HT_{2A} agonists 1-(2,5-dimethoxy-4-iodophenyl)-2-aminopropane (DOI), 3-trifluoromethyl-phenylpiperazine monohydrochloride (TFMPP), and quipazine were also effective in increasing glucocorticoid receptor binding in hippocampal culture, although not as effective as 5-HT. Selective agonists or antagonists of the 5-HT_{1A} or 5-HT_3 receptors have no effect on glucocorticoid receptor binding. Using $^{125}I_7$-amino-8-iodo-ketanserin as radioligand, we found high-affinity 5-HT_{2A} binding sites in our cultured hippocampal cells.

We then examined the nature of the secondary messenger systems involved in this serotonergic effect on glucocorticoid receptor binding. Mitchell and colleagues[56] found that low nanomolar concentrations of 5-HT (EC_{50} = 7 nM) produce a fourfold increase in cAMP levels in cultured hippocampal cells, with no effect on cGMP levels. This increase in cAMP is blocked by ketanserin and at least partially mimicked by quipazine, TFMPP, and DOI. Indeed, there is a strong correlation (+0.97) between the effects of these 5-HT receptor agonists on cAMP and glucocorticoid receptor levels.

Treatment with the stable cAMP analogue, 8-bromo-cAMP or with 10 μM forskolin produces a significant increase in glucocorticoid receptor density in cultured hippocampal neurons.[60] The effect of 8-bromo-cAMP is concentration related, and the maximal effect of 8-bromo-cAMP (~190%) is comparable to that for 5-HT (~200%). Interestingly, as with 5-HT, the effects of 8-bromo-cAMP on glucocorticoid receptor mRNA levels and receptor density are not apparent until at least four days of treatment.

Taken together, these findings suggest that changes in cAMP concentrations may mediate the effects of 5-HT on glucocorticoid receptor synthesis in hippocampal cells. We[60] also found that the cyclic nucleotide–dependent protein kinase inhibitor, H8, completely blocked the effects of 10 nM 5-HT on glucocorticoid receptor binding in hippocampal cell cultures. In contrast, the protein kinase C inhibitor, H7, had no such effect. These data suggest that activation of protein kinase A is involved in the serotonergic regulation of hippocampal glucocorticoid receptor development.

These studies involve effects on intact cells with long incubation periods, and there is ample possibility for an interaction between second messenger systems. This issue also arises because the 5-HT_{2A} receptor is linked not to cyclic nucleotide, but phospholipase C–related second messenger systems. In both *in vivo* and *in vitro* studies, 5-HT_{2A} agonists increase both diacylglycerol (DAG) levels and inositol phosphate (IP) metabolism (notably IP_1) within hippocampal membranes.[61] There are numerous examples in the literature of such "crosstalk" between second messenger systems and the stimulation of IP metabolism via phorbol esters alters cAMP levels.[62] However, in contrast to other compounds, such as glutamate or carbachol, we found that stimulation of IP metabolism by 5-HT in hippocampal slices was rather modest in animals during the first week of life.[63] Interestingly, the effect of 5-HT on IP metabolism in hippocampal slices is decreased in the handled animals on postnatal day 7, while the stimulation of DAG is slightly enhanced. However, the overall pattern of 5-HT stimulation is weak. This may be due to differences in receptor coupling at this time of life. In neonatal rat hippocampi the stimulation of IP metabolism occurred via 5-HT_{2C} and not 5-HT_{2A} receptors during the first weeks of life.[64] Since there is little 5-HT_{2C} receptor expression in dorsal hippocampus this may explain the weak stimulation of phospholipase C–related second messenger systems.

These data suggest that 5-HT directly stimulates cAMP formation in hippocampal neurons. This idea is not easy to reconcile with the involvement of a 5-HT_{2A} receptor. However, a number of 5-HT receptors have been cloned and these receptors directly stimulate adenylyl cyclase activity. These include the 5-HT_4, 5-HT_6, and 5-HT_7 receptors.[65] The mRNAs for each of these receptors is expressed in rat hippocampus. Moreover, the 5-HT_7 receptor binds ketanserin with high affinity. Interestingly, antidepressants increase glucocorticoid receptor mRNA in cortical and hippocampal cell cultures.[66,67] Both the 5-HT_6 and 5-HT_7 receptors bind various antidepressants with high affinity and the 5-HT_7 receptor shows a high affinity for ketanserin.

To examine the potential involvement of the 5-HT_7 receptor in mediating the increase in glucocorticoid receptor levels, we[68] measured receptor expression in cultured hippocampal neurons after treatment with 10 mM 8-bromo-cAMP or with various doses of the specific 5-HT_7 receptor agonist, 3-(2-aminoethyl)-1H-indole-5-carboxamide maleate (5-carboxamidotryptamine; 5-CT) for seven days. All treatments resulted in an increase in glucocorticoid receptor levels. The effect of 5-CT on glucocorticoid receptor expression was blocked by methiothepin. Likewise, 5-CT produced a significant increase in cAMP levels and the effect was blocked by methiothepin. Pindolol, which binds to the 5-HT_{1A} but not the 5-HT_7 receptor, had little effect. These results further implicate the 5-HT_7 receptor. The increase in glucocorticoid receptor expression is also mimicked with 5-methoxytryptamine (5-MeOT), an effect blocked with methiothepin as well as H8, a PKA inhibitor. Over the course of these studies we found that other serotonergic agonists (quipazine, TFMPP, DOI) could partially mimic the 5-HT effect on glucocorticoid receptor levels; this, however, was the first evidence that a more selective serotonergic agonist, 5-CT, could fully mimic the 5-HT effect. This observation is consistent with the idea that the effect of 5-HT on glucocorticoid receptor expression in hippocampal neurons is mediated by a 5-HT_7 receptor via activation of cAMP.

Activation of cAMP pathways can regulate gene transcription through effects on a number of transcription factors, including of course the cAMP-response element binding protein (CREB) through an enhanced phosphorylation of CREB. Phospho-CREB (pCREB) regulates gene transcription through pathways that involve the cofactor, CREB-binding protein (CBP). To further examine the relevant signal transduction pathway, CBP expression was investigated by Western blot analysis. Primary hippocampal cell cultures treated with 10 mM 8-bromo cAMP, 50 nM 5-CT, and 100 nM 5-HT all showed a significant increase CBP expression. Furthermore, the profile of pCREB was similar to CBP. Treatment of primary hippocampal cell cultures with 50 nM 5-CT resulted in a significant increase in phosphorylation of CREB.

In Vivo *5-HT Effects on Glucocorticoid Receptor Expression*

Our *in vivo* studies[59] of 5-HT activity provide some insight into why the hippocampus is selectively affected by handling. In rat pups handled for the first seven days of life, and sacrificed immediately following handling on postnatal day 7, 5-HT turnover was significantly increased in the hippocampus, but not in the hypothalamus or amygdala (regions where handling has no effect on glucocorticoid receptor density). These data suggest that handling selectively activates certain ascending 5-HT pathways and that this effect underlies the sensitivity of this receptor system in specific brain regions to regulation by environmental events during the first week of life.

Clearly,one concern here is the relationship between our *in vitro* results and the *in vivo* condition. Thus, it is reassuring that effects of postnatal handling of rat pups on hippocampal glucocorticoid receptor binding are blocked by concurrent administration of ketanserin.[58] Moreover, ketanserin treatment also blocked the effects of T_3 on hippocampal glucocorticoid receptor expression.[69] This finding also supports the idea that thyroid hormones mediate the handling effect by serving to increase 5-HT activity. We also examined the effects of handling on cAMP levels in hippocampal tissue in neonatal rats and found that handling stimulates a fourfold increase in cAMP levels.[70] These increases in cAMP are almost completely abolished by concurrent treatment with either ketanserin or the thyroid hormone synthesis inhibitor, PTU. Thus, to date the results from these *in vivo* studies certainly appear consistent with our earlier *in vitro* experiments.

The regulation of gene transcription by cAMP[71–78] is mediated by various transcription factors including cyclic nucleotide response element binding proteins (CREBs), cyclic nucleotide response element binding modulators (CREMs), most of which seem to be antagonists for CREBs, and the activating transcription factor family (ATF-1, ATF-2, ATF-3). In addition to the CREB/CREM-ATF family, nerve growth factor–inducible factors (NGFI-A and NGFI-B) as well as activator protein-2 (AP-2) are inducible by cAMP.[73,79] The promoter region of the human and mouse glucocorticoid receptor gene has been cloned and at least partially sequenced[80,81] and contains numerous binding sites for most of these transcription factors, providing a mechanism whereby cAMP might increase glucocorticoid receptor expression.

We[70] used a variety of techniques to study potential handling-induced changes in the expression of these transcription factors in neonatal rat hippocampus. Handling resulted in no change in NGFI-B mRNA expression, a significant (i.e., two- to three-fold) increase in AP-2 mRNA expression, and a very substantial (i.e., eight- to ten-fold) increase in NGFI-A mRNA levels. The increase in NGFI-A expression occurred across all hippocampal cell fields and in virtually every neuron. The increase in AP-2 and NGFI-A mRNAs is apparent immediately following the termination of handling, persists for at least three hours, and is associated with an increase in both AP-2 and NGFI-A immunoreactivity, indicating that the increase in mRNA expression is reflected in changes in protein levels. The handling effects on both NGFI-A and AP-2 expression are blocked by ketanserin or PTU.

The challenge at this point is to define the molecular targets for the early environmental effects. First, we are assuming that one target for regulation is the promoter region of the glucocorticoid receptor gene. We[82] identified and characterized several new glucocorticoid receptor mRNAs cloned from rat hippocampus. All encode a common protein, but differ in their 5′-leader sequences presumably as a consequence of alternative splicing of potentially 11 different exon 1 sequences. The alternate exon 1 sequences are unlikely to alter the amino acid sequence of the glucocorticoid receptor protein; there is an in-frame stop codon present immediately 5′ to the translation initiation site in exon 2, common to all the mRNA variants. From the 10 alternate exon 1 sequences we identified by 5′-RACE, four correspond to alternative exons 1 sequence previously identified in mouse—exons 1_1, 1_5, 1_9, and 1_{10}.[81,83] Most alternative exons are located in a 3-kb CpG island upstream of exon 2 that exhibits substantial promoter activity in transfected cells. Ribonuclease protection assays demonstrated significant levels of six alternative exon 1 sequences *in vivo* in the rat, with differential expression in the liver, hippocampus, and thymus presumably reflecting tissue-specific differences in promoter activity. Two of the alternative exon 1 sequences (exon 1_6 and 1_{10}) were expressed in all tissues examined, together present in 77–87% of total glucocorticoid mRNA. The remaining glucocorticoid receptor transcripts contained tissue-specific alternative first exons. Hippocampal RNA contained significant levels of the minor exon 1_5-, 1_7-, and 1_{11}-containing glucocorticoid receptor mRNA variants that were expressed at either low or undetectable levels in liver and thymus.

In transient transfection experiments, a construct encoding the whole CpG island of the glucocorticoid receptor gene, including eight of the alternate exons 1 and the splice acceptor site within the intron 5′ of exon 2, fused to a luciferase reporter gene within exon 2, exhibited substantial promoter activity in all cell lines tested. This activity results from transcripts originating at any point within the CpG island that are spliced from an appropriate donor site onto the splice acceptor site 5′ to exon 2, and represents the sum of the activity of individual promoters on the genomic DNA fragment.

Promoter activity was also associated with particular regions of the CpG island, where the fusion to luciferase was made within specific exon 1 sequences. In these cases, no splice acceptor site is available within the luciferase gene, and a transcriptional fusion is generated between the specific exon 1 and the luciferase reporter; luciferase activity therefore reflects transcription through the specific exon 1. Relative activity of these constructs in different cell types was similar with one notable exception, the exon 1_7 promoter sequence ($P1_7$). Interestingly, $P1_7$, fused to luciferase within exon 1_7, had the highest activity of any single promoter construct in B103 and C6 cells, both of which are CNS derived. The activity of this construct was low in hepatic cells, in which $P1_6$ and $P1_{10}$ had the highest activity. *In vivo*, glucocorticoid receptor mRNA transcripts containing exon 1_7 were present at significant levels in the hippocampus, but absent from the liver, suggesting that factors present in cells of CNS origin are responsible for transcription initiation at the promoter upstream of exon 1_7 in rat hippocampus.

Interestingly, tissue-specific alternative exon 1_7 usage was altered by postnatal handling that, of course, increases glucocorticoid receptor expression in the hippocampus. Handling selectively elevated glucocorticoid receptor mRNA containing exon 1_7; there was, for example, no effect on exon 1_{10}. Predictably, maternal care also affected the expression of glucocorticoid receptor splice variants. Variants containing the exon 1_7 sequence were significantly increased in the adult offspring of High LG-ABN mothers.

Serotonin appears crucial in mediating the effects of neonatal handling upon glucocorticoid receptor expression within the hippocampus. The transcription factors

NGFI-A and AP-2 are implicated in the induction of glucocorticoid receptor in the hippocampus after handling or following 5-HT treatment. A sequence in the human glucocorticoid receptor gene that binds AP-2 *in vitro*[84] is completely conserved in the rat glucocorticoid receptor gene (at 22718).[82] Additionally, within the CpG island, the glucocorticoid receptor gene contains 16 GC boxes (GGGCGG), including a sequence exactly matching the consensus binding site for the family of zinc-finger proteins that includes NGFI-A[85] immediately upstream of exon 1_7. Thus, increases in AP-2 and NGFI-A induced by neonatal handling could increase transcription from a promoter adjacent to exon 1_7, leading to increased glucocorticoid receptor mRNA. In previous studies we found that handling increased the binding of both NGFI-A and AP-2 to a promoter sequence for the human glucocorticoid receptor containing consensus sequences for both transcription factors (Weaver and colleagues, unpublished observations).

So how do these effects result in the long-term differentiation of hippocampal neurons? There are two very intriguing features of the 5-HT effect that bear directly on the question of the hippocampal cell cultures as a model for neural differentiation. First, the effects of 5-HT on glucocorticoid receptor levels in hippocampal cell cultures are restricted to the first three weeks in culture. Thus, cultures treated with 10 nM 5-HT for seven days at any time during the first three weeks in culture show a significant increase in glucocorticoid receptor density; however, the effect is lost after this point. Cultures treated with 10 nM 5-HT for seven days during the third to fourth week following plating show no increase in glucocorticoid receptor binding. Second, and most exciting, the increase in glucocorticoid receptor binding capacity following exposure to 10 nM 5-HT persists after 5-HT removal from the medium—for as long as the cultures can be studied there is a sustained increase in glucocorticoid receptor levels well past the removal of 5-HT from the medium. We have gone as long as 50 days and have seen no decrease in the magnitude of the 5-HT effect. Thus, the effect of 5-HT on glucocorticoid receptor density observed in hippocampal culture cells mimics the long-term effects of early environmental events.

Thus, we arrive at the most interesting feature of these effects: the finding that these effects persist well beyond the period of the treatment. There are at least two possible explanations for this finding. First, *in vivo* the increase in 5-HT turnover associated with the handling procedure might be accompanied by an increase in 5-HT innervation of the hippocampus, which persists throughout the life of the animal. The increased 5-HT innervation could then serve to maintain the handling effect. This possibility seems unlikely. The effect in cell cultures persists in the absence of 5-HT in the medium. Moreover, handling does not permanently alter 5-HT innervation into the dorsal hippocampus using either electrochemical[59] or immunocytohistochemical (Desjardins and Meaney, unpublished observations) measures of 5-HT content.

HOW ARE THE EFFECTS OF MATERNAL CARE ON THE OFFSPRING SUSTAINED INTO ADULTHOOD?

Epigenomic Marking of the Exon 1_7 Glucocorticoid Receptor Promoter

DNA methylation is a stable, epigenomic mark at CpG dinucleotides, which is associated with stable variations in gene transcription.[86–88] Hypomethylation of CpG

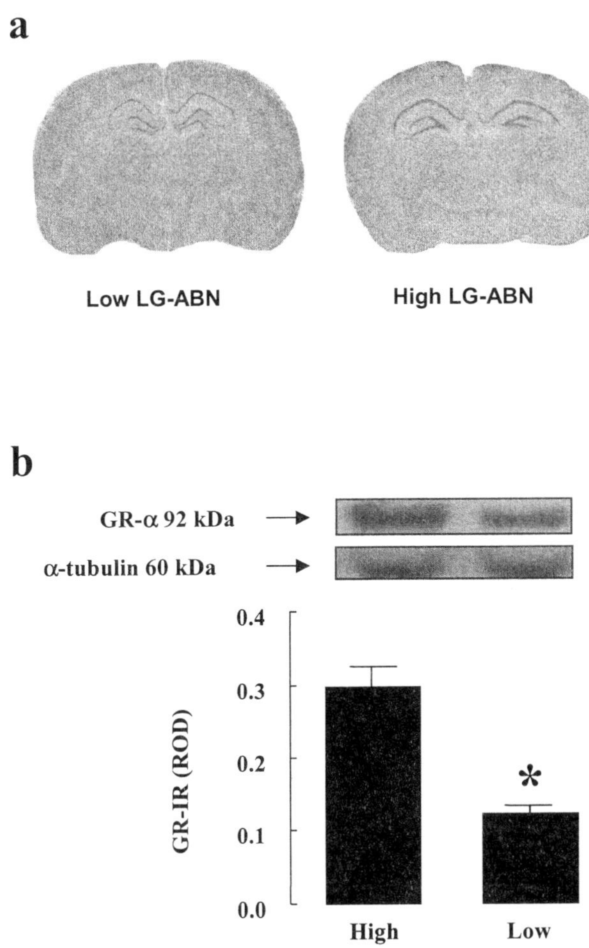

FIGURE 1. (**a**) A photomicrograph showing representative results of *in situ* hybridization studies of glucocorticoid receptor-α mRNA expression in the 90-day-old (adult) offspring of High and Low LG-ABN mothers. The results of earlier studies[46] reveal increased glucocorticoid receptor mRNA expression in the hippocampus of the adult offspring of High LG-ABN mothers. The *darkly labeled regions* in the coronal sections depicted above show the cell body regions of Ammon's horn and dentate gyrus of the dorsal hippocampus. (**b**) The *upper panel* reveals a representative Western blot illustrating glucocorticoid receptor-α and α-tubulin immunoreactivity (IR) in the 90-day-old (adult) offspring of High or Low LG-ABN mothers. Molecular weight markers (SeeBlue, Santa Cruz Biotech, Santa Cruz, CA) correspond to single major bands at 92 kDa and 60 kDa, with quantitative densitometric analysis relative optical density (ROD) of glucocorticoid receptor IR levels from samples (N = 5 animals/group). Western blot results indicate that hippocampal glucocorticoid receptor expression is greater in the adult offspring of High compared to Low LG-ABN mothers (*P <0.001).

dinucleotides of regulatory regions of genes correlates with active chromatin structure and transcriptional activity.[86,89] Thus, the methylation pattern is a stable signature of the epigenomic status of a regulatory sequence. We focused on the methylation state of the exon 1_7 glucocorticoid receptor promoter, which is activated in the hippocampus in offspring of High LG-ABN mothers. Glucocorticoid receptor gene expression is increased throughout the hippocampus in the adult offspring of High compared with Low LG-ABN mothers (FIG. 1).[40,90] We therefore examined the level of methylation across the entire exon 1_7 glucocorticoid receptor promoter sequence using the sodium bisulfite mapping technique. Sodium bisulfite treatment of DNA samples converts non-methylated cytosines to uracils; methylated cytosines are unaffected and the differences in methylation status are thus apparent on sequencing gels. In preliminary studies, we found significantly greater methylation of the exon 1_7 glucocorticoid receptor promoter sequence in the offspring of the Low LG-ABN mothers (FIG. 2). These findings are consistent with the hypothesis that maternal effects alter DNA methylation patterns in the offspring.

Two kinds of changes in DNA methylation are known to affect gene expression; regional, non–site-specific DNA methylation around a promoter[91] and site-specific methylation. To determine whether DNA methylation of specific target sites on the

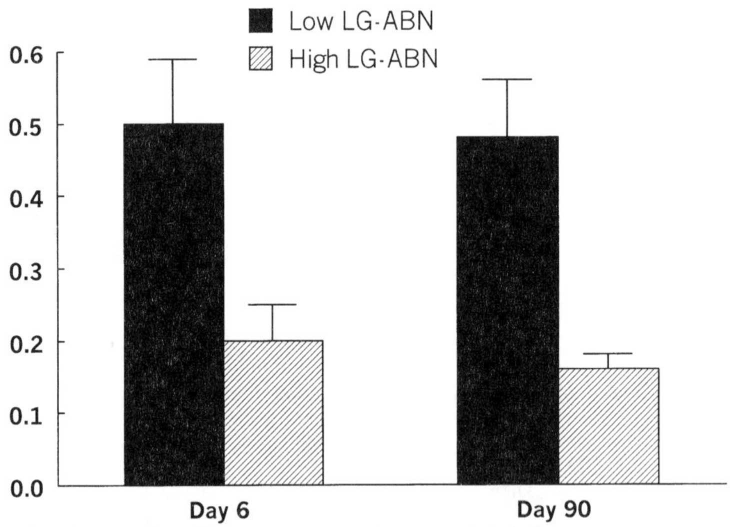

FIGURE 2. Global methylation pattern of the exon 1_7 glucocorticoid receptor promoter in 6-day-old (young pup) and 90-day-old (adult) offspring of High and Low LG mothers ($N = 4$ animals/group). The values are an average of the percentage methylation per cytosine for all 17 CpG dinucleotides in the two treatment groups ($^*P < 0.0001$). Statistical analysis with a two-way ANOVA (Group × Age) of the degree of cytosine methylation across the entire exon 1_7 glucocorticoid receptor promoter revealed a highly significant Group effect $F = 24.581$, $P < 0.0001$.

glucocorticoid receptor promoter change in response to maternal care, we mapped the differences in methylation using the sodium bisulfite mapping technique,[92,93] focusing on a region around the NGFI-A consensus sequence within the exon 1_7 promoter. The results revealed significant differences in the methylation of specific regions of the exon 1_7 glucocorticoid receptor promoter sequence. For example, the cytosine within the 5′ CpG dinucleotide (site 16) of the NGFI-A consensus sequence is always methylated in the offspring of Low LG-ABN mothers and rarely methylated in the offspring of High LG-ABN dams.

Maternal Care and Glucocorticoid Receptor Promoter Methylation

While these findings suggest that specific sites in the exon 1_7 glucocorticoid receptor promoter are differentially methylated as a function of maternal behavior, such findings are merely correlational. To directly examine the relation between maternal behavior and DNA methylation changes within the exon 1_7 glucocorticoid receptor promoter, we performed an adoption study in which the biological offspring of High or Low LG-ABN mothers were cross-fostered to either High or Low dams within 12 hours of birth.[47] Cross-fostering the biological offspring of High LG-ABN mothers or dams produced a pattern of exon 1_7 glucocorticoid receptor promoter methylation that was associated with the rearing mother. Most importantly, the cross-fostering procedure reversed the difference in methylation at specific cytosines. For example, the cytosine within the 5′ CpG dinucleotide (site 16) of the NGF I-A consensus sequence is hypomethylated following cross-fostering of offspring of Low LG-ABN to High LG-ABN dams, with no effect at the cytosine within the 3′ CpG dinucleotide. Thus, the pattern of methylation of the cytosine within the 5′ CpG dinucleotide (site 16) of the NGFI-A consensus sequence within the exon 1_7 glucocorticoid receptor promoter of the biological offspring of Low LG-ABN mothers cross-fostered to High LG-ABN dams was indistinguishable from that of the biological offspring of High LG-ABN mothers. Interestingly, cross-fostering did not have the same effect upon the methylation status of every CpG dinucleotide. For example, the CpG dinucleotide (site 12) of the AP-1 consensus sequence within the exon 1_7 glucocorticoid receptor promoter of the biological offspring of High LG-ABN mothers cross-fostered to Low LG-ABN dams remained hypomethylated; whereas the CpG dinucleotide (site 12) of the AP-1 consensus sequence within the exon 1_7 glucocorticoid receptor promoter of the biological offspring of Low LG-ABN mothers cross-fostered to High LG-ABN dams remained hypermethylated. Of course, it is tempting to speculate that when the AP-1 consensus sequence is demethylated the exon 1_7 glucocorticoid receptor promoter is transcriptionally active, whereas the methylation status of 5′ CpG dinucleotide (site 16) of the NGFI-A consensus sequence controls the relative levels of transcriptional activity, acting as a dimmer switch for transcription upon loading of methyl groups. However, NGFI-A may bind to the cognate response element and transactivate the exon 1_7 glucocorticoid receptor promoter independent of the AP-1 consensus sequence methylation status. Regardless, these findings suggest that variations in maternal care alter the methylation status within specific sites of the exon 1_7 promoter of the glucocorticoid receptor gene. This is the first demonstration of a DNA methylation pattern established through a behavioral mode of programming. Parental imprinting, a well-established paradigm of inheritance of an epigenomic mark, requires germ-line transmission.[88]

Maternal Care–Driven Demethylation of the Exon 1_7 Glucocorticoid Receptor Promoter

High and Low LG-ABN mothers differ in the frequency of pup licking/grooming and arched-back nursing only during the first week of life.[40,46] Thus, we wondered whether this period corresponds to the timing for the appearance of the difference in DNA methylation in the offspring. We used the sodium bisulfite mapping technique to map precisely the methylation status of the cytosines within the exon 1_7 glucocorticoid receptor promoter over multiple developmental time points (FIG. 3a–e). This analysis demonstrates that just before birth, on embryonic day 20, the entire region is unmethylated in both groups. Strikingly, one day following birth (postnatal day 1) the exon 1_7 glucocorticoid receptor promoter is *de novo* methylated in both groups. The 5′ and 3′ CpG sites of the exon 1_7 glucocorticoid receptor NGFI-A response element in the offspring of both High and Low LG-ABN mothers, which exhibit differential methylation later in life, are *de novo* methylated to the same extent (FIG. 3b). These data show that both the basal state of methylation and the first wave of *de novo* methylation after birth occur similarly in both groups. Whereas it is generally accepted that DNA methylation patterns are formed prenatally and that *de novo* methylation occurs early in development, there is at least one documented example of postnatal *de novo* methylation of the HoxA5 and HoxB5 genes.[94] Since similar analyses are not documented for other genes, it is unknown yet whether changes in methylation are common around birth or whether they are unique to this glucocorticoid receptor promoter. Between postnatal day 1 and postnatal day 6, the period when differences in the maternal behavior of High and Low LG-ABN dams are apparent, differences in the status of methylation of the exon 1_7 glucocorticoid receptor develop between the two groups. For example, the NGFI-A response element 5′ CpG dinucleotide (site 16) is demethylated in the High, but not in the Low LG-ABN group (FIG. 5c). This is consistent with data from the cross-fostering experiment, which illustrated that the differences between the two groups developed following birth in response to maternal behavior. The group difference in CpG dinucleotide methylation then remains consistent through to adulthood (postnatal day 90; FIG. 3c–e). Interestingly, the CpG dinucleotide (site 12) of the AP-1 consensus sequence within the exon 1_7 glucocorticoid receptor promoter is similarly hypomethylated in the High LG-ABN offspring by postnatal day 6, and this hypomethylation is sustained into adulthood. Our findings suggest that the group difference in DNA methylation occurs as a function of a maternal behavior over the first week of life. The results of earlier studies indicated that the first week of postnatal life is indeed a "critical period" for the effects of early experience on hippocampal glucocorticoid receptor expression.[95]

The results of developmental time-line study are very intriguing. From postnatal day 6, the methylation patterns for each of the 17 individual CpG sites within the exon 1_7 glucocorticoid receptor promoter do not all remain at the exactly same frequency for each developmental time-point (compare FIG. 3c–e). This is consistent with the model that methylation, like most (if not all) biological processes, is in a constant flux, but is stably maintained through a dynamic equilibrium. The developmental time-line may also help explain why the effect of maternal care on the hippocampal glucocorticoid receptor gene activity is not easily reversed when a High LG-ABN offspring is cross-fostered to a Low LG-ABN mother, in comparison to the

a 1681 ccc
1741 ctctgctagt gtgacacact t[1]**cg**[2]**cg**caact c[3]**cg**cagttgg [4]**cg**gg[5]**cg**[6]**cg**ga ccacccctg[7]**c**
1801 **g**gctctgc[8]**cg** gctggctgtc accct[9]**cg**ggg gctctggctg c[10]**cg**accca[11]**cg** ggg[12]**cg**ggct
1861 c[13]**cg**ag[14]**cg**gtt ccaagcct[15]**cg** gagtggg[16]**cgg** ggg[17]**cg**ggag ggagcctggg agaa

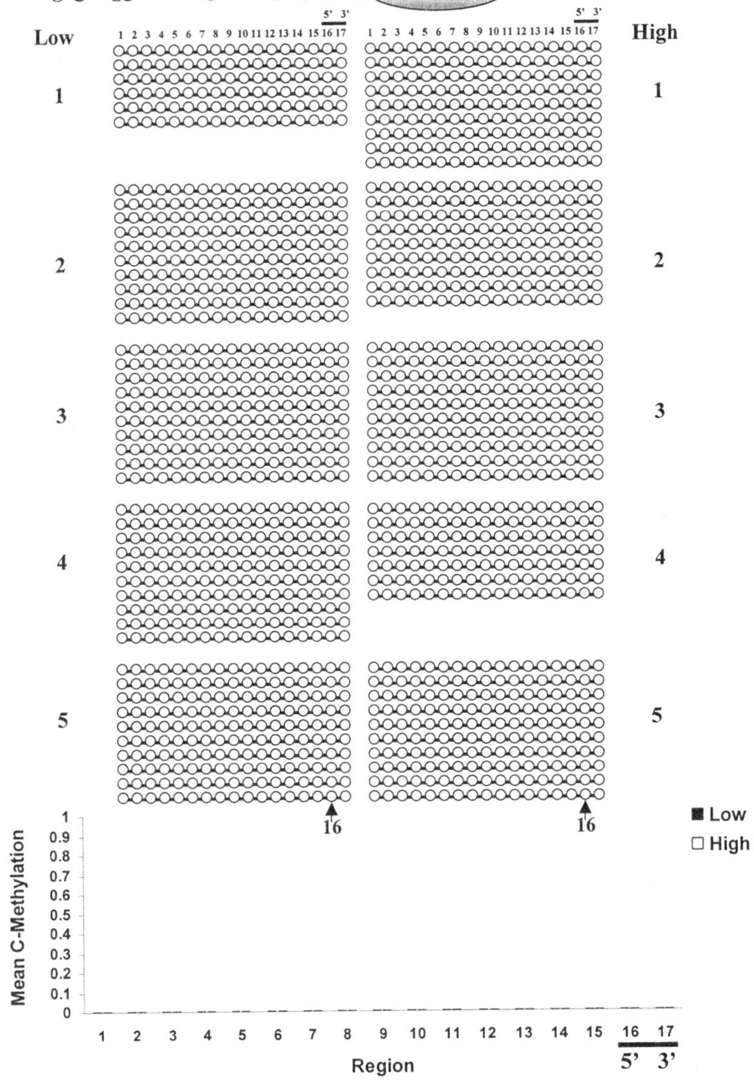

FIGURE 3. (**a**) Methylation patterns of the exon 1_7 glucocorticoid receptor promoter in the hippocampi of ED20 High and Low LG-ABN offspring ($N = 5$ animals/group). *Top panel* shows a sequence map of the exon 1_7 glucocorticoid receptor promoter including the 17 CpG dinucleotides (*highlighted in bold*) and the NGFI-A binding region (*encircled*). *Middle panel* shows bead-on-string representation of the cytosine methylation status of each of the 17 individual CpG dinucleotides of the exon 1_7 glucocorticoid receptor promoter re-

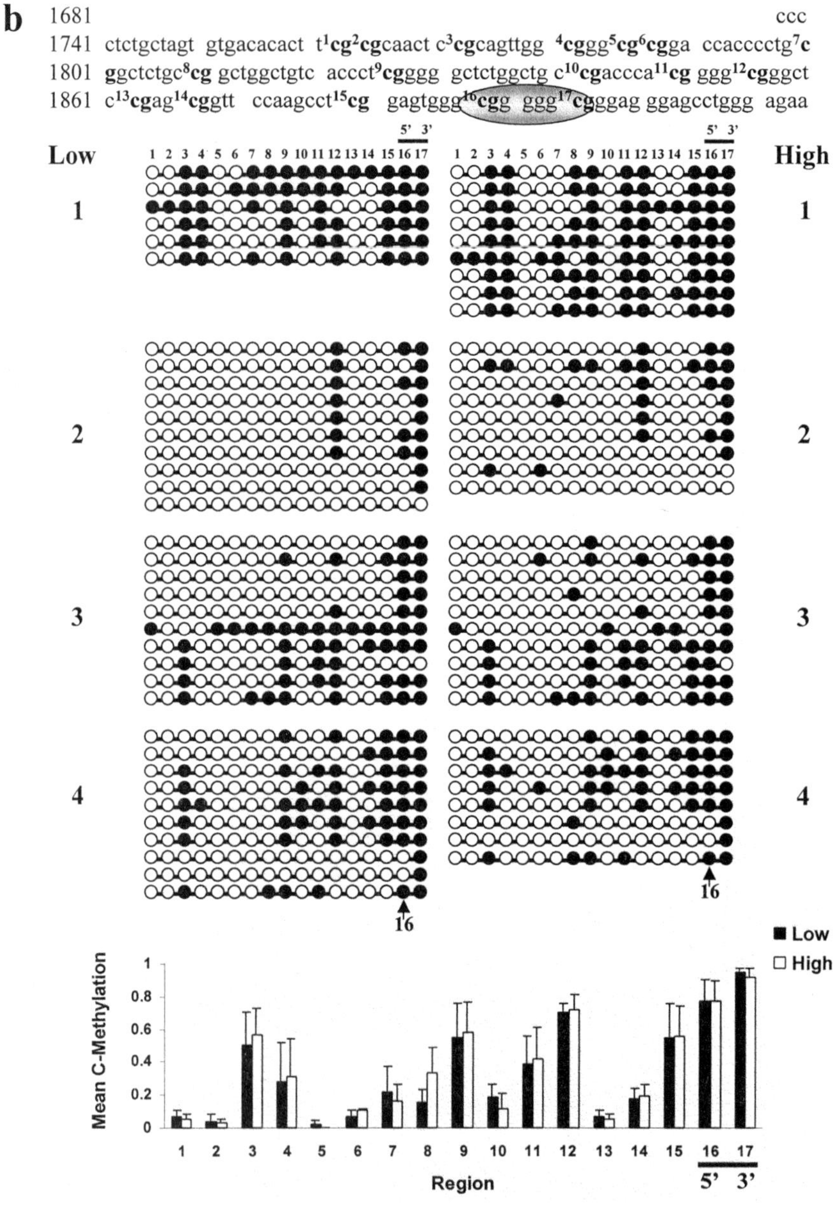

gion analyzed by sodium bisulfite mapping (6–10 clones sequenced/animal). The individual CpG dinucleotides of the exon 1_7 glucocorticoid receptor promoter region are labeled 1–17, highlighting the two CpG dinucleotides site 16 (5′) and site 17 (3′) within the NGFI-A binding region (*black line*). *Black circles* indicate methylated cytosines and *white circles* indicate non-methylated cytosines. The CpG dinucleotide at site 16 (5′) is again highlighted (*black arrow*). *Bottom panel* shows methylation analysis of the 17 CpG dinucleotides of the exon 1_7 glucocorticoid receptor promoter region from embryonic High and Low LG-ABN off-

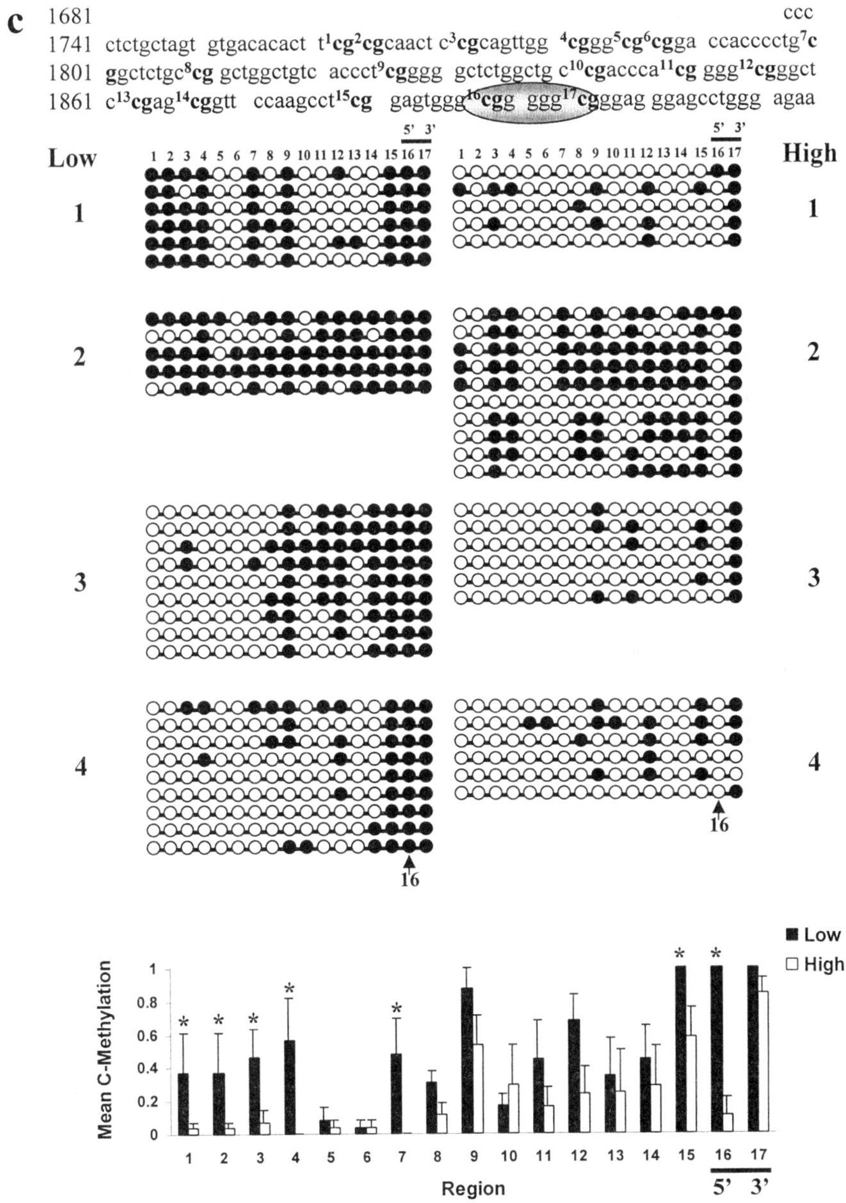

spring. (**b**) Methylation patterns of the exon 1_7 glucocorticoid receptor promoter in the hippocampi of 1-day-old (within one hour after birth) High and Low LG-ABN offspring (N = 4 animals/group). *Top panel* shows a sequence map of the exon 1_7 glucocorticoid receptor promoter including the 17 CpG dinucleotides (highlighted in bold) and the NGFI-A binding region (*encircled*). *Middle panel* shows bead-on-string representation of the cytosine methylation status of each of the 17 individual CpG dinucleotides of the exon 1_7 glucocorticoid

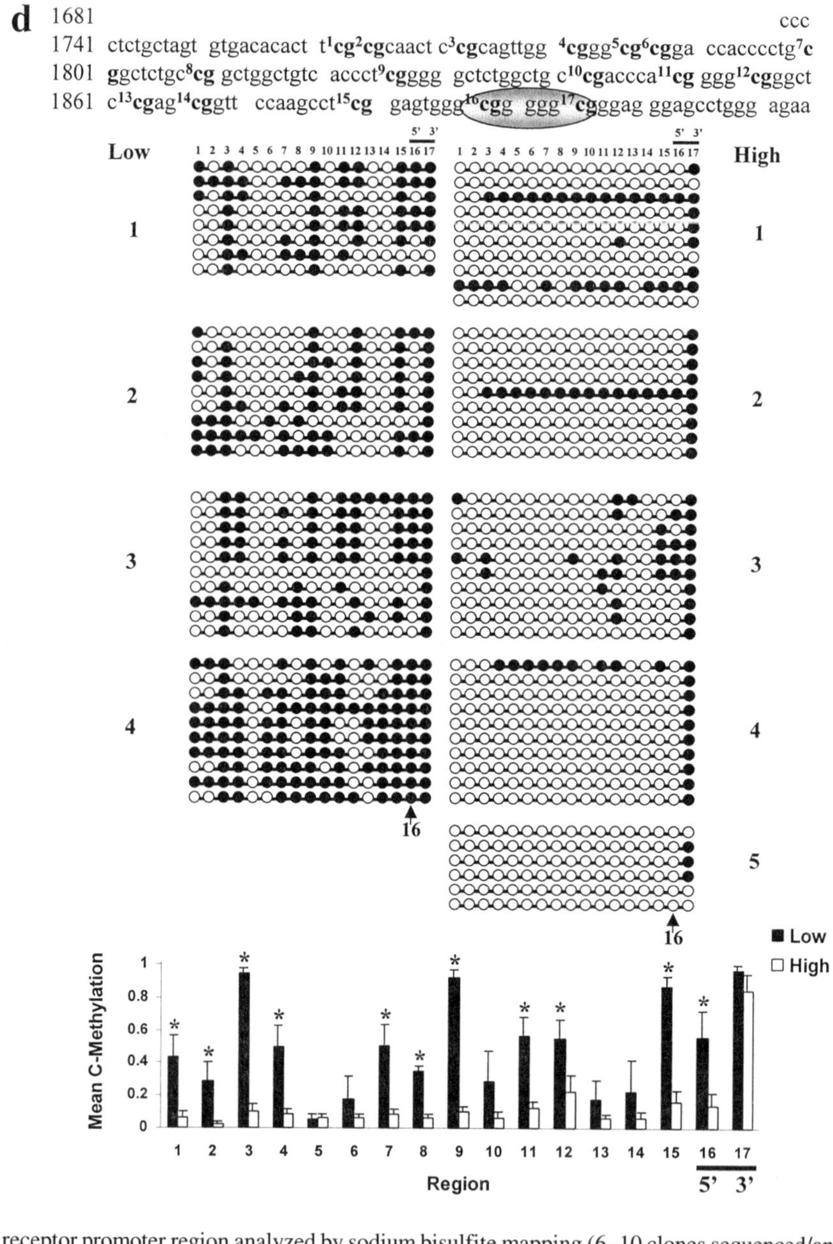

receptor promoter region analyzed by sodium bisulfite mapping (6–10 clones sequenced/animal). The individual CpG dinucleotides of the exon 1_7 glucocorticoid receptor promoter region are labeled 1–17, highlighting the two CpG dinucleotides site 16 (5′) and site 17 (3′) within the NGFI-A binding region (*black line*). *Black circles* indicate methylated cytosines and *white circles* indicate non-methylated cytosines. The CpG dinucleotide at site 16 (5′) is again highlighted (*black arrow*). *Bottom panel* shows methylation analysis of the 17 CpG

e

1681 ccc
1741 ctctgctagt gtgacacact t[1]**cg**[2]**cg**caact c[3]**cg**cagttgg [4]**cg**gg[5]**cg**[6]**cg**ga ccacccctg[7]**c**
1801 **g**gctctgc[8]**cg** gctggctgtc accct[9]**cg**ggg gctctggctg c[10]**cg**accca[11]**cg** ggg[12]**cg**ggct
1861 c[13]**cg**ag[14]**cg**gtt ccaagcct[15]**cg** gagtggg[16]**cg**g ggg[17]**cg**ggag ggagcctggg agaa

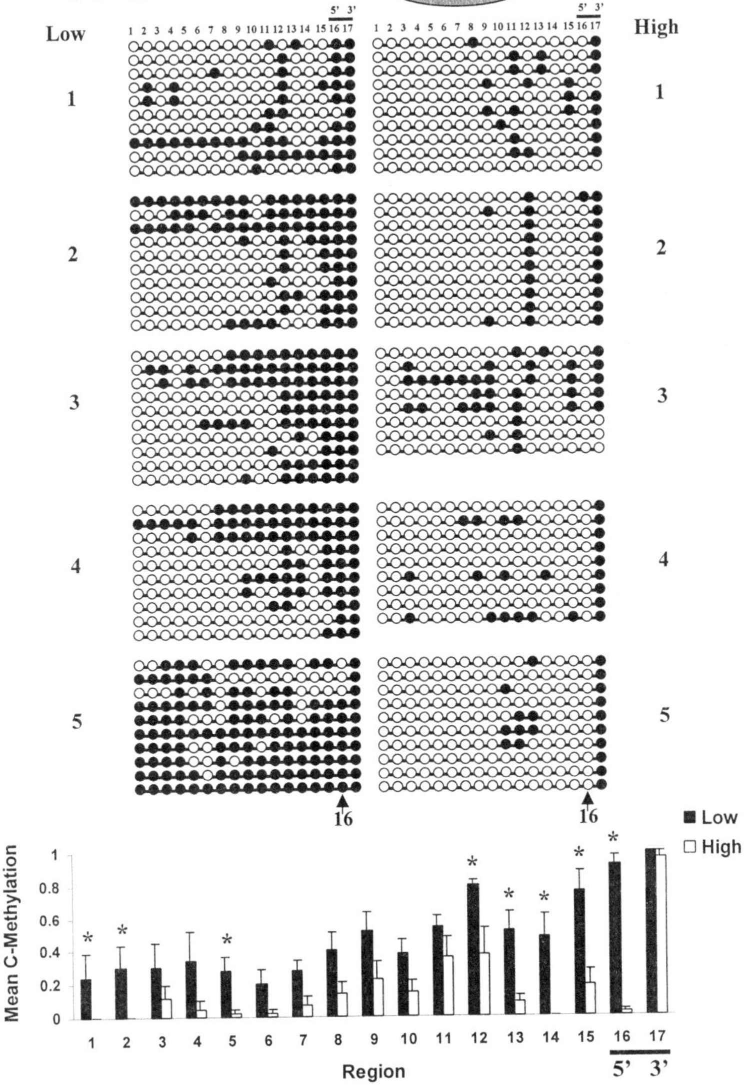

dinucleotides of the exon 1_7 glucocorticoid receptor promoter region from postnatal day-1 High and Low LG offspring. The two-way ANOVA (Group × Region) revealed a highly significant effect of Region [F =10.337, P <0.0001]. (**c**) Methylation patterns of the exon 1_7 glucocorticoid receptor promoter in the hippocampi of 6-day-old (young pup) High and Low LG-ABN offspring (N = 4 animals/group). *Top panel* shows a sequence map of the exon 1_7 glucocorticoid receptor promoter including the 17 CpG dinucleotides (*highlighted in bold*) substantial increase in hippocampal glucocorticoid receptor gene activity observed when a Low LG-ABN offspring is cross-fostered to a High LG-ABN mother. DNA methylation is a thermodynamically slower process in comparison to active demethylation. This implies that the hippocampal exon 1_7 glucocorticoid receptor promoter, within the Low LG-ABN mother's biological offspring, becomes stripped of CpG methylation by activity-dependent processive demethylation, resulting from the in-

---

and the NGFI-A binding region (*encircled*). *Middle panel* shows bead-on-string representation of the cytosine methylation status of each of the 17 individual CpG dinucleotides of the exon 1_7 glucocorticoid receptor promoter region analyzed by sodium bisulfite mapping (5–10 clones sequenced/animal). The individual CpG dinucleotides of the exon 1_7 glucocorticoid receptor promoter region are labeled 1–17, highlighting the two CpG dinucleotides site 16 (5′) and site 17 (3′) within the NGFI-A binding region (*black line*). *Black circles* indicate methylated cytosines and *white circles* indicate non-methylated cytosines. The CpG dinucleotide at site 16 (5′) is again highlighted (*black arrow*). *Bottom panel* shows methylation analysis of the 17 CpG dinucleotides of the exon 1_7 glucocorticoid receptor promoter region of the glucocorticoid receptor from postnatal day-6 High and Low LG-ABN offspring $*P < 0.05$; $*P < 0.0001$ for site 16 (5′) lying within the NGFI-A binding region. The two-way ANOVA (Group × Region) revealed a highly significant effect of both Group [F = 32.569, $P < 0.0001$] and Region [F = 5.353, $P < 0.0001$]. The Group × Region interaction effect was not significant [F = 1.265, $P = 0.057$]. (**d**) Methylation patterns of the exon 1_7 glucocorticoid receptor promoter in the hippocampi of 21-day-old (weaning age) High and Low LG-ABN offspring ($N = 4$–5 animals/group). *Top panel* shows a sequence map of the exon 1_7 glucocorticoid receptor promoter including the 17 CpG dinucleotides (*highlighted in bold*) and the NGFI-A binding region (*encircled*). *Middle panel* shows bead-on-string representation of the cytosine methylation status of each of the 17 individual CpG dinucleotides of the exon 1_7 glucocorticoid receptor promoter region analyzed by sodium bisulfite mapping (6–10 clones sequenced/animal). The individual CpG dinucleotides of the exon 1_7 glucocorticoid receptor promoter region are labeled 1–17, highlighting the two CpG dinucleotides site 16 (5′) and site 17 (3′) within the NGFI-A binding region (*black line*). *Black circles* indicate methylated cytosines and *white circles* indicate non-methylated cytosines. The CpG dinucleotide at site 16 (5′) is again highlighted (*black arrow*). *Bottom panel* shows methylation analysis of the 17 CpG dinucleotides of the exon 1_7 glucocorticoid receptor promoter region from postnatal day 21 High and Low LG-ABN offspring ($*P < 0.05$). The two-way ANOVA (Group × Region) revealed a highly significant effect of both Group [F = 150.450, $P < 0.0001$] and Region [F = 12.474, $P < 0.0001$], as well as a significant Group × Region interaction effect [F = 4.223, $P < 0.0001$]. (**e**) Methylation patterns of the exon 1_7 glucocorticoid receptor promoter in the hippocampi of 90-day-old (adult) High and Low LG-ABN offspring ($N = 5$ animals/group). *Top panel* shows a sequence map of the exon 1_7 glucocorticoid receptor promoter including the 17 CpG dinucleotides (*highlighted in bold*) and the NGFI-A binding site (*encircled*). *Middle panel* shows bead-on-string representation of the cytosine methylation status of each of the 17 individual CpG dinucleotides of the exon 1_7 glucocorticoid receptor promoter region analyzed by sodium bisulfite mapping (8–10 clones sequenced/animal). The individual CpG dinucleotides of the exon 1_7 glucocorticoid receptor promoter region are labeled 1–17, highlighting the two CpG dinucleotides site 16 (5′) and site 17 (3′) within the NGFI-A binding region (*black line*). *Black circles* indicate methylated cytosines and *white circles* indicate non-methylated cytosines. The CpG dinucleotide at site 16 (5′) is again highlighted (*black arrow*). *Bottom panel* shows methylation analysis of the 17 CpG dinucleotides of the exon 1_7 glucocorticoid receptor promoter region from postnatal day-90 High and Low LG-ABN offspring $*P < 0.05$; $*P < 0.0001$ for site 16 (5′) lying within the NGFI-A binding region. The two-way ANOVA (Group × Region) revealed a highly significant effect of Group [F = 104.782, $P < 0.0001$] and Region [F = 11.443, $P < 0.0001$], as well as a significant Group × Region interaction effect [F=2.321, $P < 0.01$].

tense tactile stimulation exerted toward the pups by the High LG-ABN foster mother. Whereas the hippocampal exon 1_7 glucocorticoid receptor promoter within the High LG-ABN mother's biological offspring passively gains CpG methylation (assuming some loss of methylation through maternal care by the biological High LG-ABN parent prior to cross-fostering), through a lack of stimulation of processive demethylation by the Low LG-ABN adoptive parent. The fact that High and Low LG-ABN offspring differ on epigenomic control of gene activity leaves the reader to speculate upon why the Low LG-ABN offspring are so plastic in response to the foster, High LG-ABN mother.

Site-Specific Methylation of the Cytosine within the 5′ CpG Dinucleotide (Site 16) of the NGFI-A Response Element Blocks Transcription Factor Binding

The next question concerns the functional importance of such differences in methylation. DNA methylation affects gene expression either by attracting methylated DNA binding proteins to a densely methylated region of a gene[96] or by site-specific interference with the binding of a transcription factor to its recognition element.[97] Our data showing site-specific demethylation of the cytosine within the 5′ CpG dinucleotide (site 16) of the NGFI-A response element (FIG. 3c) is consistent with the hypothesis that methylation at this site interferes with the binding of NGFI-A protein to its binding site. To address this question, we determined the *in vitro* binding of increasing concentrations of purified recombinant NGFI-A protein[98] to its response element under different states of methylation using the electrophilic mobility shift assay (EMSA) technique with four ^{32}P-labeled synthetic oligonucleotide sequences (FIG. 4a) bearing the NGFI-A binding site that was either (1) non-methylated, (2) methylated in the 3′ CpG site, (3) methylated in the 5′ CpG site, (4) methylated in both sites, or (5) mutated at the two CpGs with an adenosine replacing the cytosines. NGFI-A formed a protein-DNA complex with the non-methylated oligonucleotide (FIG. 4b, lanes 2–4), while the protein was unable to form a complex with either a fully methylated sequence or a sequence that was methylated at the 5′ CpG site (FIG. 4b, lanes 10–12, 14–16). Partial activity was seen with the sequence methylated at the 3′ CpG site (FIG. 4b, lanes 6–8). The specificity of the protein-DNA interaction is indicated by the fact that the recombinant protein fails to form a complex with the mutated NGFI-A response element, even at high protein concentrations (36 pM) (FIG. 4b, lanes 18–20). This difference in binding was confirmed by competition experiments (FIG. 4c). NGFI-A recombinant protein was incubated with a labeled, non-methylated oligonucleotide in the presence of increasing concentrations of non-labeled oligonucleotides containing the NGFI-A consensus sequence that were either 3′ CpG methylated, 5′ CpG methylated, methylated at both sites, or mutated at the two CpGs with an adenosine replacing the cytosines. As expected, the non-methylated oligonucleotide competed with the labeled oligonucleotide protein-DNA complex (FIG. 4c, lane 7). The specificity of the protein-DNA interaction is indicated by the fact that the mutated oligonucleotide is unable to compete away the labeled oligonucleotide protein-DNA complex (FIG. 4c lanes 17–19). Neither the oligonucleotide methylated in both the 3′ and 5′ CpGs nor the 5′ CpG methylated oligonucleotide were able to compete (FIG. 4c, lanes 11–16). Importantly, the 3′ CpG methylated oligonucleotide, which mimics the sequence observed in the off-

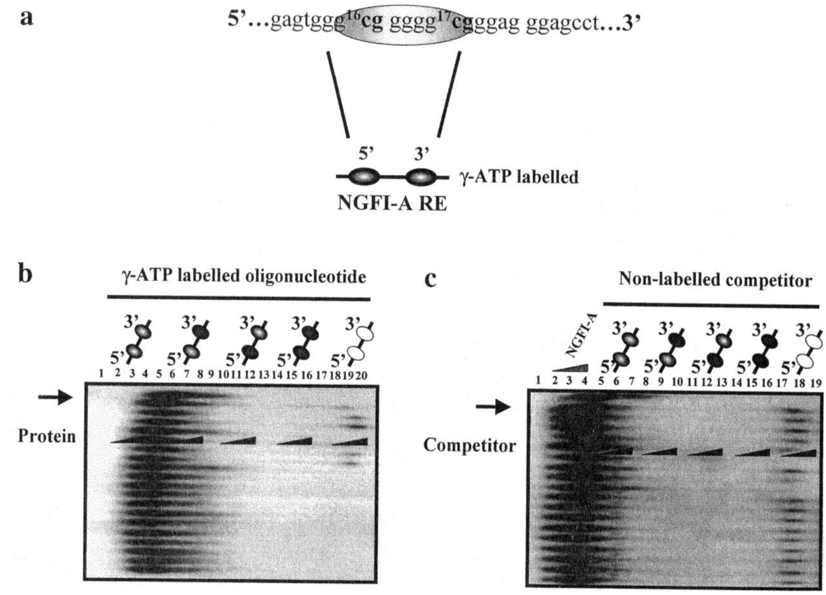

FIGURE 4. (**a**) The exon 1_7 glucocorticoid receptor promoter sequence with the NGFI-A binding region (*encircled*). Beneath is a bead-on-string representation of a synthesized radio-labeled oligonucleotide probe, highlighting the two CpG dinucleotides [*ovals* represent the cytosines at site 16 (5′) and site 17 (3′)] within the NGFI-A binding region (response element, RE). (**b**) EMSA analysis of protein-DNA complex formation between recombinant purified NGFI-A protein and radiolabeled oligonucleotide (**a**) bearing the NGFI-A response element containing differentially methylated cytosines within the 5′ CpG (site 16) and 3′ CpG (site 17) dinucleotides. Non-methylated cytosines are represented by *gray ovals*, methylated cytosines are shown as *black ovals*, and *white ovals* are mutated CpG dinucleotides with an adenosine replacing the cytosine. The presence of increasing amounts of purified NGFI-A protein (9 pM, 18 pM, or 36 pM) is indicated by the *black triangle*. The *black arrow* indicates the shift in labeled oligonucleotide mobility. Lane 1: free oligonucleotide non-methylated at either dinucleotide. Lanes 2–4: non-methylated oligonucleotide with an increasing amount of NGFI-A. Lane 5: free oligonucleotide methylated at the 3′ CpG dinucleotide. Lanes 6–8: oligonucleotide methylated at the 3′ CpG dinucleotide with an increasing amount of NGFI-A. Lane 9: free oligonucleotide methylated at the 5′ CpG dinucleotide. Lanes 10–12: oligonucleotide methylated at the 5′ CpG dinucleotide with an increasing amount of NGFI-A. Lane 13: free oligonucleotide methylated at both 5′ and 3′ CpG dinucleotides. Lanes 14–16: oligonucleotide methylated at both 5′ and 3′ CpG dinucleotides with an increasing amount of NGFI-A. Lane 17: free non-methylated oligonucleotide mutated with an adenosine replacing the cytosine in both the 5′ and 3′ CpG dinucleotides. Lanes 18–20: mutated non-methylated oligonucleotide with increasing amount of NGFI-A. Note, methylation of the cytosine within the 3′ CpG dinucleotide reduced binding at the higher levels of NGFI-A protein, while methylation of the cytosine within the 5′ CpG dinucleotide completely eliminated protein binding to the NGFI-A binding region (response element, RE). (**c**) EMSA analysis of competition of protein-DNA complex formation between NGFI-A protein and a radiolabeled oligonucleotide probe containing the NGFI-A response element (RE) (**a**) by an excess of non-labeled oligonucleotides containing differentially methylated cytosines within the 5′ and 3′ CpG dinucleotides of the NGFI-A response element (RE). Non-methylated cytosines are represented by *gray ovals*, methylated cytosines are

spring of High LG-ABN mothers, exhibited substantial competition (FIG. 4c, lanes 8–10). The results indicate that while methylation of the cytosine within the 5′ CpG dinucleotide (site 16) reduced NGFI-A protein binding to the same extent as methylation in both CpG sites, methylation of the cytosine within the 3′ CpG dinucleotide (site 17) only partially reduced NGFI-A protein binding (FIG. 4a, b). These data support the hypothesis that methylation of the cytosine within the 5′ CpG dinucleotide (site 16) in the NGFI-A response element of the exon 1_7 glucocorticoid receptor promoter region in the offspring of Low LG-ABN mothers inhibits NGFI-A protein binding, potentially explaining the reduced glucocorticoid receptor gene transcription in the offspring of the Low LG-ABN mothers.

Forming a Mechanism for Environment-Driven Gene-Imprinting of Behavior

The defining question of early experience studies concerns the mechanism by which environmental effects occurring in early development are sustained into adulthood (i.e., "environmental programming" effects). The offspring of High LG-ABN mothers exhibit increased hippocampal glucocorticoid receptor expression from the exon 1_7 promoter and dampened HPA responses to stress that persists into adulthood. We propose that the differential epigenomic status of the exon 1_7 glucocorticoid receptor promoter in the offspring of High LG-ABN mothers serves as a mechanism by which maternal behavior can sustain maternal effects into adulthood (see FIG. 5 for our proposed model). We show that forms of maternal care that increase tactile stimulation of the neonate (i.e., licking/grooming and arched-back nursing) result in a functional demethylation of a number of sites in the exon 1_7 glucocorticoid receptor promoter as well as increased acetylation of H3 histones and increased occupancy of the transcription factor NGFI-A on the exon 1_7 glucocorticoid receptor promoter, which is stably maintained into adulthood.[99]

---

shown as *black ovals*, and *white ovals* are mutated CpG dinucleotides with an adenosine replacing the cytosine. The presence of increasing amounts of purified NGFI-A protein (9 pM, 18 pM, or 36 pM) is indicated by the *grey triangle*. The presence of increasing amount of non-labeled oligonucleotide competitor (1:100 fold, 1:500 fold, or 1:1,000 fold) is indicated by the *black triangle*. The *black arrow* indicates the shift in labeled oligonucleotide mobility. Lane 1: free labeled non-methylated oligonucleotide. Lanes 2–4: labeled non-methylated oligonucleotide with increasing amount of NGFI-A. Lanes 5–7: labeled non-methylated oligonucleotide with increasing amount of non-labeled non-methylated oligonucleotide competitor. Lanes 8–10: labeled non-methylated oligonucleotide with increasing amount of non-labeled oligonucleotide competitor methylated at the 3′ CpG dinucleotide. Lanes 11–13: labeled non-methylated oligonucleotide with increasing amount of non-labeled oligonucleotide competitor methylated at the 5′ CpG dinucleotide. Lanes 14–16: labeled non-methylated oligonucleotide with increasing amount of non-labeled oligonucleotide competitor methylated at both 5′ and 3′ CpG dinucleotides. Lanes 17–19: labeled non-methylated oligonucleotide with increasing amount of non-labeled oligonucleotide competitor non-methylated but mutated with an adenosine replacing the cytosine in both the 5′ and 3′ CpG sites. Note that only the oligonucleotides with a non-methylated cytosine at the 5′ CpG dinucleotide of the NGFI-A binding region (response element, RE) showed effective competition. Methylation of the cytosine within the 5′ CpG dinucleotide completely eliminated the ability of the non-labeled oligonucleotides to compete for NGFI-A protein binding to the radiolabeled, non-methylated oligonucleotide sequence containing the NGFI-A binding region.

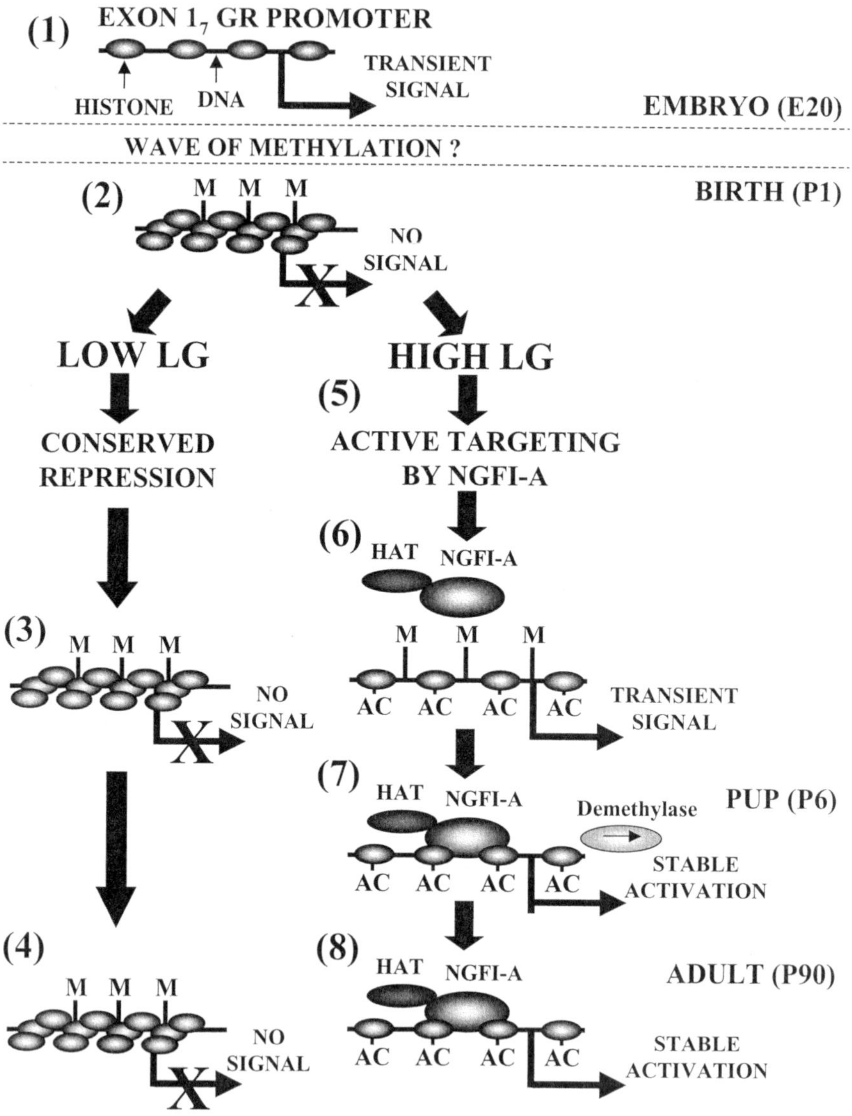

FIGURE 5. A model for environmental gene programming. (**1**) Prior to parturition the hippocampal exon 1_7 glucocorticoid receptor is entirely non-methylated (FIG. 3a). (**2**) Following birth, the hippocampal exon 1_7 glucocorticoid receptor becomes hypermethylated (FIG. 3b) and associated with tightly packed histones to form inactive chromatin. (**3**) In the absence of High tactile stimulation, the hippocampal exon 1_7 glucocorticoid receptor of the Low LG-ABN offspring remains hypermethylated and associated with the inactive chromatin, which endured into adulthood (**4**) (compare FIG. 3c–e). (**5**) High licking/grooming dams stimulate and maintain increased levels of activity-dependent gene expression in the offspring. The activity-dependent transcription factor NGFI-A actively targets its cognate response element within the hypermethylated hippocampal exon 1_7 glucocorticoid receptor.

Note that the effect of maternal care on glucocorticoid receptor expression is subtler (FIG. 1a, b) than the more pronounced effect on the methylation status of the 5′ CpG dinucleotide (site 16). However, in previous studies[82] we found evidence for the use of at least three promoters regulating hippocampal glucocorticoid receptor expression, suggesting that exon 1_7 is but one of the regulatory sequences determining glucocorticoid receptor expression within the hippocampus.

Environmental Variability Meets Epigenomic Predictability

Further studies are required to determine how maternal behavior alters the epigenomic status of the exon 1_7 glucocorticoid receptor promoter. In addition, the exact causal relationship between DNA methylation and other changes in the epigenomic status described here, such as altered histone acetylation and NGFI-A binding, remains unclear. Regardless of these as-yet-unanswered questions, our findings provide the first evidence that maternal behavior, early after birth, stably alters the epigenome of the offspring, providing a mechanism for the long-term effects of early experience on gene expression in the adult. These studies offer an opportunity to clearly define the nature of gene–environment interactions during development and how such effects result in the sustained "environmental programming" of gene expression and function over the life-span. Finally, it is important to note that maternal effects on the expression of defensive responses, such as increased HPA activity, are a common theme in biology,[7,100,101] such that the magnitude of the maternal influence on the development of HPA and behavioral responses to stress in the rat should not be surprising. Maternal effects on defensive responses to threat are apparent in plants, insects, and reptiles. Such effects commonly follow from the exposure of the mother to the same or similar forms of threat and may represent examples where the environmental experience of the mother is translated through an epigenetic mechanism of inheritance into phenotypic variation in the offspring. Indeed, maternal effects could result in the transmission of adaptive responses across generations.[7,100,101] Epigenomic modifications of targeted regulatory sequences in response to even reasonably subtle variations in environmental conditions might serve as a major source of epigenetic variation in gene expression and function and ultimately as a process mediating such maternal effects. We propose that epigenomic changes serve as an intermediate process that imprints dynamic environmental experiences on the fixed genome resulting in stable alterations in phenotype.

(**6**) Transcription factors commonly recruit histone modifying proteins, such as histone acetylase transferase (HAT). DNA methylation is associated with changes in chromatin activity states that gate accessibility of promoters to transcription factors through effects on histone acetylation (AC), a marker of active chromatin. Acetylation of the histone tails neutralizes the positively charged histones, which disrupts histone binding to negatively charged DNA and thus promotes transcription factor binding, resulting in transient exon 1_7 glucocorticoid receptor promoter activity. (**7**) Following processive DNA demethylation (compare FIG. 3b and c), the transcription factor NGFI-A can firmly bind to the demethylated response element (FIG. 4b, c) within the hippocampal exon 1_7 glucocorticoid receptor and allow stable transcription. The hippocampal exon 1_7 glucocorticoid receptor of the High LG-ABN offspring remains hypomethylated into adulthood (**8**), allowing stable transcription that is sustained throughout life (FIG. 1a, b).

REFERENCES

1. DENENBERG, V.H. 1964. Critical periods, stimulus input, and emotional reactivity: A theory of infantile stimulation. Psychol. Rev. **71:** 335–351.
2. DENENBERG, V.H. *et al.* 1967. Increased adrenocortical activity in the neonatal rat following handling. Endocrinology **81:** 1047–1052.
3. LEVINE, S. 1966. Infantile stimulation and adaptation to stress. Res. Publ. Assoc. Res. Nerv. Ment. Dis. **43:** 280–291.
4. LEVINE, S. 1957. Infantile experience and resistance to physiological stress. Science **126:** 405.
5. LEVINE, S. 1962. Plasma-free corticosteroid response to electric shock in rats stimulated in infancy. Science **135:** 795–796.
6. ZARROW, M.X., P.S. CAMPBELL & V.H. DENENBERG. 1972. Handling in infancy: increased levels of the hypothalamic corticotropin releasing factor (CRF) following exposure to a novel situation. Proc. Soc. Exp. Biol. Med. **141:** 356–358.
7. MEANEY, M.J. 2001. Maternal care, gene expression, and the transmission of individual differences in stress reactivity across generations. Annu. Rev. Neurosci. **24:** 1161–1192.
8. CHROUSOS, G.P. & P.W. GOLD. 1992. The concepts of stress and stress system disorders. Overview of physical and behavioral homeostasis. J. Am. Med. Assoc. **267:** 1244–1252.
9. AKANA, S.F. *et al.* 1992. Feedback and facilitation in the adrenocortical system: unmasking facilitation by partial inhibition of the glucocorticoid response to prior stress. Endocrinology **131:** 57–68.
10. MCEWEN, B.S. & E. STELLAR. 1993. Stress and the individual. Mechanisms leading to disease. Arch. Intern. Med. **153:** 2093–2101.
11. SECKL, J.R. & M.J. MEANEY. 1993. Early life events and later development of ischaemic heart disease. Lancet **342:** 1236.
12. ADER, R. 1970. The effects of early experience on the adrenocortical response to different magnitudes of stimulation. Physiol. Behav. **5:** 837–839.
13. MEANEY, M.J. *et al.* 1989. Neonatal handling alters adrenocortical negative feedback sensitivity and hippocampal type II glucocorticoid receptor binding in the rat. Neuroendocrinology **50:** 597–604.
14. VIAU, V. *et al.* 1993. Increased plasma ACTH responses to stress in nonhandled compared with handled rats require basal levels of corticosterone and are associated with increased levels of ACTH secretagogues in the median eminence. J. Neurosci. **13:** 1097–1105.
15. BHATNAGAR, S. & M.J. MEANEY. 1995. Hypothalamic-pituitary-adrenal function in chronic intermittently cold-stressed neonatally handled and non handled rats. J. Neuroendocrinol. **7:** 97–108.
16. MEANEY, M.J. *et al.* 1991. Postnatal handling attenuates certain neuroendocrine, anatomical, and cognitive dysfunctions associated with aging in female rats. Neurobiol. Aging **12:** 31–38.
17. MEANEY, M.J. *et al.* 1992. Basal ACTH, corticosterone and corticosterone-binding globulin levels over the diurnal cycle, and age-related changes in hippocampal type I and type II corticosteroid receptor binding capacity in young and aged, handled and nonhandled rats. Neuroendocrinology **55:** 204–213.
18. LANDFIELD, P.W. & T.A. PITLER. 1984. Prolonged Ca^{2+}-dependent after hyperpolarizations in hippocampal neurons of aged rats. Science **226:** 1089–1092.
19. SAPOLSKY, R.M., L.C. KREY & B.S. MCEWEN. 1984. Glucocorticoid-sensitive hippocampal neurons are involved in terminating the adrenocortical stress response. Proc. Natl. Acad. Sci. USA **81:** 6174–6177.
20. KERR, D.S. *et al.* 1989. Corticosteroid modulation of hippocampal potentials: increased effect with aging. Science **245:** 1505–1509.
21. ISSA, A.M. *et al.* 1990. Hypothalamic-pituitary-adrenal activity in aged, cognitively impaired and cognitively unimpaired rats. J. Neurosci. **10:** 3247–3254.
22. BHATNAGAR, S., N. SHANKS & M.J. MEANEY. 1996. Plaque-forming cell responses and antibody titers following injection of sheep red blood cells in nonstressed, acute,

and/or chronically stressed handled and nonhandled animals. Dev. Psychobiol. **29:** 171–181.
23. LABAN, O. *et al.* 1995. Experimental allergic encephalomyelitis in adult DA rats subjected to neonatal handling or gentling. Brain Res. **676:** 133–140.
24. MASON, D. 1991. Genetic variation in the stress response: susceptibility to experimental allergic encephalomyelitis and implications for human inflammatory disease. Immunol. Today **12:** 57–60.
25. PLOTSKY, P.M. & M.J. MEANEY. 1993. Early, postnatal experience alters hypothalamic corticotropin-releasing factor (CRF) mRNA, median eminence CRF content and stress-induced release in adult rats. Brain Res. Mol. Brain. Res. **18:** 195–200.
26. FRANCIS, D., P.M. PLOTSKY & M.J. MEANEY. 2004. Handling increases CRH mRNA expression in selected hypothalamic neuronal populations. Submitted for publication.
27. PAGE, M.E. & R.J. VALENTINO. 1994. Locus coeruleus activation by physiological challenges. Brain Res. Bull. **35:** 557–560.
28. PEARSON, D. *et al.* 1997.The effect of postnatal environment on stress-induced changes in hippocampal FOS-like immunoreactivity in adult rats. Soc. Neurosci. Abstr. **23:** 1849.
29. LIU, D. *et al.* 2000. Influence of neonatal rearing conditions on stress-induced adrenocorticotropin responses and norepinephrine release in the hypothalamic paraventricular nucleus. J. Neuroendocrinol. **12:** 5–12.
30. DE KLOET, E.R. 2000. Stress in the brain. Eur. J. Pharmacol. **405:** 187–198.
31. MEANEY, M.J. *et al.* 1985. Early postnatal handling alters glucocorticoid receptor concentrations in selected brain regions. Behav. Neurosci. **99:** 765–770.
32. SARRIEAU, A., S. SHARMA & M.J. MEANEY. 1988. Postnatal development and environmental regulation of hippocampal glucocorticoid and mineralocorticoid receptors. Brain Res. **471:** 158–162.
33. O'DONNELL, D. *et al.* 1994. Postnatal handling alters glucocorticoid, but not mineralocorticoid messenger RNA expression in the hippocampus of adult rats. Brain Res. Mol. Brain Res. **26:** 242–248.
34. JACOBSON, L. & R. SAPOLSKY. 1991. The role of the hippocampus in feedback regulation of the hypothalamic-pituitary-adrenocortical axis. Endocr. Rev. **12:** 118–134.
35. DIORIO, D., V. VIAU & M.J. MEANEY. 1993. The role of the medial prefrontal cortex (cingulate gyrus) in the regulation of hypothalamic-pituitary-adrenal responses to stress. J. Neurosci. **13:** 3839–3847.
36. DE KLOET, E.R. *et al.* 1998. Brain corticosteroid receptor balance in health and disease. Endocr. Rev. **19:** 269–301.
37. OWENS, M.J., M.A. VARGAS & C.B. NEMEROFF. 1993. The effects of alprazolam on corticotropin-releasing factor neurons in the rat brain: implications for a role for CRF in the pathogenesis of anxiety disorders. J. Psychiatr. Res. **27:** 209–220.
38. DE BOER, S.F., J.L. KATZ & R.J. VALENTINO. 1992. Common mechanisms underlying the proconflict effects of corticotropin-releasing factor, a benzodiazepine inverse agonist and electric foot-shock. J. Pharmacol. Exp. Ther. **262:** 335–342.
39. CALDJI, C. *et al.* 2000. The effects of early rearing environment on the development of GABAA and central benzodiazepine receptor levels and novelty-induced fearfulness in the rat. Neuropsychopharmacology **22:** 219–229.
40. CALDJI, C. *et al.* 1998. Maternal care during infancy regulates the development of neural systems mediating the expression of fearfulness in the rat. Proc. Natl. Acad. Sci. USA **95:** 5335–5340.
41. WILSON, M.A. 1996. GABA physiology: modulation by benzodiazepines and hormones. Crit. Rev. Neurobiol. **10:** 1–37.
42. MEHTA, A.K. & M.K. TICKU. 1999. An update on GABAA receptors. Brain Res. Brain Res. Rev. **29:** 196–217.
43. NUTT, D.J. *et al.* 1992. Clinical correlates of benzodiazepine receptor function. Clin. Neuropharmacol. **15:** 679A–680A.
44. HENNESSY, M.B., J. LI & S. LEVINE. 1980. Infant responsiveness to maternal cues in mice of 2 inbred lines. Dev. Psychobiol. **13:** 77–84.

45. LEE, M.H. & D.I. WILLIAMS. 1975. Long term changes in nest condition and pup grouping following handling of rat litters. Dev. Psychobiol. **8:** 91–95.
46. LIU, D. *et al.* 1997. Maternal care, hippocampal glucocorticoid receptors, and hypothalamic-pituitary-adrenal responses to stress. Science **277:** 1659–1662.
47. FRANCIS, D. *et al.* 1999. Nongenomic transmission across generations of maternal behavior and stress responses in the rat. Science **286:** 1155–1158.
48. STERN, J.M. 1997. Offspring-induced nurturance: animal-human parallels. Dev. Psychobiol. **31:**19–37.
49. CHAMPAGNE, F.A. *et al.* 2003. Variations in maternal care in the rat as a mediating influence for the effects of environment on development. Physiol. Behav. **79:** 359–371.
50. PURVES, D., W.D. SNIDER & J.T. VOYVODIC. 1988. Trophic regulation of nerve cell morphology and innervation in the autonomic nervous system. Nature **336:** 123–128.
51. LEVINE, S. & G.W. LEWIS. 1959. The relative importance of experimenter contact in an effect produced by extra-stimulation in infancy. J. Comp. Physiol. Psychol. **52:** 368–369.
52. MEANEY, M.J. & D.H. AITKEN. 1985. The effects of early postnatal handling on hippocampal glucocorticoid receptor concentrations: temporal parameters. Brain Res. **354:** 301–304.
53. MEANEY, M.J. *et al.* 1994. Corticosteroid receptors in rat brain and pituitary during development and hypothalamic-pituitary-adrenal (HPA) function. *In* Receptors and the Developing Nervous System. P.A.Z. McLaughlin, Ed.: 163–202. Chapman and Hall. London.
54. MEANEY, M.J., D.H. AITKEN & R.M. SAPOLSKY. 1987. Thyroid hormones influence the development of hippocampal glucocorticoid receptors in the rat: a mechanism for the effects of postnatal handling on the development of the adrenocortical stress response. Neuroendocrinology **45:** 278–283.
55. BANKER, G.A. & W.M. COWAN. 1977. Rat hippocampal neurons in dispersed cell culture. Brain Res. **126:** 397–342.
56. MITCHELL, J.B. *et al.* 1990. Serotonin regulates type II corticosteroid receptor binding in hippocampal cell cultures. J. Neurosci. **10:** 1745–1752.
57. SAVARD, P. *et al.* 1984. Effect of neonatal hypothyroidism on the serotonin system of the rat brain. Brain Res. **292:** 99–108.
58. MITCHELL, J.B., L.J. INY & M.J. MEANEY. 1990. The role of serotonin in the development and environmental regulation of type II corticosteroid receptor binding in rat hippocampus. Brain Res. Dev. Brain Res. **55:** 231–235.
59. SMYTHE, J.W., W.B. ROWE & M.J. MEANEY. 1994. Neonatal handling alters serotonin (5-HT) turnover and 5-HT2 receptor binding in selected brain regions: relationship to the handling effect on glucocorticoid receptor expression. Brain Res. Dev. Brain Res. **80:** 183–189.
60. MITCHELL, J.B. *et al.* 1992. Serotonergic regulation of type II corticosteroid receptor binding in hippocampal cell cultures: evidence for the importance of serotonin-induced changes in cAMP levels. Neuroscience **48:** 631–639.
61. SANDERS-BUSH, E., M. TSUTSUMI & K.D. BURRIS. 1990. Serotonin receptors and phosphatidylinositol turnover. Ann. N.Y. Acad. Sci. **600:** 224–235; discussion 235–236.
62. YOSHIMASA, T. *et al.* 1987. Cross-talk between cellular signalling pathways suggested by phorbol-ester-induced adenylate cyclase phosphorylation. Nature **327:** 67–70.
63. PARENT, M. *et al.* 2001. Analysis of amino acids and catecholamines, 5-hydroxytryptamine and their metabolites in brain areas in the rat using in vivo microdialysis.Methods **23:** 11–20.
64. IKE, J., H. CANTON & E. SANDERS-BUSH. 1995. Developmental switch in the hippocampal serotonin receptor linked to phosphoinositide hydrolysis. Brain Res. **678:** 49–54.
65. LUCAS, J.J. & R. HEN. 1995. New players in the 5-HT receptor field: genes and knockouts. Trends Pharmacol. Sci. **16:** 246–252.
66. PEPIN, M.C., S. BEAULIEU & N. BARDEN. 1989. Antidepressants regulate glucocorticoid receptor messenger RNA concentrations in primary neuronal cultures. Brain Res. Mol. Brain Res. **6:** 77–83.
67. PEPIN, M.C., M.V. GOVINDAN & N. BARDEN. 1992. Increased glucocorticoid receptor gene promoter activity after antidepressant treatment. Mol. Pharmacol. **41:** 1016–1022.

68. LAPLANTE, P., J. DIORIO & M.J. MEANEY. 2002. Serotonin regulates hippocampal glucocorticoid receptor expression via a 5-HT7 receptor. Brain Res. Dev. Brain Res. **139:** 199–203.
69. FRANCIS, D. *et al.* 1996. The role of early environmental events in regulating neuroendocrine development. Moms, pups, stress, and glucocorticoid receptors. Ann. N.Y. Acad. Sci. **794:** 136–152.
70. MEANEY, M.J. *et al.* 2000. Postnatal handling increases the expression of cAMP-inducible transcription factors in the rat hippocampus: the effects of thyroid hormones and serotonin. J. Neurosci. **20:** 3926–3935.
71. DE GROOT, R.P. & P. SASSONE-CORSI. 1993. Hormonal control of gene expression: multiplicity and versatility of cyclic adenosine 3′,5′-monophosphate-responsive nuclear regulators. Mol. Endocrinol. **7:** 145–153.
72. HABENER, J.F. 1990. Cyclic AMP response element binding proteins: a cornucopia of transcription factors. Mol. Endocrinol. **4:** 1087–1094.
73. IMAGAWA, M., R. CHIU & M. KARIN. 1987. Transcription factor AP-2 mediates induction by two different signal-transduction pathways: protein kinase C and cAMP. Cell **51:** 251–260.
74. MONTMINY, M.R., G.A. GONZALEZ & K.K. YAMAMOTO. 1990. Regulation of cAMP-inducible genes by CREB. Trends Neurosci. **13:** 184–188.
75. SHENG, M. & M.E. GREENBERG. 1990. The regulation and function of c-fos and other immediate early genes in the nervous system. Neuron **4:** 477–485.
76. VALLEJO, M. 1994. Transcriptional control of gene expression by cAMP-response element binding proteins. J. Neuroendocrinol. **6:** 587–596.
77. WALTON, K.M. & R.P. REHFUSS. 1990. Molecular mechanisms of cAMP-regulated gene expression. Mol. Neurobiol. **4:** 197–210.
78. YAMAMOTO, K.K. *et al.* 1988. Phosphorylation-induced binding and transcriptional efficacy of nuclear factor CREB. Nature **334:** 494–498.
79. VACCARINO, F.M. *et al.* 1993. Induction of immediate early genes by cyclic AMP in primary cultures of neurons from rat cerebral cortex. Brain Res. Mol. Brain Res. **19:** 76–82.
80. LECLERC, S. *et al.* 1991. Purification of a human glucocorticoid receptor gene promoter-binding protein. Production of polyclonal antibodies against the purified factor. J. Biol. Chem. **266:** 8711–8719.
81. STRAHLE, U. *et al.* 1992. At least three promoters direct expression of the mouse glucocorticoid receptor gene. Proc. Natl. Acad. Sci. USA **89:** 6731–6735.
82. MCCORMICK, J.A. *et al.* 2000. 5′-heterogeneity of glucocorticoid receptor messenger RNA is tissue specific: differential regulation of variant transcripts by early-life events. Mol. Endocrinol. **14:** 506–517.
83. CHEN, F., C.S. WATSON & B. GAMETCHU. 1999. Multiple glucocorticoid receptor transcripts in membrane glucocorticoid receptor-enriched S-49 mouse lymphoma cells. J. Cell. Biochem. **74:** 418–429.
84. NOBUKUNI, Y. *et al.* 1995. Characterization of the human glucocorticoid receptor promoter. Biochemistry **34:** 8207–8214.
85. CROSBY, S.D. *et al.* 1991. The early response gene NGFI-C encodes a zinc finger transcriptional activator and is a member of the GCGGGGGCG (GSG) element-binding protein family. Mol. Cell. Biol. **11:** 3835–3841.
86. RAZIN, A. 1998. CpG methylation, chromatin structure and gene silencing—a three-way connection. EMBO J. **17:** 4905–4908.
87. SZYF, M. 2001. Towards a pharmacology of DNA methylation. Trends Pharmacol. Sci. **22:** 350–354.
88. SAPIENZA, C. 1990. Parental imprinting of genes. Sci. Am. **263:** 52–60.
89. RAZIN, A. & H. CEDAR. 1977. Distribution of 5-methylcytosine in chromatin. Proc. Natl. Acad. Sci. USA **74:** 2725–2728.
90. MYERS, M.M. *et al.* 1989. Relationships between maternal behavior of SHR and WKY dams and adult blood pressures of cross-fostered F1 pups. Dev. Psychobiol. **22:** 55–67.
91. KESHET, I., J. YISRAELI & H. CEDAR. 1985. Effect of regional DNA methylation on gene expression. Proc. Natl. Acad. Sci. USA **82:** 2560–2564.
92. CLARK, S.J. *et al.* 1994. High sensitivity mapping of methylated cytosines. Nucleic Acids Res. **22:** 2990–2997.

93. FROMMER, M. *et al.* 1992. A genomic sequencing protocol that yields a positive display of 5-methylcytosine residues in individual DNA strands. Proc. Natl. Acad. Sci. USA **89:** 1827–1831.
94. HERSHKO, A.Y. *et al.* 2003. Methylation of HoxA5 and HoxB5 and its relevance to expression during mouse development. Gene **302:** 65–72.
95. MEANEY, M.J. *et al.* 1996. Early environmental regulation of forebrain glucocorticoid receptor gene expression: implications for adrenocortical responses to stress. Dev. Neurosci. **18:** 49–72.
96. NAN, X. *et al.* 1998. Transcriptional repression by the methyl-CpG-binding protein MeCP2 involves a histone deacetylase complex. Nature **393:** 386–389.
97. COMB, M. & H.M. GOODMAN. 1990. CpG methylation inhibits proenkephalin gene expression and binding of the transcription factor AP-2. Nucleic Acids Res. **18:** 3975–3982.
98. MILBRANDT, J. 1987. A nerve growth factor-induced gene encodes a possible transcriptional regulatory factor. Science **238:** 797–799.
99. WEAVER, I.C. *et al.* Epigenomic programming by maternal behavior. Nat. Neurosci.
100. AGRAWAL, A.A. 2001. Phenotypic plasticity in the interactions and evolution of species. Science **294:** 321–326.
101. ROSSITER, M.C. 1999. Maternal Effects as Adaptations. T.A.F. Mousseau, Ed. Oxford University Press. London.

Subnuclear Trafficking and Gene Targeting by Steroid Receptors

AKHILESH K. NAGAICH, GEETHA V. RAYASAM, ELISABETH D. MARTINEZ, MATTHIAS BECKER, YI QIU, THOMAS A. JOHNSON, CEM ELBI, TERACE M. FLETCHER,[a] SAM JOHN, AND GORDON L. HAGER

Laboratory of Receptor Biology and Gene Expression, National Cancer Institute, National Institutes of Health, Bethesda, Maryland 20892-5055, USA

ABSTRACT: Through the use of novel imaging techniques, we have observed direct steroid receptor binding to a tandem array of a hormone-responsive promoter in living cells. We found that the glucocorticoid receptor (GR) exchanges rapidly with regulatory elements in the continued presence of ligand. We have also reconstituted a GR-dependent nucleoprotein transition with chromatin assembled on promoter DNA, and we discovered that GR is actively displaced from the chromatin template during the chromatin remodeling process. Using high-intensity UV laser crosslinking, we have observed highly periodic interactions of GR with promoter chromatin. These periodic binding events are dependent on GR-directed hSWI/SNF remodeling of the template and require the presence of ATP. Both the *in vitro* and *in vivo* results are consistent with a dynamic model ("hit-and-run") in which GR first binds to chromatin after ligand activation, recruits a remodeling activity, and is simultaneously lost from the template. We also find that receptor mobility in the nucleoplasm is strongly enhanced by molecular chaperones. These observations indicate that multiple mechanisms are involved in transient receptor interactions with nucleoplasmic targets.

KEYWORDS: glucocorticoid receptor; nuclear receptor; chromatin; progesterone receptor; laser crosslinking; chaperone; FRAP; protein dynamics

Nuclear receptors exert their physiological effects largely through the modulation of gene expression rates for responsive genes in target tissues. Classically, mechanisms of transcription regulation have been modeled in terms of interactions between soluble members of the regulatory apparatus and the DNA template. We now know that a key feature of receptor action lies in the epigenetic modification of the nucleoprotein template. Receptors recruit many activities that change local chromatin structure, and these alterations in turn are central to the resulting activation or suppression of gene activity. These alterations include primary sequence tags (acetylation, phos-

Address for correspondence: Gordon L. Hager, Laboratory of Receptor Biology and Gene Expression, Bldg. 41, B602, 41 Library Drive, National Cancer Institute, NIH, Bethesda, MD 20892-5055. Voice: 301-496-9867; fax: 301-496-4951.

hagerg@exchange.nih.gov

[a]Current address: University of Miami School of Medicine, Dept. of Biochemistry and Molecular Biology, 1011 NW 15th Street, Gautier Bldg., Rm. 210 (R629), Miami, FL 33136.

Ann. N.Y. Acad. Sci. 1024: 213–220 (2004). © 2004 New York Academy of Sciences.
doi: 10.1196/annals.1321.002

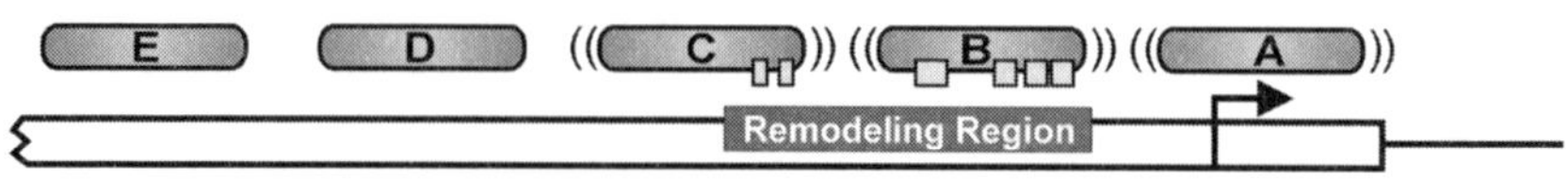

FIGURE 1. MMTV chromatin structure. The MMTV promoter is organized in a series of nucleosome families, covering 1.2 kb of the promoter regulatory region. Each family consists of a cluster of nucleosomes nonrandomly positioned on the promoter DNA. Hormone response elements (*square gray boxes*) for GR, PR, and AR are associated with the B and C nucleosomes. Hormone induction of the promoter gives rise to a receptor-dependent reorganization of the nucleosome B family and the right half of the C nucleosome region.

phorylation, methylation, sumolation) on histone tails and other chromatin-bound proteins, as well as the rearrangement of nucleosome structures (often referred to as nucleosome mobilization). Although a large literature now exists describing many of these primary sequence modifications, less is known concerning the actual structural changes to local promoter chromatin, and how these changes impact gene expression.

We have exploited one of the best-characterized systems, the MMTV promoter,[1] as a model to understand mechanisms of chromatin reorganization. The MMTV chromatin transition has been extensively characterized *in vivo*.[1–8] The regulatory elements for this promoter are organized in a series of six nucleosome families (A–F), and a specific reorganization of the B/C region (FIG. 1) occurs in response to activation by one of several receptors, including the glucocorticoid receptor (GR), the progesterone receptor (PR), or the androgen receptor (AR). These receptors bind to their cognate response elements in promoter chromatin and initiate a complex series of nucleoprotein modifications that result in activation of the promoter. A thorough grasp of the processes and components involved in chromatin reorganization is clearly central to a complete understanding of the mechanism of gene regulation by this receptor family.

RECEPTOR EXCHANGE ON CHROMATIN *IN VITRO*

To explore and understand the molecular events involved in this chromatin transition, we have developed an *in vitro* system that accurately recapitulates the nucleosome positioning described *in vivo*, and the hormone-dependent transition in structure.[9] Addition of purified GR to this system results in a dramatic increase in restriction enzyme access to all regions of the B nucleosome family, but only to the 3′ half of the C nucleosome cluster.[9,10] This is precisely the same structural transition that occurs *in vivo*.[4]

Results from the enzyme access experiments[10] presented a logical conundrum. Although receptor-dependent remodeling creates a zone of increased sensitivity *in vivo*, access to the DNA in the region of the receptor binding sites was decreased in the absence of ATP and remodeling complex.[10] Parallel experiments with receptor and naked DNA indicated that the receptor itself would block enzyme access to the DNA by simple steric hindrance (FIG. 2). These findings imply that the receptor must be lost from the template during the remodeling process (FIG. 2), and they led to the first suggestion of a "hit-and-run" mechanism.[10]

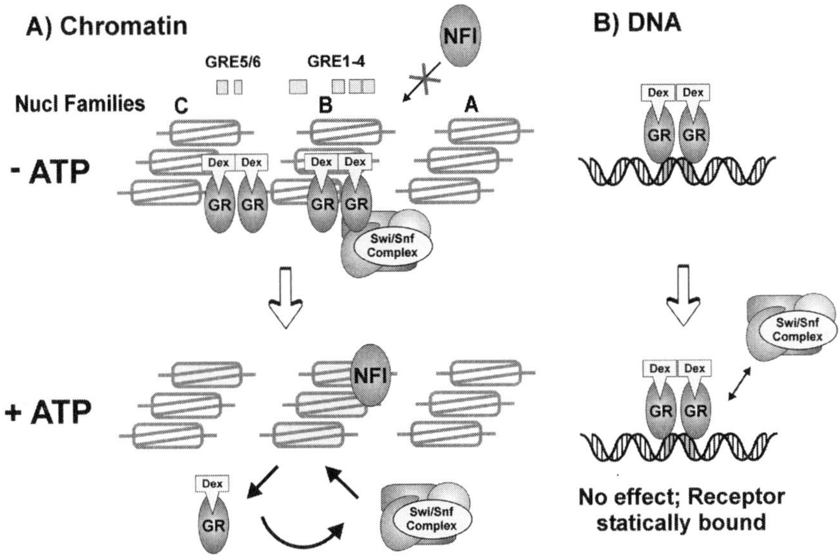

FIGURE 2. Chromatin remodeling causes loss of the receptor. (**A**) During the receptor-dependent gene activation, the SWI/SNF remodeling complex is recruited to the promoter, causing a nucleoprotein reorganization localized to the B/C region. During this process, the receptor is actively displaced from the template. (**B**) On pure DNA templates, the receptor recognizes and binds to the hormone-response elements. However, no displacement of receptor is observed in the presence of SWI/SNF.

It became clear at this point that classic assays for chromatin transitions (DNase I access, restriction enzyme access, MPE chemical cleavage) were inadequate to monitor the receptor-induced events at this promoter. We therefore embarked on a program to develop technology that could follow the events in real time. Laser UV crosslinking has been utilized by a small group of investigators[11,12] to study protein–DNA interactions. The use of a laser UV source exhibits several advantages over conventional low-intensity light sources. The high photon flux delivered by the laser generates radical cations of nucleic acid bases via a biphotonic mechanism.[13,14] The high quantum yield of laser-induced radical cations leads to an efficiency of crosslinking that exceeds that obtained with conventional UV light sources by at least two orders of magnitude. Finally, the crosslinking reaction itself is completed in less than 1 μs; this avoids the possibility of artifactual crosslinking of UV-damaged molecules and permits trapping of rapid dynamic changes in protein–DNA interactions.

The application of this new approach to analyze the interaction of GR and the SWI/SNF complex with the template led to startling findings.[15,16] GR interactions with the template during the remodeling process were found to be highly transient and periodic (see Fig. 1 of Ref. 16). A sharp peak of laser-detected binding was observed at 5 min after initiation of the reaction, followed by an equally rapid loss of the receptor. This cycle was then repeated periodically, with a cycle time of 5 min.

A similar cycle of binding was observed for the SWI/SNF complex, although the detailed binding profile was different. There appeared to be a loss of SWI/SNF interaction as GR binding increased, with a return to the basal level of interaction as GR left the template. Laser-detected interactions of core histones with the template were also periodic, but more complex.[15,16] H2A and H2B each manifested a sharp peak of interaction, but out of phase with each other.

These findings have led us to propose the following model for GR and remodeling complex interaction with the template[17] (see Fig. 7 from Ref. 16). We suggest that the rapid binding of GR results from the initial recruitment of the SWI/SNF complex. At this stage, nucleosome remodeling "opens" the structure and increases the number of available GR binding sites (there are a total of six potential binding sites in the B/C region[9]). We suggest that this local perturbed chromatin state is transient, leading to subsequent loss of the remodeling complex. (Note in Fig. 2 from Ref. 16 that SWI/SNF binding is significantly reduced after the initial GR loading.) Progression of the remodeling process would lead in turn to collapse of the high-energy state and to return of the local chromatin domain to the ground state. Because this state is incompatible with binding of multiple GR homodimers, GR would be rapidly lost. Thus, GR is actively ejected from the chromatin structure as a direct result of the progression of the remodeling process.

RAPID EXCHANGE ON REGULATORY ELEMENTS *IN VIVO*

The implications of the findings discussed above, if generalized to other members of the receptor superfamily and other transcription factors, are quite profound. Since the elaboration of the general receptor/DNA regulatory element model 30 years ago, it has been a central tenet in endocrinology that hormone-activated receptors bind stably to their regulatory sites and nucleate the formation of large multiprotein complexes. In contrast to this view, we recently discovered that receptors exchange rapidly with regulatory sites in living cells. This breakthrough required the development of methodology that permits direct visualization of gene expression in real time. We first characterized a murine cell line[3] (3134) in which a steroid receptor-regulated gene is amplified at one site on chromosome 4. An array containing approximately 200 copies of a perfect head-to-tail 9-kb repeat that includes the complete steroid-responsive MMTV promoter and reporter structure is inserted near the centromere (FIG. 3). Each gene copy contains binding sites for six to eight receptors; thus, the array contains an aggregate capacity for approximately 1,000 receptors. We then introduced cell transcription factors tagged with the green fluorescence protein (GFP). Because the amplified gene array is localized at one site in the nucleus, it is possible to directly observe binding of GFP-labeled factors to the promoter repeated within the array (FIG. 3).[18] Thus, for the first time we could see the process of transcription activation in real time.

This system opened the opportunity to study, with photobleaching technologies, the movement of transcription-associated factors in real time. In the most common approach, fluorescence recovery after photobleaching (FRAP), fluorescent proteins associated with a known target are rapidly bleached with an intense laser beam, which can be focused on a small target within the nucleus and whose energy is tuned to the excitation wavelength of the fluorescence tag. If proteins subjected to such a

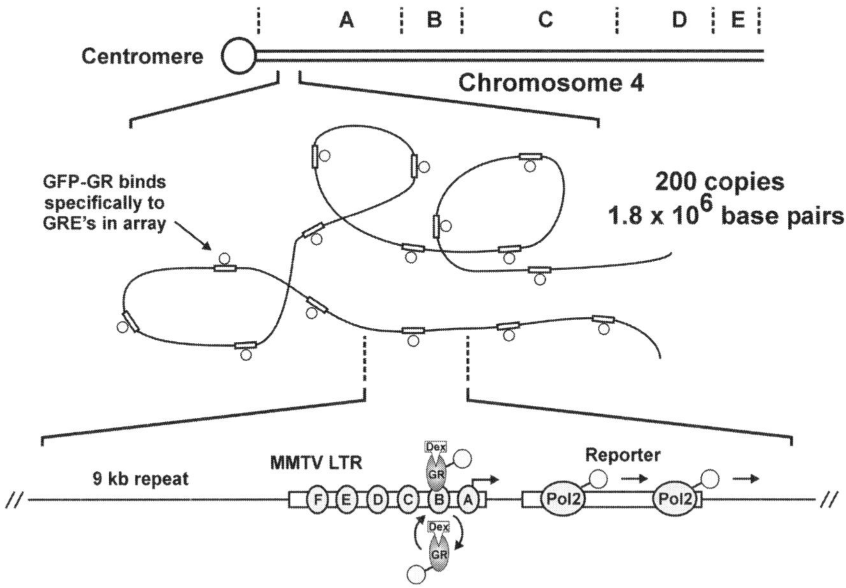

FIGURE 3. Visualization of receptor binding to response elements in living cells. A tandem array of MMTV/reporter elements was characterized near the centromere of murine chromosome 4. This array contains 200 copies of a head-to-tail 9-kb repeat, spanning 1.8×10^6 base pairs of DNA. GFP-labeled receptors and RNA pol II can be observed to bind directly to this array in living cells.

bleaching protocol are stably bound to their target, the fluorescence signal will not recover and the target will remain bleached. If the factors are in a dynamic equilibrium, however, the fluorescence will recover with kinetics determined by the proteins' rate of exchange.

Studies with factors that bind to the amplified gene revealed two general types of behavior. RNA Pol II, which is responsible for polymerization of the RNA transcript, recovers slowly after the bleach pulse, requiring 13 min to regain the original fluorescence (FIGS. 3 and 4)[19] because all of the elongating polymerases on each gene copy must complete their RNA chains and be released from the templates before they can be replaced by new, unbleached GFP-Pol II molecules. However, experiments with GFP-glucocorticoid receptor, which binds to the regulatory region and activates transcription from the gene, led to a highly unexpected finding (FIGS. 3 and 4).[18] The receptor and its coactivator GRIP-1 are present only briefly, with a FRAP recovery period of less than 30 s. Thus, the receptor must exchange rapidly and constantly with the regulatory element.

These findings contrast markedly with the view of hormone action that has dominated endocrinology since the original characterization of ligand-binding nuclear receptors. In this classic view, receptors remain bound to their cognate genes essentially as long as hormone is present in the cell, and the receptors serve as platforms for the accumulation of large multiprotein complexes. These "coactivator" complex-

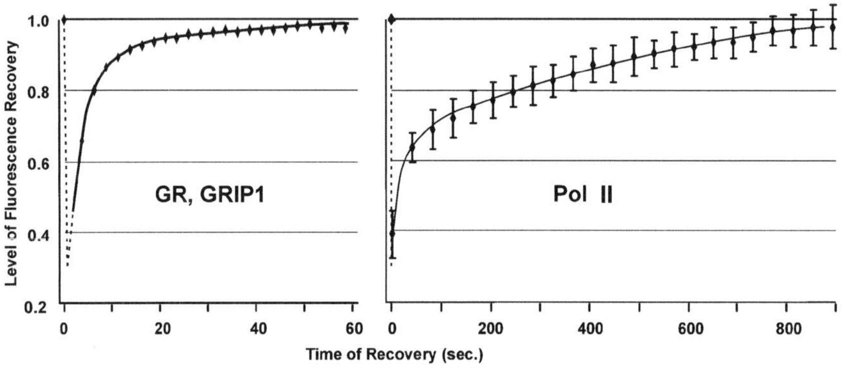

FIGURE 4. Photobleaching analysis of transcription factors bound to the MMTV array. The dynamic interaction of GFP-labeled GR, GRIP1, and RNA Pol II with the MMTV array was studied by fluorescence recovery after photobleaching (FRAP) analysis. Both GR and the coactivator GRIP1 exchange rapidly with the array, with a residence time of less than 30 s. In contrast, GFP-Pol II is resident on the array for an extended period, requiring 13 min for complete recovery after photobleaching.

es in turn harbor the many activities required for gene activation. However, our findings indicate that the receptors are not "static" on the template, but are moving rapidly on and off their regulatory elements. While initially controversial, recent findings with other receptors confirm this result. We have examined residence times for both the PR [20] and the AR on the MMTV array (Elbi and Hager, unpublished data), and found exchange rates equivalent to that for GR. Mancini and colleagues have observed binding of the estrogen receptor to an amplified prolactin regulatory element in living cells (M. Mancini, personal communication), and also find very brief residence times (10 s). Thus, findings from multiple systems now indicate that this rapid exchange is a common property of the steroid/nuclear receptor family.

CHAPERONE-MEDIATED NUCLEAR MOBILITY

As one approach to analyze potential mechanisms of nuclear receptor mobility, we established a permeablized cell system. Our goal was to translate tagged receptor cDNAs in reticulocyte lysates and to observe the behavior of the resulting GFP-labeled proteins after translocation into nuclei of the permeablized cells. After permeablization of cells, we showed that receptor mobility was lost (for both PR and GR).[21] Much to our surprise, addition of the reticulocyte lysate to these cells restored a large fraction of the receptor movement that was normally found in living cells. We then found that recovery of receptor mobility by lysate addition was inhibited by geldanamycin, a specific inhibitor of hsp90. This finding led to the demonstration that mobility of both receptors (GR and PR) could be restored by the addition of highly purified molecular chaperones.[21,22] In contrast, the equally mobile HP1α (*in vivo*) was refractory to addition of chaperones. These findings introduce a new paradigm for chaperone involvement in nuclear protein mobility.

We currently entertain two general models for chaperone action on nuclear protein mobility. Many of the steroid/nuclear receptors exist in inactive chaperone complexes in the absence of ligand.[23] Treatment of cells with hormone causes the disruption of these complexes, with the formation of an active receptor; this is often referred to as the "chaperone cycle." One tenet of this model is that these receptors cannot rebind ligand without reforming the chaperone complex. Thus, nuclear events involving ligand exchange (loss from the template, interaction with cofactor) could require a constant chaperone involvement to rebind hormone. Under this model, a chaperone role might be restricted to factors (nuclear receptors) that have this unusual requirement for ligand cycling.

Alternatively, the role of chaperones could be much more general. Although not frequently discussed, many nuclear proteins, including steroid receptors, exist in the cell with extensively unfolded regions.[24–26] Induced folding of the GR AF1 domain has been shown to occur upon binding to a GRE,[26] and induced folding of the AR AF1 has been shown upon binding of a cofactor.[24] In principle, the extensive domain reorganizations that accompany these events could require participation of chaperone refolding activities.

CONCLUSION

In summary, it seems that our view of receptor function is in need of major revision. The nucleus is a highly dynamic environment. In addition to passive diffusion, there appear to be multiple ATP-dependent mechanisms (chromatin remodeling, chaperone action) involved in active receptor movement. We currently favor a model[17,27] in which ligand-activated receptors exist in the nucleoplasm in complexes with many different coregulators. All of the complexes can interact transiently and randomly with hormone response elements specific for a given receptor. A productive interaction occurs only when the appropriate complex interacts at the regulatory site. As activities are recruited to the element that alter the local chromatin domain, the promoter is continuously modified and passes through a series of states. Which receptor-coregulator complex will be productive at a given site is determined by the status of the promoter at the time of the receptor–template interaction.

REFERENCES

1. HAGER, G.L. 2001. Understanding nuclear receptor function: From DNA to chromatin to the interphase nucleus. Prog. Nucleic Acids Res. Mol. Biol. **66:** 279–305.
2. HAGER, G.L. & G. FRAGOSO. 1999. Analysis of nucleosome positioning in mammalian cells. Methods Enzymol. **304:** 626–638.
3. KRAMER, P., G. FRAGOSO, W.D. PENNIE, *et al.* 1999. Transcriptional state of the mouse mammary tumor virus promoter can effect topological domain size in vivo. J. Biol. Chem. **274:** 28590–28597.
4. FRAGOSO, G., W.D. PENNIE, S. JOHN & G.L. HAGER. 1998. The position and length of the steroid-dependent hypersensitive region in the mouse mammary tumor virus long terminal repeat are invariant despite multiple nucleosome B frames. Mol. Cell. Biol. **18:** 3633–3644.

5. FRAGOSO, G. & G.L. HAGER. 1997. Analysis of in vivo nucleosome positions by determination of nucleosome-linker boundaries in crosslinked chromatin. Methods **11:** 246–252.
6. FRAGOSO, G., S. JOHN, M.S. ROBERTS & G.L. HAGER. 1995. Nucleosome positioning on the MMTV LTR results from the frequency-biased occupancy of multiple frames Genes Dev. **9:** 1933–1947.
7. ROBERTS, M.S., G. FRAGOSO & G.L. HAGER. 1995. Nucleosomes reconstituted in vitro on mouse mammary tumor virus B region DNA occupy multiple translational and rotational frames. Biochemistry **34:** 12470–12480.
8. SMITH, C.L. & G.L. HAGER. 1997. Transcriptional regulation of mammalian genes in vivo: a tale of two templates. J. Biol. Chem. **272:** 27493–27496.
9. FLETCHER, T.M., B.-W. RYU, C.T. BAUMANN, *et al.* 2000. Structure and dynamic properties of the glucocorticoid receptor-induced chromatin transition at the MMTV promoter. Mol. Cell. Biol. **20:** 6466–6475.
10. FLETCHER, T.M., N. XIAO, G. MAUTINO, *et al.* 2002. ATP-dependent mobilization of the glucocorticoid receptor during chromatin remodeling Mol. Cell. Biol. **22:** 3255–3263.
11. HOCKENSMITH, J.W., W.L. KUBASEK, E.M. EVERTSZ, *et al.* 1993. Laser cross-linking of proteins to nucleic acids. II. Interactions of the bacteriophage T4 DNA replication polymerase accessory proteins complex with DNA. J. Biol. Chem. **268:** 15721–15730.
12. DMITROV, S.I. & T. MOSS. 2001. UV laser-induced protein-DNA crosslinking. Methods Mol. Biol. **148:** 395–402.
13. DOUKI, T., D. ANGELOV & J. CADET. 2001. UV laser photolysis of DNA: effect of duplex stability on charge-transfer efficiency. J. Am. Chem. Soc. **123:** 11360–11366.
14. HOCKENSMITH, J.W., W.L. KUBASEK, W.R. VORACHEK, *et al.* 1991. Laser cross-linking of protein-nucleic acid complexes. Methods Enzymol. **208:** 211–236.
15. NAGAICH, A.K. & G.L. HAGER. 2004. Rapid detection of protein/DNA interactions by UV laser crosslinking. Science STKE. In press.
16. NAGAICH, A.K., D.A. WALKER, R.G. WOLFORD & G.L. HAGER. 2004. Rapid periodic binding and displacement of the glucocorticoid receptor during chromatin remodeling. Mol. Cell **14:** 163–174.
17. HAGER, G.L., C.C. ELBI & M. BECKER. 2002. Protein dynamics in the nuclear compartment. Curr. Opin. Genet. Dev. **12:** 137–141.
18. MCNALLY, J.G., W.G. MUELLER, D. WALKER, *et al.* 2000. The glucocorticoid receptor: rapid exchange with regulatory sites in living cells. Science **287:** 1262–1265.
19. BECKER, M., C.T. BAUMANN, S. JOHN, *et al.* 2002. Dynamic behavior of transcription factors on a natural promoter in living cells. EMBO Rep. **3:** 1188–1194.
20. RAYASAM, G.V., C.C. ELBI, D.A. WALKER, *et al.* 2004. Dynamic exchange of the progesterone receptor on gene targets. Submitted for publication.
21. ELBI, C.C., D.A. WALKER, G. ROMERO, *et al.* 2004. Molecular chaperones function as steroid receptor nuclear mobility factors. Proc. Natl. Acad. Sci. USA **101:** 2876–2881.
22. ELBI, C.C., D.A. WALKER, M. LEWIS, *et al.* 2004. A novel in situ assay for the identification and characterization of soluble nuclear mobility factors Science STKE. In press.
23. PRATT, W.B. 1998. The hsp90-based chaperone system: involvement in signal transduction from a variety of hormone and growth factor receptors. Proc. Soc. Exp. Biol. Med. **217:** 420–434.
24. KUMAR, R., R. BETNEY, J. LI, *et al.* 2004. Induced alpha-helix structure in AF1 of the androgen receptor upon binding transcription factor TFIIF Biochemistry **43:** 3008–3013.
25. KUMAR, R. & E.B. THOMPSON. 2003. Transactivation functions of the N-terminal domains of nuclear hormone receptors: protein folding and coactivator interactions Mol. Endocrinol. **17:** 1–10.
26. KUMAR, R., J.C. LEE, D.W. BOLEN & E.B. THOMPSON. 2001. The conformation of the glucocorticoid receptor af1/tau1 domain induced by osmolyte binds co-regulatory proteins J. Biol Chem. **276:** 18146–18152.
27. HAGER, G.L., A.K. NAGAICH, T.A. JOHNSON, *et al.* 2004. Dynamics of nuclear receptor movement and transcription Biochim. Biophys. Acta **1677:** 46–51.

Index of Contributors